SHAPE CASTING:
5th International Symposium 2014

Cover Photograph: Image of a sand inclusion on the fracture surface of an A356-T6 alloy castings subjected to hot isostatic pressing (HIP). Further details can be found in "Influence of Oxide Film Filling Defects on Fatigue Properties of. Cast Al-7Si Mg Alloy", Ph.D. Thesis by C.W.M. Nyahumwa (1997), The University of Birmingham, UK.

TMS2014

143rd Annual Meeting & Exhibition

SHAPE CASTING:
5th International Symposium 2014

Proceedings of a symposium sponsored by
the Aluminum Committee of the Light Metals Division
and the Solidification Committee of the
Materials Processing & Manufacturing Division of TMS
(The Minerals, Metals & Materials Society)

held during

TMS2014
143rd Annual Meeting & Exhibition

February 16-20, 2014
San Diego Convention Center
San Diego, California, USA

Edited by:

Murat Tiryakioğlu
John Campbell
Glenn Byczynski

Editors
Murat Tiryakioğlu
John Campbell

Glenn Byczynski

ISBN 978-3-319-48594-2
ISBN 978-3-319-48130-2 (eBook)
DOI 10.1007/978-3-319-48130-2

Chemistry and Materials Science: Professional

Published by Springer International Publishers, Switzerland, 2016
Reprint of the original edition published by John Wiley & Sons, Inc., 2014, 978-1-118-88818-6

Printed on acid-free paper

This Springer imprint is published by Springer Nature
The registered company is Springer International Publishing AG
The registered company address is: Gewerbestrasse 11, 6330 Cham, Switzerland

TABLE OF CONTENTS
Shape Casting: 5th International Symposium 2014

Process Design and Innovation

Solidification and Microstructure I

Mechanical Properties

Solidification and Microstructure II

PREFACE

Significant progress has been made in the quality and reliability of shape castings in the past several decades through a better understanding of the nature of structural defects, how they form, and how to avoid them. Light-alloy castings are now replacing complex assemblies by offering weight savings and significant reduction in tooling, assembly, and quality costs. These advances have been made possible by research on the physics behind simulation codes, the quality of molten metal, the hydraulics of mold filling, and the nature of bifilms and their effect on mechanical properties.

The first Shape Casting symposium held at the TMS 2005 Annual Meeting & Exhibition initiated a forum where researchers and foundry engineers could exchange their latest findings to improve the quality and reliability of shape castings. Since then, three other Shape Casting symposia followed this trend. Leading-edge technologies and the latest innovations in casting process design and quality improvements relative to shape casting were explored through presentations by researchers from around the world and are documented in the articles in this book.

We would like to give special thanks to Dr. Paul N. Crepeau, who served as the co-editor of the first four volumes in this series. We also welcome Dr. Glenn Byczynski as the third co-editor of this volume.

Murat Tiryakioğlu

John Campbell

Glenn Byczynski

SYMPOSIUM ORGANIZERS/EDITORS

Murat Tiryakioğlu is the director of the School of Engineering and a Professor of Mechanical Engineering at the University of North Florida. He received his B.Sc. in Mechanical Engineering from Boğaziçi University (1990), M.S. (1991) and Ph.D. (1993) in Engineering Management from the University of Missouri-Rolla, and another Ph.D in Metallurgy and Materials (2002) from the University of Birmingham, England.

Dr. Tiryakioğlu grew up around his family's foundry, which continues to thrive in Istanbul, Turkey. This has led to research interests in process design for high quality castings, aluminum heat treatment modeling and optimization, process–structure–property relationships in metals, statistical modeling, and quality and reliability improvement. He has written a book chapter, published more than 100 technical papers, and edited eight books on these topics.

Dr. Tiryakioğlu was awarded the inaugural John Campbell Medal from the Institute of Cast Metals Engineers (UK), and was named the Eugene Merchant Outstanding Young Manufacturing Engineer by the Society of Manufacturing Engineers. He was also selected as a TMS Young Leader in Light Metals. He is a member of TMS and ASQ, and is an ASQ Certified Quality Engineer.

John Campbell was educated at Cambridge, Sheffield and Birmingham and first trained as a physicist, then becoming a metallurgical engineer. He has spent most of his life in foundries, where he thinks he has made more aluminum alloy cylinder head castings than he has had hot dinners.

He developed the Cosworth Casting Process, then a novel process, making Al alloy cylinder heads and blocks for Formula One racing engines using counter-gravity filling of molds by electromagnetic pumps. Later still, after 15 years as professor of casting technology at the University of Birmingham, UK, he worked for over ten years with his partner in the United States, John Grassi, to set up a new casting operation. This new operation, Alotech, helped to develop the new and exciting Ablation Casting Process.

He continues to work in the industry in different parts of the world. He introduces new casting concepts to greatly improve quality and reduce costs, and he redesigns foundries to adopt new melting and counter-gravity techniques for the production of fault-free castings. These techniques promise to herald a revolution in engineering attainment.

Glenn Byczynski is the Technical Manager of Nemak Canadian Operations. He received his Ph.D. in metallurgy and materials science from the University of Birmingham in the UK in 2002. His master's (materials science in 1997) and bachelor's (mechanical engineering in 1994) degrees were received at the University of Windsor. In his 20-year career in metal casting he has held several R&D and engineering positions within Nemak and Ford Motor Company including R&D Manager of Nemak Europe based in Germany (2007–2009), Manager of the Nemak Engineering Centre (Windsor), and Engineering Manager at Nemak's Windsor Aluminum Plant. He was chairman of the Detroit-Windsor Chapter of the American Foundry Society (AFS) in 2006–2007, is a director and regional chairman of the Foundry Educational Foundation, and is a registered professional engineer in the province of Ontario.

SHAPE CASTING:
5th International Symposium 2014

Process Design and Innovation

Shape Casting: 5th International Symposium 2014
Edited by: Murat Tiryakioğlu, John Campbell, and Glenn Byczynski
TMS (The Minerals, Metals & Materials Society), 2014

A Draft Melting Procedure for Al Alloys

J. Campbell
Emeritus Professor of Casting Technology
Department of Metallurgy and Materials
University of Birmingham, UK

Keywords: Melting Procedure; liquid metal quality; bifilm; degassing.

Abstract

Current melting practices for Al alloys are in general unsatisfactory because it is clear that properties of Al alloy castings are not under control. In particular the variability of tensile ductility appears to indicate the great variability of oxide bifilm contents. A melting and casting procedure is outlined in which (i) the primary oxide skins from the surface of charge materials are first separated from the melt, followed by (ii) procedures for reducing the population of oxide bifilms in suspension from the interiors of charge materials. (iii) Hydrogen gas is reduced passively, avoiding potentially damaging bubbling techniques. (iv) Finally, techniques for the transfer of good quality metal into molds without re-introduction of damage are briefly mentioned.

Introduction

The Al alloy casting industry spends significant time and money on each of their melts in an attempt to 'clean up' the melt, or sometimes merely to 'degass' the melt. The extent to which these current procedures are unreliable, and sometimes completely fail, is clear from the survey by Tiryakioglu [1] revealing the scatter in tensile elongation results sourced mainly from US aerospace foundries who would, of course, be expected to use the best procedures available to the Al alloy casting industry (Figure 1).

It sobering to reflect that most other Al foundries will perform even less well. Figure 2 shows results from an Al alloy foundry who wishes to remain anonymous. They had been rigorous in attempting to increase the quality of their melts and the properties of their castings. When, on recommendation, they introduced the comparative Quality Index test [2] they were amazed to discover that they had achieved, on average, only 25 % of the possible level of tensile elongation. This prompted them to reconsider their melting and melt treatment procedures, which was an incentive to write this paper.

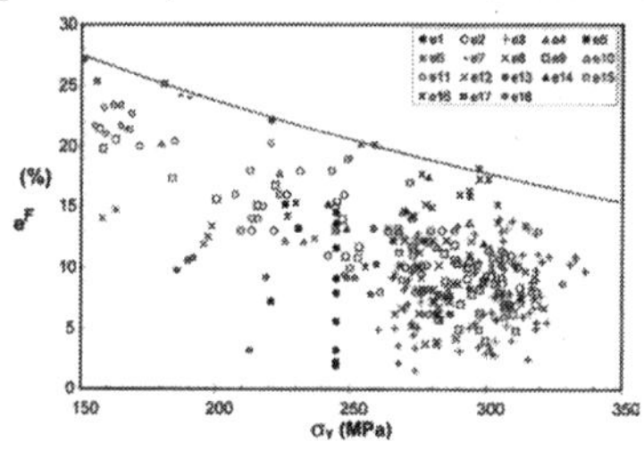

Fig 1. Scattered tensile elongation results mainly from US Aerospace foundries [1].

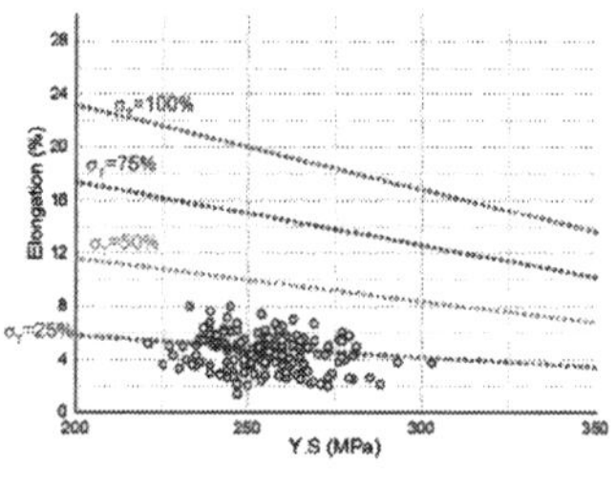

Fig 2. A typical Al foundry obtaining, on average, only 25 % of potential properties.

The foundry discovering its 25 % achievement is to be congratulated on their use of the new Quality Index which revealed to them the precise extent to which their melt quality fell short of what should have been attainable. It is to be noted with regret that most founders have no idea what their quality is really like, and no idea what they can achieve with good practice. The problem arises from oxides in the melt; these are all in the form of bifilms – folded over oxide films which act as cracks to reduce mechanical properties. Traditionally oxides have not been monitored nor treatments devised for their reduction, although their action as cracks in the liquid, inherited by the solid, means that the oxides effectively control the attainment of good properties.

My initial suggestions to this foundry included the avoidance of the addition of flux on the melt which may raise hydrogen levels unnecessarily (flux may have benefits such as the reduction of suspended oxides by a mopping up type of mechanism, but this has never been proven, and in the absence of convincing data to indicate useful benefits, the current decision was to discard it. This decision could be reversed at some future time if evidence to support its usefulness can be demonstrated.) In addition their argon degassing for lengthy periods measured in hours would almost certainly be damaging because of increased oxide bifilm content of the melt. Rotary degassing was also not necessarily recommended because of its operational uncertainties as discussed below, and in any case the hydrogen is easily reduced later without such threats to quality as is explained in due course.

The Rotary Degassing Technique

Most of the current efforts to increase quality involve rotary degassing; the dispersing of a cloud of fine bubbles of inert gas from a spinner submerged in, and vigorously stirring the melt. The technique is named a degassing technique, and researchers have been preoccupied by measurements and predictions of degassing efficiency, but it seems likely to this author that its degassing effect is of secondary importance to its effect in reducing the population of large oxide bifilms. So far this important and critical aspect of the process appears to have been overlooked.

In its role as a treatment for the reduction in oxide films, however, the rotary process may not be completely satisfactory. This is partly because the rising bubbles can, in general, only float out films larger than themselves as illustrated in Figure 3. For this reason the process appears to be highly effective in removing the large oxide skins, often the size of newspapers, which were the original skins on the surface of the charge materials such as ingots and foundry returns such as running and feeding systems and scrap castings. These large, floppy, planar defects are easily impacted by the bubbles and transported to the surface of the melt from where they can be skimmed off. The removal of these massive defects is an important feature of the rotary technique, and explains the generally much improved quality of melts following treatment.

It is noteworthy that these large oxide skins are always spinels ($Al_2O_3.MgO$) [2] as would be expected from the relatively long cooling times for cast products when originally cast, especially in sand molds, and the long reheating and melting time in air when remelted, giving plenty of time for Mg to diffuse into the surface oxide converting the originally pure alumina to a spinel.

Turning now to the case of those films in the liquid metal which are smaller than the bubbles provided by the spinner, as the bubbles rise the films will tend to move around the flow lines of the bubble, never touching the bubble, despite being apparently directly in the flow path of the bubble as seen in Figure 3. It is clear therefore that on average the bubble technique can only eliminate oxides larger than the average bubble size. Since the bubbles are approximately several millimeters in diameter, typically 5 mm, this leaves relatively untouched a large population of oxide bifilms in the melt of significant size.

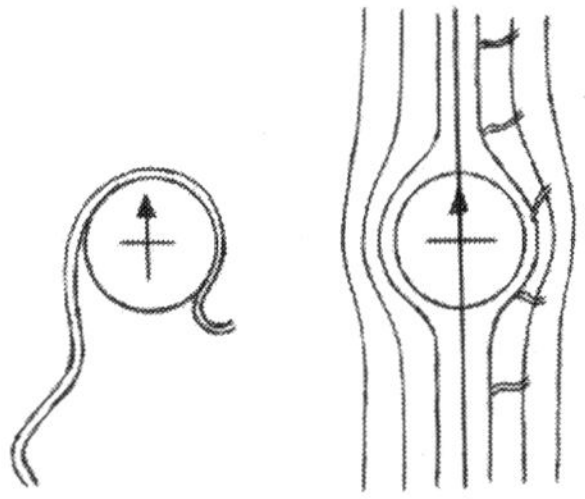

Fig 3. Bifilms in the paths of rising bubbles (a) the bifilm larger than the bubble is impacted and floated out; (b) the bifilm smaller than the bubble tends to follow the flow lines around the bubble, not contacting the bubble, and is therefore not floated out.

Furthermore, as the bubbles break at the surface of the melt their freshly opened surfaces are oxidized by exposure to air. As the opened and oxidized surface closes again it naturally creates an oxide bifilm. Interestingly this will be a pure alumina bifilm, as corroborated by experiment which finds, as rotary degassing progresses, that the oxides in suspension change from being 100 % spinels to 100 % pure alumina [2].

Finally, at the high shaft speeds to achieve fine bubbles, a common technique is the introduction of a baffle plate in the melt to inhibit the general rotation of the liquid and avoid the development of a central vortex. This feature may take air down into the melt and thereby introduce oxides,. However, in the conditions when the baffle is working hard to hold back the general rotation of the melt, water modelling clearly reveals the consequential development of severe air entrainment in the wake of the baffle, probably countering its usefulness.

It is clear therefore that although the rotary degassing technique is valuable, it is also seems likely to be limited if very high metal quality is required; although the technique appears to remove the very damaging large spinel sheet bifilms, it simultaneously appears to create millions of confetti sized pure alumina bifilms.

Clearly therefore, if very high quality is required, some alternative technique to eliminate oxide bifilms is desirable.

The Cosworth Casting Process

To the author's knowledge, the Cosworth Casting Process was the first melting and casting process to be set up to counter porosity in castings by reducing the population of suspended oxides in the liquid metal. It was first set up in 1980 specifically to produce cylinder block and head castings for racing engines. The Al alloy melts were cleaned from oxide bifilm by a sink and float separation technique. The Process enjoyed significant success, demonstrating the ability to cast a pair of blocks, each weighing over 30 kg in a single sand mold every 45 seconds, putting its productivity well above any other block producing process. In addition the properties of the castings were unexcelled. However, despite its high achievements, it is now known never to have achieved its full potential because the fundamental requirements of the process were never fully met [4], hampering properties of products because of inadequate cleanness of the melt. This less than optimum performance resulted because the contribution of the massive oxide films which formed the skins of the charge materials had been overlooked. In addition, the design of the holding furnace was not optimum for effective sedimentation. These shortcomings are the subject of current developments by the author.

The specification for the Process is simple. Avoid any turbulence or other disturbance of the metal. This was designed to be achieved simply by melting the interior of charge materials but separating and discarding the outer oxide skin, ensuring that the liquid alloy from that point on in the process never poured downhill, but travelled only horizontally. The interior bifilms in the alloy were allowed to sink, possibly aided by precipitation of heavy intermetallics. Finally the alloy was transferred into the mold via an uphill process in which the velocity of the melt was always controlled to be below the critical velocity 0.5 m/s to ensure that any surface oxide was not entrained by surface turbulence.

This logic of this specification is most effectively translated into hardware using a commercially available dry hearth melter in tandem with an electrically heated reverbatory holding furnace. Preferably a casting station is provided separate from the holder (the use of a pump in a pump well of a holder is at risk from the holder developing a leak, reducing temperature in the pump well to the point that casting production cannot be continued. A separate casting furnace, most conveniently a crucible furnace, can have independent heating for casting temperature control, and is easily swapped for a spare furnace in the rare case of total failure of the furnace). A counter-current of dry nitrogen flowing through the holder is a long understood technique for passively degassing the melt without the trauma of turbulence caused by bubbles. Effectively everything works by itself in this system; there are practically no moving parts, and very little operator involvement.

In general, however, most foundries do not currently enjoy the luxury of such a high volume continuous melting and treatment system, if only because of the number of different alloys their customers demand. Batch melting in crucible furnaces is therefore appropriate, but is, of course, significantly more difficult to achieve a consistently good level of quality control. The following procedure is designed to attain the best results possible from this less favorable starting point.

A suggested draft General Procedure

A melt treatment to take advantage of the most recent thinking and experience in Al foundries using crucible furnaces is outlined below. At this time it is proposed as a draft, which can be revised and added to as the technology is developed.

1. After the melting of the charge, add alloying elements, perhaps stirring for half a minute to ensure that all alloys are in solution and to ensure homogeneity.

2. Eliminate the primary oxide skins resulting from the surface of the original charge materials the surfaces of the alloying additions. These are large oxide skins, some

probably approaching the size of newspapers. Flushing the melt with fine bubbles of argon from a stationary lance with a porous refractory tip, or from a rotary 'degasser', is generally helpful to lift out these large skins. After 5 minutes the fine bubbles should have raised all these oxides to the surface and should be skimmed off. If there is any sign of additional skins coming to the surface after the termination of the flushing action, additional treatment with the fine bubbles for a further 5 minutes and skimming off again may be helpful. It is possible that these times are excessive, but shorter times, for instance 1 minute, may be insufficient for the shaft and rotor to degass its contained hydrogen, so there may be some danger of raised gas levels. This danger is worth exploring, since short times would definitely reduce the dangers of the creation of large populations of pure alumina bifilms.

3. Eliminate the dispersion of finer oxide bifilms in the centres of the charge materials and alloying additions, and those introduced by rotary degassing, as far as possible. At this stage the melt is probably choked with a snow-storm of fine oxide bifilms which will degrade properties. Furthermore, the bifilms are nearly neutral buoyancy and, with the assistance of the thermal convection present in all conventional furnaces, will stay in suspension for hours or days. The rate of sedimentation can be encouraged by the addition of a heavy solute such as the grain refiner Ti+B. An addition of Ti-rich material needs to be gently stirred in to the melt to homogenize it. However, after that, the melt should not be stirred or disturbed in any way again. This is because Ti-rich compounds precipitate out on the favoured substrates provided by the oxide bifilms. This causes them to sink to the bottom like stones leaving the main body of the melt clean. The melt can remain clean while the bottom sediment is not disturbed.

   (It is a widespread fallacy among metallurgists and foundry engineers that the melt must be stirred vigorously immediately prior to pouring so as to obtain as much of the Ti-rich compounds transferred into the casting as possible, maximizing grain refinement. This treatment has almost no effect on grain refinement, but because the bifilms are now also carried over into the casting, also reduces properties and increases the susceptibility to hot tearing, cracking and porosity.)

4. Reduce hydrogen by passive degassing. This means, instead of active degassing by bubbling, doing nothing for a further 30 minutes (or perhaps more with a fuel-fired furnace – this needs to be checked by RPT or other device capable of quantifying hydrogen content as a function of time). After this 30 minute period in which the melt should be left alone, not stirred or disturbed in any way, the hydrogen should have evaporated from the melt, equilibrating with its environment. This assumes that the environment is fairly free from moisture. The environment may be a slight problem with a fuel-fired furnace. It might also be a problem with an induction furnace in which the water-cooled induction coils are dripping with condensed water. Dryness of all furnace systems is achieved by maintaining the furnaces hot 365 days of the year. Provided the furnace casing and refractories, and the melt crucible, are all fairly free from moisture, the level of equilibration achievable should sufficiently low for practically all casting purposes. Also during this 30 minute quiescent dwell time most of any remaining fine oxide bifilms from the original population will have settled out.

5. The metal should be delivered into the lowest point of the casting without turbulence. The velocity of metal delivery into the mold cavity should be controlled to be less

than 1 m/s and preferably less than 0.5 m/s. If it is faster the melt will fountain, and so fold in new bifilms which will impair the casting, effectively undoing the good of the metal/bifilm separation techniques employed so far.

6. Provide feed metal from feeders on top of the casting to continue to pressurise and feed the casting during freezing.

7. If uphill filling against gravity is employed, a slide gate or other device to cut off the metal supply from the furnace is necessary to allow the mould to be lifted clear from the furnace as soon as possible. (If the mold is retained on the casting station, with continued furnace pressurisation to feed the casting during freezing, the grave danger is introduced from the development of convection problems. These lead to what appears to be shrinkage porosity which, frankly, cannot usually be cured. Although it is good to fill uphill, it is dangerous to feed uphill. The fundamentally correct practice is to *fill uphill* against gravity, but *feed downhill* with the help of gravity. Removing the filled mold from the casting station prior to the freezing of the casting greatly increases productivity, of course, compared to the operation of most low pressure casting units.)

Assessment of Cleaning Effectiveness

The RPT technique is now sufficiently well developed to quantify the presence of relatively large oxide bifilms in Al alloy melts [5].

However, it is becoming clear that the test is not sensitive to very small bifilms, of size probably of the order of 10 μm or less. Such fine crack-like defects are revealed at the much higher stresses involved in tensile fracture, and so are clear on a tensile fracture surface. For instance in an Al-Si alloy, each fractured Si particle can be assumed to have fractured because of the presence of a bifilm. (Si is strong and would not normally be expected for fracture since solidification would have produced it free from cracks, and therefore enjoying its maximum crack resistance). Thus every ductile dimple on a fracture surface denotes the presence of a bifilm which has caused the inclusion or Si particle to 'fracture' i.e. simply come apart at the bifilm in its center or appear to decohere from the matrix as a result of a bifilm along one of its sides.

Conclusions

1. An optimum continuous melting and casting system based on a development of the original Cosworth Casting Process uses three separate specialized furnaces working in series.

2. A melting and casting procedure for general use in existing foundries with crucible furnaces is outlined in draft form.

3. Hydrogen reduction is recommended to be carried out passively to avoid damage to the melt.

References

1. M. Tiryakioglu, J. Campbell, N. D. Alexopoulos; "On the ductility of cast Al-7Si-Mg alloys." *Metall. & Mater. Trans. A* 2009 vol.40A (April) pp 1000 – 1007.
2. M. Tiryakioglu, J. Campbell; Mater Sci Technol 2009 vol 25 (No. 6) pp. 784-789
3. Phil Enright, NTec Limited. Personal communication 2009.
4. J. Campbell; "Quality Castings – A personal history of the Cosworth Casting Process" to be published.
5. D. Dispinar and J. Campbell; Reduced Pressure Test (RPT) for Bifilm Assessment; a Draft Procedure; TMS Annual Congress, San Diego, February 2014, V Shape Casting Symposium.

Shape Casting: 5th International Symposium 2014
Edited by: Murat Tiryakioğlu, John Campbell, and Glenn Byczynski
TMS (The Minerals, Metals & Materials Society), 2014

Effect of Casting Condition in Semi-Solid Aluminum Alloy Injection Process on Distribution of Defects and Density

Yuichiro Murakami[1], Kenji Miwa[2], Masayuki Kito[3], Takashi Honda[3], Naoyuki Kanetake[4],Shuji Tada[1]

[1]National Institute of Advanced Industrial Science and Technology (AIST);
2266-98, Anagahora, Shimo-Shidami, Moriyama-ku; Nagoya, Aichi, 463-8560, Japan
[2]Aichi Science and Technology Foundation;1-157-1, Onda-cho; Kariya, Aichi, 448-0013, Japan
[3]Aisan Industry Co., Ltd.;1-1-1 Kyowa-cho; Obu, Aichi, 474-8588, Japan
[4]Nagoya University; Furo-cho, Chikusa-ku; Nagoya, Aichi, 464-8603, Japan

Keywords: Semi-solid Process, Aluminum Alloy, Casting defect, Density, Shear Rate

Abstract

Semi-solid process is expected as near net shape method with high quality. In this study, the semi-solid aluminum slurry was injected into a metallic mold with spiral shape cavity via some gates of several thicknesses. The effects of injection speed and gate thickness on the distribution of casting defects and density were investigated. Most of the casting defects appeared at the center of the specimen. The casting defects were generated most frequently on the tip of the spiral specimen. The amount of casting defects was decreased by semi-solid injection process than general high pressure die casting. The density of specimen made by the semi-solid injection process was approximately the same as high pressure die cast at near the gate, and increased in the other area.

Introduction

The development and diffusion of energy and environment technologies has become increasingly important in recent years. Particularly, CO_2 emission from transportation sector is accounting for 23% of the world total, hence the improvement of fuel efficiency is urgent issues for the automobile industry. For the solution, reduction of vehicle weight by the use of light metals is one highly effective means. Additionally, the range of application of in-vehicle electronics and electro actuation is extending in recent years; also amounts of heat generation of them are increasing with an enhanced performance. Aluminum alloys have low density and high heat conductivity, hence increasing the use of the aluminum alloys is expected to increase a radiation performance and to decrease vehicle weight.

The production of light metal parts using aluminum is mainly performed by high pressure die casting (HDPC), which directly fabricate the required shape of the liquid state. However in this process, liquid metal which has low viscosity is injected into the mold with high velocity, hence components molded by die-casting exhibit low engineering performance due to the existence of inherent defect such as porosity, hot cracks and oxide inclusions, and heat treatment is impossible.

Semi-solid processing is the fabrication in solid-liquid co-existing state, and semi-solid material has a greater viscosity than that of the liquid. Hence the semi-solid products have the potential of decreasing gas defects than HPDC because of reducing air entrainment. Additionally, solidification shrinkage would be reduced in the semi-solid alloy, which is expected to reduce shrinkage porosity and improve dimensional accuracy. This process has been studied since 1970s

[1,2], and recently it has been developed to fabricate high-quality aluminum alloy [3-5] and magnesium alloy [6-9] products.

On the other hand, semi-solid metal has a lower fluidity than the liquid metal. For the solution to this problem, we have developed the method of increasing fluidity by applying shear stress[10- 12]. However, the slurry is applied not only high shear stress but also high gate velocity and piston speed in this process, then these factors have the risk for increase of air entrainments. Whereat, the effects of gate velocity and shear velocity on casting defects and density of products were investigated in this study.

Experimental procedure

The alloy used for the experiments was an Al-7wt%Si-0.3wt%Mg alloy (AC4CH in the JIS, equivalent to A356 in the ASTM standard). The semi-solid slurry was prepared by a nano-cast method [13]. In this method, molten AC4CH alloy was poured into a stainless steel cup with the room temperature. In this method, molten AC4CH alloy was poured into a stainless steel cup with the room temperature, therefore the molten AC4CH alloy was drawn the heat by the cup. By this means, the temperature of the molten AC4CH alloy was decreased to the semi-solid temperature, moreover the electromagnetic stirring was applied during cooling. For this reason, the slurry with dispersed small and fine solid particles can be obtained. In this study, the poured temperature of AC4CH alloy was 700°C, and the pouring weight was about 220g. The dimension of the stainless steel cup is 90 mm in height, 38.7 mm in inner diameter and 2.0 mm in thickness. The electromagnetic stirring was applied to the vertical direction for 5 sec, then to the rotational direction for 10 sec.

In this study, a 135t HPDC machine was used. The specimens were cast by injection into a permanent mold with spiral cavity, shown in Figure 1. The permanent mold had spiral cavities of 5.7 mm in width, 1350 mm in length and 4.0 mm in thickness, also it had a gate of 5 mm in width and gate velocity was controlled by changing thickness 1.0 mm, 2.2 mm, 3.1 mm and 4.0 mm, respectively. The piston speed (injection velocity) was set to 0.1 or 0.35 m/s.

The microstructure observation was carried out at the surface perpendicular to the flow

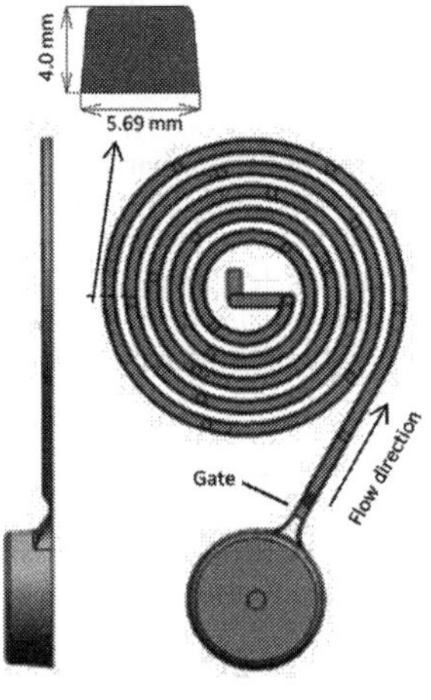

Figure 1. Appearance image of spiral cavity of permanent mold

direction with respect to each 100 mm or 50 mm in length. The specimens were polished by grinding with SiC water proof paper, followed by polishing with diamond paste, then, the specimens were etched with 0.5% HF solution. The microstructures of these specimens were observed by optical microscopy. The density of specimens was measured by the Archimedes method with respect to each 100 mm in length.

Results and discussion

Microstructure observations

Figure 2 shows typical microstructures of the specimen perpendicular to flow direction. These micrographs show that the AC4CH aluminum alloys had a microstructure with a dispersion of the primary α-Al particles in the matrix (eutectic α-Al and Si). It is considered that the primary α-Al particles were the solid phase and the matrix was the liquid phase when the slurry was injected. The black areas in this micrograph were casting defects. Microstructures of the vertical section of the specimens near the gate are shown in Figure 3 and microstructures of the vertical section of the specimens at the middle part of the total flow length are shown in Figure 4. Under the condition of forming at liquid state (HPDC), the casting defects appeared at the center of the specimen. On the other hand, in specimen formed in the semi-solid state, some casting defects were detected between alpha particles, but these casting defects became smaller than HPDC. Also the casting defects were located at the center part of the specimen. Both the distribution and size of casting defects have the same tendency near the gate and the middle part of the specimen. Figure 5 shows microstructures of the vertical section of the specimens near the end. The specimens had an increased number of casting defects than the other part (Figure 3and Figure 4), and the casting defects were located not only the center of the specimen. Additionally, area of casting defects was increased in all conditions. However, size and the number of casting defects were decreased by semi-solid forming.

The slurry or the melt was considered to be solidified from the mold surface in a plane perpendicular to flow direction. For this reason, the defects caused by shrinkage was thought to be located at the center of the specimen. In the micrographs of specimen near the gate and middle of flow length, most of casting defects were located at the center. Then, casting defects near the gate and middle of flow length were considered to be caused by solidification shrinkage. On the other hand, the casting defects were dispersed at the end of the specimen. Therefore, the casting defects in the end of the specimen were thought to be caused by air entrainment.

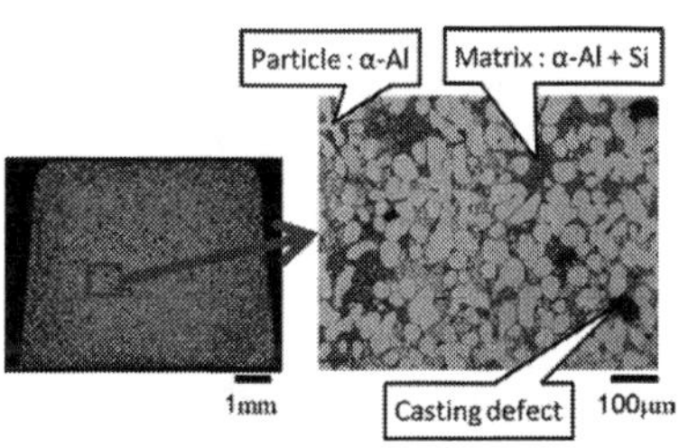

Figure 2. Typical microstructures of the specimen perpendicular to flow direction.

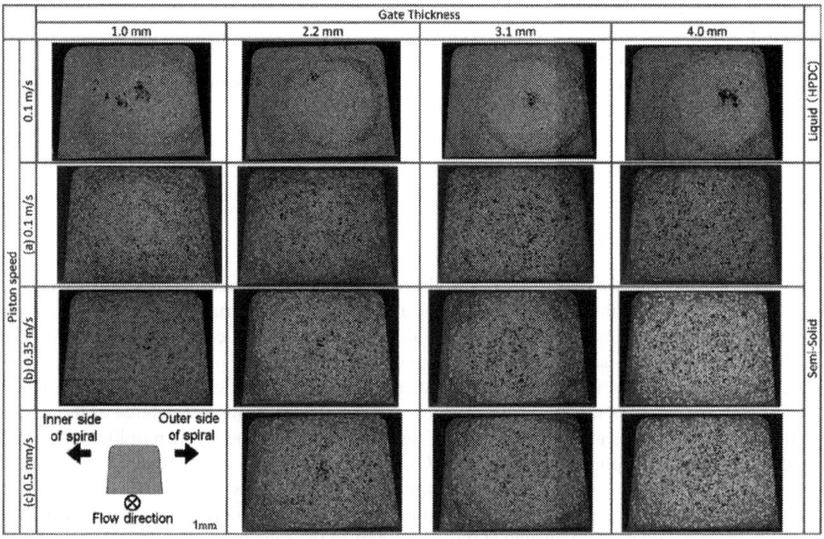

Figure 3 Microstructures of the vertical section of the specimens near the gate. Observation points were 50 mm from the gate (injection speed 0.1 m/s) or 100 mm from the gate (injection speed 0.35 m/s or 0.5 m/s)

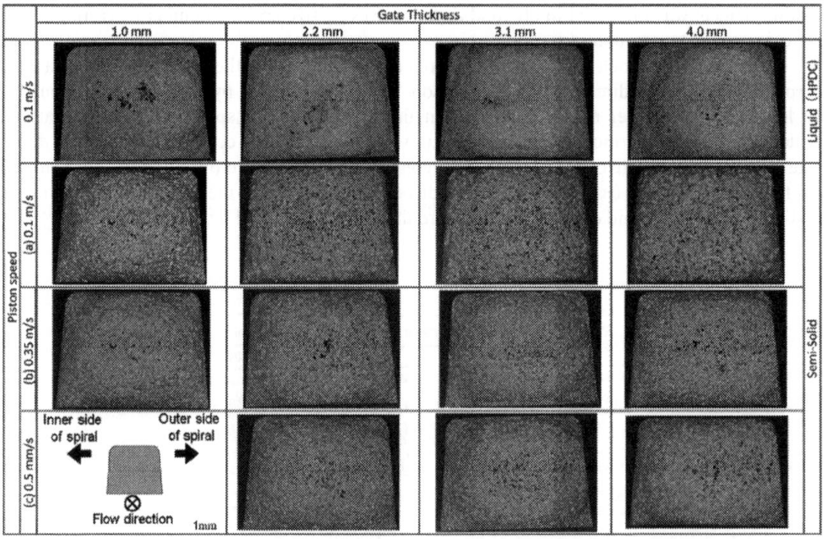

Figure 4 Microstructures of the vertical section of the specimens at the middle part of the total flow length. These photographs were taken at the points of 50% for their total fluidity length.

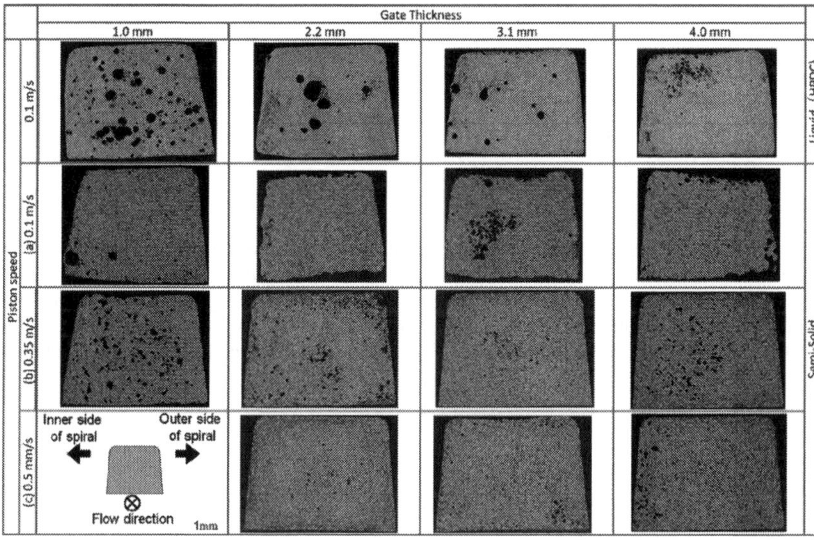

Figure 5 Microstructures of the vertical section of the specimens near the end. These photographs were taken at the points of the end for their total fluidity length.

Distribution of casting defects and effect of gate velocity

Areal fractions were calculated as below with a distance of 100 mm between each micrographs. Area of each casting defect A_{di} and cross-sectional area of the sample A_s were calculated by image analysis. Next, areal fraction f_d were calculated as follows.

$$f_d = \frac{\Sigma A_{di}}{A_d}$$

Figure 6 shows a comparison of the distribution of the areal fraction of casting defects between semi-solid forming and HPDC in the same conditions. Until 60% of flow length, the areal fraction of casting defects were smaller than 0.01 by semi-solid state forming . On the other hand, the areal fraction was slightly increased in HDPC specimen. Additionally, the areal fractions of casting defects were increased at the end of the specimen in all conditions. The areal fractions of the specimens made by HPDC were greater than semi-solid forming at same gate thickness.

The flow length of semi-solid forming is smaller than HPDC at the same conditions. The comparison of the areal fraction of casting defects between semi-solid forming with piston speed of 0.35 m/s and HPDC with piston speed of 0.1 m/s is shown in Figure 7. Tendencies of distribution of the areal fraction of casting defects were same in semi-solid forming between piston speed 0.1 m/s and 0.35 m/s, and until 60% of flow length, the areal fraction was smaller than HPDC. However, the areal fraction increased at the end of specimen by increasing piston speed.

As the discussion in the preceding section, the casting defects near the gate and the middle of specimen were considered to be caused by solidification shrinkage, and the defects at the end of the specimen were thought to be caused by air entrainment. The results indicated that both of the causes, solidification shrinkage and air entrainment were decreased by using semi-solid forming. However, increasing piston speed has possibilities to increase air entrainment even if it is semi-solid forming.

Distribution of density

The density of specimen measured by Archimedes method was shown in Figure 8 (gate thickness 1.0 mm) and Figure 9 (gate thickness 4.0 mm). The density of the specimens was higher at near the gate and decreased at farther than 60% of flow length. This means the same tendency of the distribution of casting defects, but the difference between semi-solid casting and HPDC is greater in the top of the specimen. This result suggested that the porosities which could

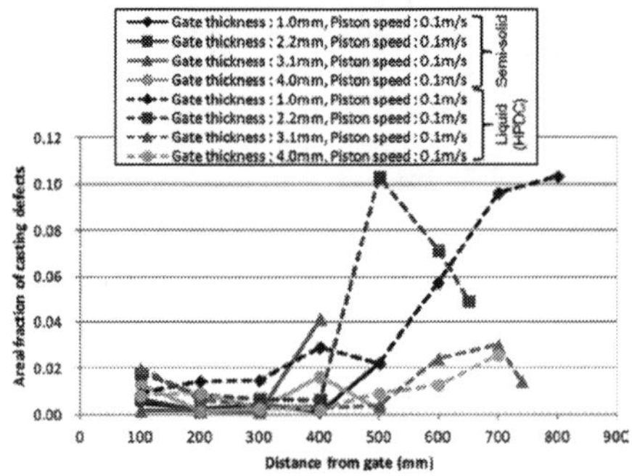

Figure 6 Distribution of areal fraction of casting defects at piston speed 0.1 m/s

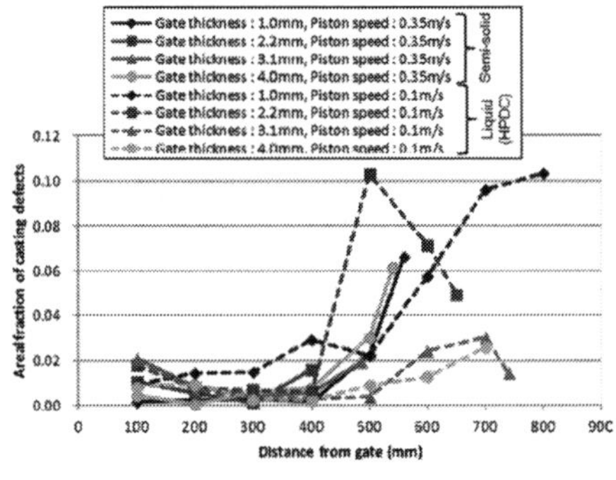

Figure 7 Distribution of areal fraction of casting defects at piston speed 0.1 m/s (liquid state) and piston speed 0.35 m/s (semi-solid forming)

not detect by optical microscopy existed, and they were mostly located at the end of the specimen. Additionally, they were decreased by the use of semi-solid forming. The densities at near the gate were little difference between HPDC and semi-solid forming. On the other hand, the density in semi-solid specimen at middle of flow length is higher than HDPC. Moreover the density of the specimen at the gate thickness of 4.0 mm was greater than the gate thickness of 1.0 mm. These results indicated that the amount of air entrainment was decreased by the use of semi-solid forming. Then decreasing the gate thickness leads increasing gate velocity, so this was considered to be the cause of the increasing air entrainment.

Note that, in this mold which has a spiral cavity, the effect of intensification pressure is considered to be smaller than the normal HPDC mold. Because the slurry was not fully filled in the cavity, the top of the slurry while the injection was thought to be free.

Conclusions

The distribution of casting defects on the plane perpendicular to the flow direction in the spiral shape AC4CH (A356) specimens were investigated. Also, the distribution of the areal fraction of casting defects in the flow direction was calculated. Additionally, the density distribution of the spiral specimen in flow direction was measured, and the following conclusions have been derived.

The most of casting defects were located in center of the specimen from gate to middle on flow direction. On the other hand, the casting defects were dispersed at the end of specimen and both

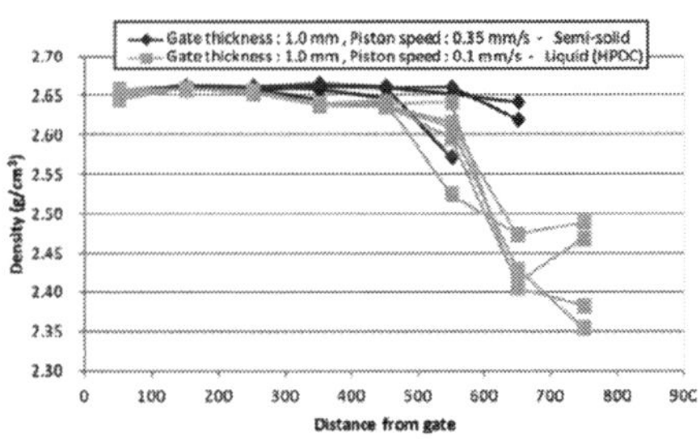

Figure 8 Distribution of density of specimens at gate thickness of 1.0 mm

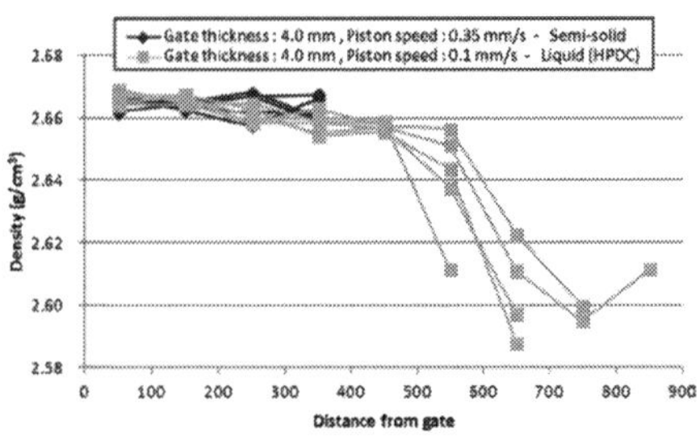

Figure 9 Distribution of density of specimens at gate thickness of 4.0 mm

amount and area of casting defects increased. By application of semi-solid casting, areal fraction of casting defects of specimen increased and density decreased, respectively.

References

1. M.C Flemings and R. Meharabian, "Casting Semi-Solid Metals," *AFS Transactions*, 102 (1973), 81-88.

2. M.C Flemings, "Solidification processing," *Metallurgical and Materials Transaction*, 5B (1974), 2121-2134

3. W.G. Cho and C.G. Kang, "Mechanical properties and their microstructure evaluation in the thixoforming process of semi-solid aluminum alloys," *Journal of Materials Processing Technology*, 105 (2000), 269-277

4.D. G. Eskin, and L. Katgerman. "Mechanical properties in the semi-solid state and hot tearing of aluminium alloys." *Progress in Materials Science,* 49 (2004), 629-711.

5. S. Nafisi and R. Ghomashchi, "Grain refining of conventional and semi-solid A356 Al–Si alloy," *Journal of Materials Processing Technology*, 174 (2006), 371-383

6. F. Czerwinski, "Size evolution of the unmelted phase during injection molding of semisolid magnesium alloys," *Scripta materialia* 48 (2003) 327-331.

7. N. Omura, Y. Murakami, M.G. LI, T. Tamura and K. Miwa, "Effect of Volume Fraction Solid and Injection Speed on Mechanical Properties in New Type Semi-solid Injection Process," *Solid State Phenomena*, 141-143 (2008) 761-766

8. Y. Murakami, K. Miwa, N. Omura and S. Tada, "Effects of Injection Speed and Fraction Solid on Tensile Strength in Semisolid Injection Molding of AZ91D Magnesium Alloy," *Materials Transactions*, 53 (2012) 1775-1781

9. Y. Murakami, K. Miwa, N. Kanetake and S. Tada, "Effect of Casting Defects Distribution on the Beginning of Tensile Fracture in Semi-solid Injected Magnesium AZ91D Alloy," *Magnesium Technology 2013*, (2013) 271-274

10. Y. Murakami, N. Omura, M.G. Li, T. Tamura, S. Tada and K. Miwa "Microstructures and Casting Defects of Magnesium Alloy Made by a New Type of Semisolid Injection Process," *Magnesium Technology 2011*, (2011) 107-112

11. Y. Murakami, N. Omura, M.G. Li, T. Tamura and K. Miwa "Effect of Injection Speed on Microstructure of AZ91D Magnesium Alloy in Semi-Solid Injection Process," *Materials Transactions*, 53 (2012) , 1094 - 1099

12. Y. Murakami, K. Miwa, M. Kito, T. Honda, N. Kanetake and S. Tada, "Effects of Injection Conditions in the Semi-Solid Injection Process on the Fluidity of JIS AC4CH Aluminum Alloy," *Materials Transactions*, 54 (2013) , 1788-1794

13. M. Itamura, C. P. Hong and J. M. Kim , "Development of New Semi-Solid Die Casting Method," *Journal of Japan Foundry Engineering Society* 77 (2005), 537-541 (in Japanese)

Shape Casting: 5th International Symposium 2014
Edited by: Murat Tiryakioğlu, John Campbell, and Glenn Byczynski
TMS (The Minerals, Metals & Materials Society), 2014

RELIABILITY-BASED CASTING PROCESS DESIGN OPTIMIZATION

Richard Hardin[1], K.K. Choi[1], and Christoph Beckermann[1]

[1]University of Iowa, Department of Mechanical and Industrial Engineering, 3131 SC, Iowa City, IA 52242-1527

Keywords: Casting Process Design, Optimization, Reliability-Based Design Optimization

Abstract

Optimum casting designs are unreliable without consideration of the statistical and physical uncertainties in the casting process. In the present research, casting simulation is integrated with a general purpose reliability-based design optimization (RBDO) software tool previously developed at the University of Iowa. The RBDO methodology considers uncertainties in both the input variables as well as in the model itself. The output consists not only of a reliable optimum design but also of the knowledge of the confidence level in this design. An example is presented where the design of a riser is optimized while considering uncertainties in the fill level, riser diameter, and the riser pipe depth prediction. It is shown that the present reliability-based method provides a much different optimum design than a traditional deterministic approach.

Introduction

Casting process simulation has become an invaluable tool in the production of economical and high performance cast components. Its application by experienced and knowledgeable operators leads to reduced castings defects, casting yield improvement, and reduced trial and error iteration in development of a casting's rigging. Increasingly casting simulation is being used as a collaborative tool between component designers and casting producers to reduce lead times, to develop casting friendly component designs, and to produce better castings. The majority of casting simulation is being used in a purely deterministic approach, replacing iterative trial-and-error process development on the shop floor with iterations on the computer. In this purely deterministic approach, the experience and knowledge of the engineer operating the software determines to a great extent that the software is used effectively, and that the casting process developed is the best it can be.

To maximize the effectiveness of casting simulation and improve the likelihood of an operator achieving an optimal solution, automatic optimization algorithms for casting process development are being researched [1-4], and commercial software such as MAGMAfrontier [5-8], OPTICast [9] and AutoCAST-X [3,10,11] have been developed. The most common application found in all of these automatic process optimization tools is the solution to the problem of casting feeding system optimization. In optimization of the casting feeding system, optimal sizes and locations of feeders are determined such that casting yield (ratio of mass of casting produced to mass of metal poured) is maximized and the desired quality level is met, which is typically defined as an absence of, or low level of, shrinkage porosity in the casting. Despite the power and promise of these developments in casting process optimization there are major shortcomings in these purely deterministic optimization approaches: neither the reliabilities of the casting production process nor the reliabilities of the casting model are considered. Uncertainties in the casting process conditions and variables, and in the casting model parameters and properties must be considered since these will affect the feasibility of the optimized solution. Casting process optimization where the feasibility of the process solution is not considered is termed Deterministic Design Optimization (DDO). In DDO, the probability of success or failure of the solution is not known, and based on the software operators' experience they must judge the feasibility of the optimized solution and make adjustments if needed. Here the authors present an optimization study for a casting feeding system using a general purpose *Iowa Reliability-Based Design Optimization* (*I-RBDO*) software [12-15]. By considering uncertainties in the casting production process and models in the *I-RBDO* software, not only does the software tool provide a reliable optimal casting process design, but it also provides a measure of the design's confidence level. The resulting reliability-based design optimization (RBDO) solution for the optimal casting feeding system will be compared with a DDO solution and a solution as might be developed by a casting simulation operator based on a riser piping safety margin.

Optimization Methods

Whether by DDO or RBDO methods, casting process design optimization inherently involves multiple variables, multi-objectives and multi-constraints. Many of the objectives conflict, such as achieving a porosity-free casting with the smallest size feeders. As such, the most successful multi-objective optimization algorithms developed for casting process optimization with conflicting objects have been the multi-objective evolutionary algorithms (MOEA) [1] or multi-objective genetic algorithms (MOGA) such as modeFRONTIER [16], which has been implemented in the casting process optimization software MAGMAfrontier [5-8]. As has been demonstrated in [6], given casting process and model variables (i.e. initial rigging, metal and mold properties, heat transfer coefficients (HTCs), pouring temperature and time, etc.), and given process constraints (i.e. porosity level, other defects, customer requirements, alloy, etc.), an optimized process can be determined to meet required objectives (i.e. maximum casting yield, required mechanical properties, etc.).

For reliability-based casting process optimization, both the multi-objective aspect of the problem and the uncertainties in casting process and model variables must be defined. In the *I-RBDO* software used in the current work, uncertainties and variations in the casting process variables and the casting modeling software variables and parameters are described via statistical distributions. Normal, Lognormal, Weibull, Gamma, Gumbel, and Extreme I and Extreme II distributions may be used in the *I-RBDO* software. Assuming the variations in the variables follow normal distributions, a standard deviation is sufficient to define the variability about a mean value. The desired confidence level, the probability that the optimized casting process solution will be successful, is also defined in the *I-RBDO* software. For example, a 95% confidence level for the RBDO solution would mean there is a 5% probability of failure.

Description of Example for Casting Process Feeding System Optimization

The example case study application presented here to compare the DDO and RBDO methods for casting feeding system design is shown in Figure 1. The casting is a 600mm long by 100 mm wide by 50 mm thick bar. A cylindrical-shaped riser is assumed as shown in green in the figure. The average porosity within the casting must less than 0.1%; this is the constraint. The riser volume is to be minimized; this is the objective function. Hence the casting yield is to be maximized while keeping the average porosity in the casting to a low level. The two design variables to be optimized are the radius R and height H of the riser. Since a sample-based optimization method is used, the sensitivities of the riser volume to the radius and height are not needed by the software. For the RBDO analysis the uncertainties in *R* and *H* must be defined. In this example problem, it is assumed that there is much less control in the process of filling the riser to a given height than to form the riser radial dimension during the molding process. The distributions for *R* and *H* to be used in the RBDO analysis are shown schematically in Figure 1. Here the uncertainty in *R* is defined by assuming it follows a normal distribution with a standard deviation of 3 mm, and to define the uncertainty in *H* a standard deviation of 10 mm is used. The results from the DDO and RBDO cases will be compared to a case termed here as "typical practice", where the riser diameter is set to the plate width and the riser height was determined using a 10 mm safety margin, defined as the distance between the end of the riser pipe and the casting cope surface.

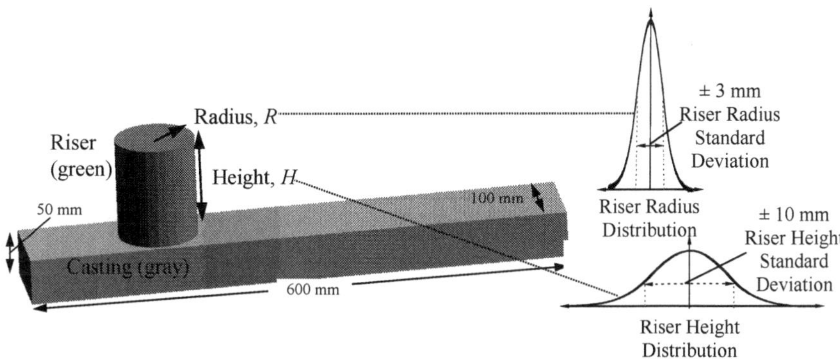

Figure 1. Casting (shown in gray) with dimensions and riser (shown in green) used in the casting feeding system design case study. Distributions for the riser radius and height assumed for the RBDO analysis are also shown.

The typical practice cases were run first to determine the shortest riser height that satisfies the margin of safety condition. Once *R* and *H* for the "Typical Practice" case were found, those dimensions were used as the starting point for the DDO analysis. For the DDO and RBDO analyses, search ranges for *R* and *H* were defined in the software as: 30 to 65 mm for *R*, and 60 to 190 mm for *H*. The DDO analysis was run first using the *I-RBDO* software and the commercial casting simulation package *MAGMAsoft* in an iterative fashion. Output from the *I-RBDO* software are the values of *R* and *H* that are to be simulated in the casting model. When the total casting porosity is determined from the casting simulation, for a given set of values of *R* and *H*, the porosity and riser volume are passed back to *I-RBDO*. Porosity and riser volume are the performance measure responses to the requested sets of variables *R* and *H*. The sequential quadratic programming algorithm was used in the optimization analysis for both the DDO and RBDO methods. The *I-RBDO* software uses normalized tolerances, and these were set to values recommended by its developers. In the DDO analyses the tolerances for the objective function and variables were set to 0.001, and the constraints to 0.05. In the RBDO analysis all tolerances were 0.05 as will be discussed below. The RBDO analysis used a sampling-based method with a dynamic kriging surrogate model.

Casting Process Feeding System Optimization Results

In Figure 2, sections are shown through the mid-width section of the casting and riser to visualize the feeding "shrinkage pipe" for the three approaches used to design the risers. The shrinkage pipe is the gray (empty) v-shaped region in the riser that transitions through the blue (porous) region to the white (sound) region. A low carbon steel with properties from the MAGMAsoft database was the cast metal used in these studies. This v-shaped shrinkage pipe forms as metal is drained from the riser to make up for solidification shrinkage in the casting. In Figure 2(a) the result is shown for the typical practice case with riser diameter equal to the plate width and using a 10 mm safety margin to determine the riser height. The resulting riser dimensions, riser volumes, average porosity predicted and casting yield are given for each feeding system design method in Figure 2.

A summary of results from the casting feeding system design studies for all three design methods is provided in Figure 3(a) for the riser dimensions and aspect ratios, and in Figure 3(b) for the casting yield and probability of failure results. The casting yield for the typical process safety-margin design approach is about 75% and its probability of failure was found to be 5.6%, as seen in Figure 3(b). This casting yield is relatively high for the steel foundry industry, where typical yields are in the 50% to 60% range. Because of this, the increase in casting yield for the optimized casting process is not as dramatic as it would be in applying optimization to most industrial casting feeding systems. Bearing this in mind, the optimized casting process solution

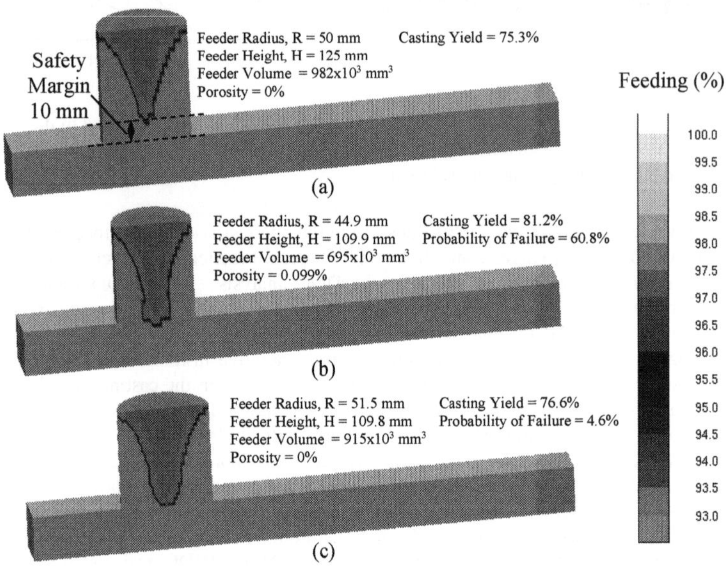

Figure 2. Mid-width sections showing riser pipes via the *MAGMAsoft* feeding percentage result for (a) the case run using a 10 mm safety margin to determine the riser height with riser diameter equal to plate width, (b) the resulting riser piping and results for the DDO analysis, and (c) the results and predicted riser piping for the RBDO case.

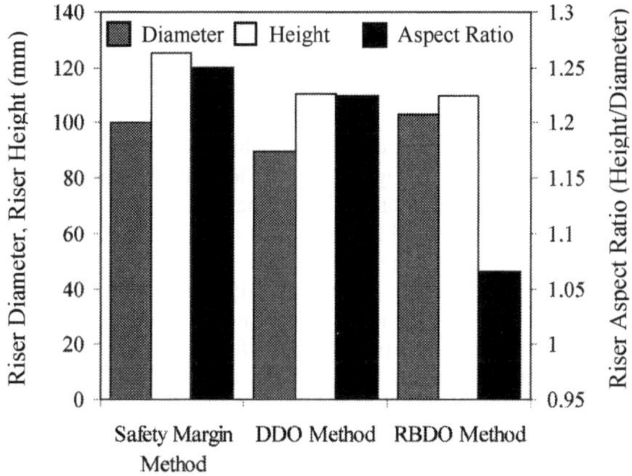

(a)

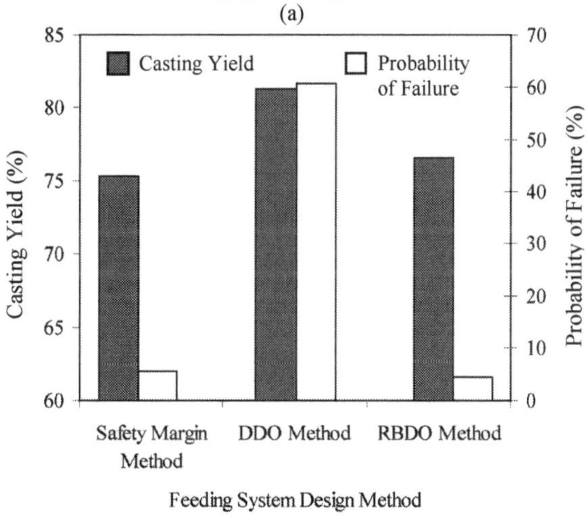

(b)

Figure 3. Summary results from casting feeding system design studies for the three design methods (a) riser dimensions and aspect ratio and (b) casting yield and probability of failure of design to meet the porosity constraint.

from the DDO method is shown in Figure 2(b) and in the summary Figure 3 by the middle bars. The DDO solution required 115 runs of the casting simulation software to determine the solution. In the DDO result, both R and H are markedly reduced from the safety margin method; R is reduced from 50 to 44.9 mm, and H is reduced from 125 to 109.9 mm. The DDO solution maximizes the casting yield shown in Figure 3(b) while keeping porosity in the casting to just less than 0.1%. The DDO solution has an average porosity of 0.099% whereas the safety margin method had 0%. The casting yield in the DDO solution is 81.2%, about a 6% increase over the safety margin method as seen in Figure 3(b). Qualitatively speaking, the shrinkage pipe of the DDO solution feeder in Figure 2(b) appears much flatter at the bottom than the safety margin one, and it has no margin for error that it might extend down into the casting and violate the constraint. Therefore it is not surprising that the probability of failure for the DDO solution was determined to be 60.8% as shown in Figure 3(b). The RBDO solution riser pipe and results shown in Figure 2(c) required an additional 125 casting simulations. The probability of failure of the RBDO solution is only 4.6%, which is much lower than that of the DDO solution and slightly lower than that of the safety margin method. The large probability of failure for the DDO solution is not surprising. DDO solutions are typically found to have a probability of failure in the neighborhood of 50% when analyzed using the *I-RBDO* software, according to its developers. The small difference in the probability of failure between the RBDO and the safety margin case is insignificant from a practical point of view. For the RBDO solution the casting yield decreases to 76.6%, which is 4.6% less than the DDO solution, and is 1.3% higher than the safety margin solution as seen in Figure 3(b). These results indicate that the 10 mm safety margin design approach gives a reasonably safe design, but that it is less economical than the RBDO solution. Clearly from Figure 3(b), the DDO method is offering a dramatic increase in casting yield (or decrease in riser volume), but it is not feasible.

Note that in Figures 2 and 3(a) for the RBDO solution the *H* is nearly identical to that in the DDO solution (109.8 versus 109.9 mm for RBDO and DDO, respectively), and *R* is increased from 44.9 mm in the DDO solution to 51.5 mm in the RBDO solution. The uncertainty for *H* is much larger than that for *R*, and to prevent the shrinkage from piping into the casting one might wrongly think that *H* should be increased to be sure it is large enough to prevent this. However, foundries know from experience that the radial dimension should be increased to prevent piping into the casting. The resulting RBDO solution is seen to agree with foundry practice and achieves its solution by increasing *R* to a large enough value that the solution is insensitive to a large change in *H*.

The aspect ratio (AR) of a riser is its height divided by its diameter. For top risers used in steel casting (the type of riser examined here) the AR is recommended to be at least 1, and for side risers (with contact to the side rather than the top of a casting) as large as 1.5. If the AR exceeds 1.5 there is no benefit to the riser's feeding effectiveness arising from the additional height and the additional wasted metal in the riser is uneconomical. In addition, secondary under-riser shrinkage may form in steel casting with ARs greater than 1.5. The AR results of this study are shown in Figure 3(a). The safety-margin approach gave an AR of 1.25, the DDO result was only marginally smaller with an AR of 1.22, while the AR for the RBDO solution was 1.07. As a result from the RBDO analysis, ARs closer to 1.1 for casting feeding systems with top risers can be used and appear to be more efficient and reliable.

The sets of *R* and *H* values requested by the *I-RBDO* software in the solution process are presented in Figure 4. Here, values for *R* and *H* selected for casting simulation runs by the *I-RBDO* software are plotted for the DDO analysis in Figure 4(a), and for the RBDO analysis in Figure 4(b). Lines in Figure 4 indicate the optimal solution values for *H* and *R* found for each variable and optimization method. It is evident from the sets of *R* and *H* values in Figure 4(a) that the software performs the DDO analysis searching throughout the allowable ranges of the variables until it homes in on the solution. Note, there are many simulation runs in the neighborhood of the DDO solution, indicating the tolerances in *I-RBDO* could probably have

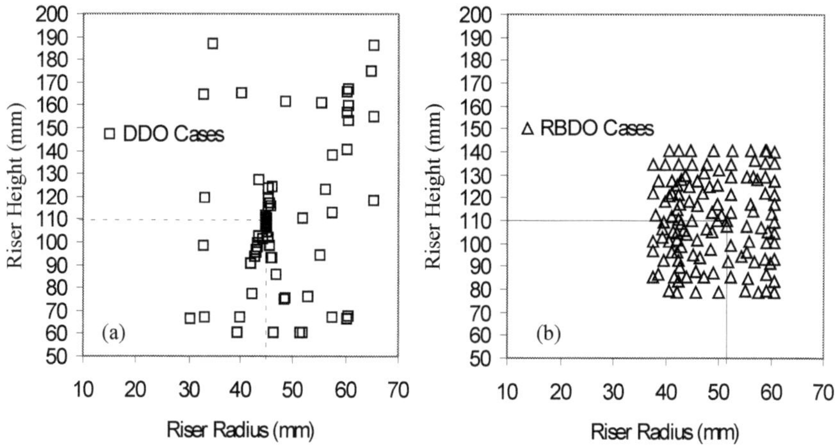

Figure 4. Values for *R* and *H* selected for casting simulation runs by the *I-RBDO* software as part of (a) the DDO analysis, and (b) the RBDO analysis. Lines indicate the optimal solutions found for each variable and optimization method.

been relaxed. For this reason, the RBDO tolerances for the objective function and variables were increased to 0.05 from the 0.001 tolerances used in the DDO analyses. This demonstrates the importance of gaining experience with the *I-RBDO* software and its parameters for a given problem in using it efficiently. In Figure 4(b) the RBDO sets of values appear to be more uniformly distributed in the search window. This is due to the construction of the surrogate model used in the reliability analysis.

Conclusions

Application of optimization methods to casting process design provides more than just optimal solutions. It provides an overview of possible solutions, some of which might be novel and innovative. It gives foundry engineers insight into the sensitivity and stability in both the actual process, and process models, to variables and parameters. Here reliability-based design optimization and casting simulation are integrated to go a step further in the development of optimization methods by including uncertainties in process and model variables, and determining an optimal solution with a known probability of success. For a typical approach to riser design using a safety margin of 10 mm, the probability of failure was 5.6% based on assumed uncertainties in riser height and radius. This probability of failure was found to be slightly greater than that from the RBDO method, which was 4.6%. The safety margin design approach gives a reasonably safe design, but is less economical than the RBDO solution, which had a 3% casting yield improvement over the safety margin approach. It has been demonstrated that a purely deterministic optimal solution offers a remarkable 6% increase in casting yield over typical design practice, but had an unacceptable 61% probability of failure. The advantages of the RBDO method are that its output consists not only of a reliable optimum design but also of the knowledge of the confidence level in this design. An additional insight from this study is that the RBDO solution determines, on its own, the practice followed by most foundries that increasing the radial dimension of risers, rather than the height, is the most reliable way to resize a riser that is not feeding a casting adequately.

References

1. J. Kor, X. Chen, Z. Sun, H. Hu, "Casting Design through Multi-Objective Optimization," *Proc. of the 18th IFAC World Congress 2011*, (2011), 6481-6486.
2. D.M. Liao, J.X. Zhou, L.L. Chen, X. Shen, and X.D. Gong, "Casting CAD/CAE Automatic Optimal Riser Design Technology," *Int. Journal of Cast Metals Research*, 24 (2011), 247-252.
3. M. Sutaria, D. Joshi, M. Jagdishwar and B. Ravi, "Automatic Optimization of Casting Feeders Using Feed-Paths Generated by VEM," ASME 2011 Int. Mech. Eng. Cong. and Expo., *Volume 3: Design and Manufacturing* (2011), 137-146.
4. R.S. Ransing, M.P. Sood, and W.K. Pao, "Computer Implementation of Heuvers' Circle Method for Thermal Optimisation in Castings," *Int. Journal of Cast Metals Research*, 18 (2005), 119-128.
5. MAGMA Gießereitechnologie GmbH, *MAGMAfrontier Module Manuals: Part One, Part Two and Reference*, (Aachen Germany, 2005).
6. I. Hahn and J.C. Sturm, "Autonomous Optimization of Casting Processes and Designs," World Foundry Congress, Hangzhou, China October 16-20 (2010), 1-20.
7. I. Hahn and G. Hartmann, "Automatic Computerized Optimization in Die Casting Processes," *Casting Plant & Technology*, No. 4 (2008), 2–14.
8. C. Tutum and J. Hattel, "State-of-the-Art Multi-Objective Optimisation of Manufacturing Processes Based on Thermo-Mechanical Simulations," Chapter 3 in *Multi-objective Evolutionary Optimisation for Product Design and Manufacturing*, L. Wang et al. (eds.), (London England: Springer-Verlag, 2011), 71-133.
9. Finite Solutions Inc., *OPTICast Training Course Workbook*, Version 8.1.1 (2011).
10. Advanced Reasoning Technologies, *AutoCAST Software User Manual and Case Studies*, http://www.autocast.co.in, (2009).
11. B. Ravi, "Casting Simulation and Optimisation: Benefits, Bottlenecks, and Best Practices," *Indian Foundry Journal*, 38 (2008), 47-52.
12. L. Zhao, K.K. Choi, and I. Lee, "Metamodeling Method Using Dynamic Kriging for Design Optimization," *AIAA Journal*, 49 (2011), 2034-2046.
13. H. Song, K.K. Choi, and D. Lamb, "A Study on Improving Accuracy of Kriging Models by Using Correlation Model Selection and Penalized Log-Likelihood Function," 10th World Congress on Structural and Multidisciplinary Optimization, May 20-24 Orlando, Florida, (2013).
14. I. Lee, K.K. Choi, Y. Noh, L. Zhao, and D. Gorsich, "Sampling-Based Stochastic Sensitivity Analysis Using Score Functions for RBDO Problems with Correlated Random Variables," *ASME Journal of Mechanical Design*, 133 (2011) , 021003.
15. I. Lee, K.K. Choi, and L. Zhao, "Sampling-Based RBDO Using the Stochastic Sensitivity Analysis and Dynamic Kriging Method," *Structural and Multidisciplinary Optimization*, 44 (2011), 299-317.
16. E. Rigoni and S. Poles, "NBI and MOGA-II, Two Complementary Algorithms for Multi-Objective Optimization," *Practical Approaches to Multiobjective Optimization*, (Dagstuhl Germany: Internationales Begegnungs- und Forschungszentrum für Informatik (IBFI), 2005), 1–22.

Shape Casting: 5th International Symposium 2014
Edited by: Murat Tiryakioğlu, John Campbell, and Glenn Byczynski
TMS (The Minerals, Metals & Materials Society), 2014

INFLUENCE OF PROCESS PARAMETERS ON BLISTERING DURING T6 HEAT TREATMENT OF SEMI-SOLID CASTINGS

Youfeng He, Xiaojing Xu, Fan Zhang, Daquan Li, Stephen Midson & Qiang Zhu

General Research Institute for Non-Ferrous Metals, No.2 Xinjiekou Wai Street, Beijing 100088, China

Keywords: Blister, Semi-solid thixo-casting, T6 heat treatment

Abstract

High pressures applied to the castings during solidification mean that semi-solid castings tend to be prone to surface blistering during subsequent T6 heat treatment. It is believed that the blistering originates from subsurface defects present in the semi-solid castings, which expand when exposed to high temperatures during the solution heat treatment. Despite the significance of blistering to the commercial development of the semi-solid casting process, there have only been limited quantitative studies of the impact of process parameters on blistering. This paper, therefore, will report on a study to examine the impact of a number of process parameters including intensification pressure, plunger velocity and solid fraction of the feed material on the blistering of semi-solid castings during T6 heat treatment. The location and average size of blisters formed at each condition have been measured and related to the casting conditions.

Introduction

Unlike all other casting processes, semi-solid casting utilizes a feed material that is about 50% solid and 50% liquid. This highly viscous feed material provides a greater level of control during die filling, which, together with the extremely high intensification (feeding) pressures applied as the casting solidifies, provide many benefits compared with fully liquid casting processes such as sand, permanent mold and die casting. These benefits include lower residual porosity levels, improved surface appearance, improved net-shape capability, faster cycle rates, and the ability to generate improved mechanical properties.

To maximize mechanical properties, however, it is necessary to T6 heat treat the semi-solid castings, but surface blistering can be a problem when the castings are exposed to high temperatures during solution heat treatment. So, semi-solid castings are often used in the T5 temper rather than T6 temper in order to avoid significant blistering on the surfaces of the castings, which limits one of the major advantages of the semi-solid casting process - the ability to generate better mechanical properties than competing casting processes.

It is believed that there are several possible causes for the formation of the blisters, including excessive die and plunger lubricants, and turbulent die filling. Midson [1] proposed a mechanism for blister formation in semi-solid castings. He noted that surface blistering is generally not a problem for components produced by gravity casting processes such as sand or permanent mold casting, as these castings are poured and solidify close to atmospheric pressure. Semi-solid castings, however, are produced using pressures close to 1000 bar. Midson suggested that during casting it is possible to trap defects (entrapped air or lubricants) below the casting's surface, and any gas entrapped in the casting will also be pressurized close to 1000 bar. During subsequent solution heat treatment, these gases will expand (while the strength of the aluminum weakens), and blisters can form at the surface of the castings. According to the combined gas law (see Equation1), p_1 is the pressure of the gas or air trapped in a defect, V_1 is its volume and T_1 is its temperature. Semi-solid castings are typically produced in a die preheated to a temperature of about 250°C at an intensification pressure of 1000 bar. During subsequent solution heat treatment, the casting is heated to a higher temperature, resulted in the increasing pressure of gas or air trapped within the defect. At high temperature, the gas or air pressure within a defect close to the surface of the casting will exceed the strength of the aluminum and the defect will expand, forming a blister.

$$\frac{p_1V_1}{T_1} = \frac{p_2V_2}{T_2} \qquad \text{Equation 1}$$

Despite the importance of minimizing blistering for the commercial development of semi-solid castings, there have only been a few published reports discussing the impact of process conditions on blistering. Kopper [2] reviewed the impact of different types of plunger lubricants on the blistering of semi-solid castings during T6 heat treatment, while He et al. [3] evaluated the impact of both plunger lubricants and die lubricants on blister formation, and also provided quantitative measurements of the size of the blisters.

The goal of this project was to examine the influence of several semi-solid process parameters on the propensity for blistering, including intensification pressure, plunger velocity and solid fraction of the feed material. Both the size and position of blisters formed on the surfaces of semi-solid castings after solution heat treatment were analyzed using quantitative measurements. In addition, a more detailed examination of the inside surfaces of the blisters was performed using SEM and EDS, with the objective of identifying the source of the blisters.

Experimental Procedures

The trials described in this study were performed on the step casting shown in Figure.1a. The step castings were produced using the thixocasting semi-solid (billet) process, and were cast using the 340-ton Buhler horizontal cold chamber die casting machine with a large-diameter shot cylinder modified for semi-solid casting described in previous papers [4,5]. The 319S alloy (Al-6%Si-3%Cu-0.35%Mg) feed material was produced using a commercial continuous casting process, with electromagnetic stirring used to generate the globular semi-solid structure. Slugs cut from this pre-cast feed material were reheated to the semi-solid temperature range using a 10-coil carousel-style induction heater (operating at about 1,000 Hz). The configuration of the runner and gating system used to produce the step castings is shown schematically in Figure1b.

The process parameters evaluated in this study included intensification pressure, plunger velocity and solid fraction of the feed material and the detailed testing parameters are listed in Table 1.

Table 1 Detailed testing parameters

Intensification pressure (bar)	415, 900, 1120
Plunger velocity (m/s)	0.12, 0.25, 0.50
Slug temperature (°C)	580, 590

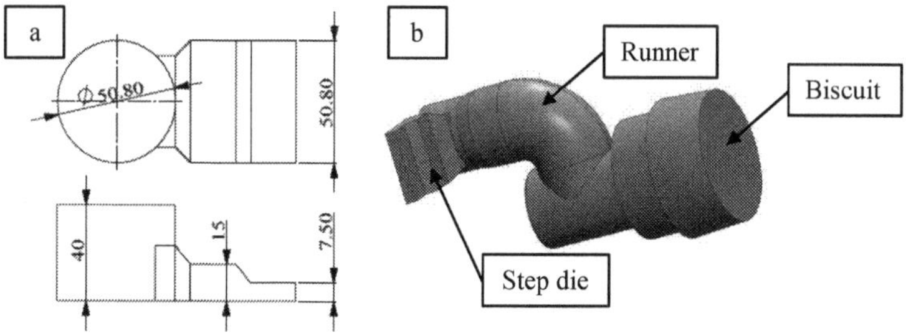

Figure 1 Semi-solid step casting using thixo casting process
a) The dimension of step casting; b) Schematic drawing of runner and gating system

All the castings were solution heat treated at 500°C for two hours and water quenched. Note that <u>all</u> the blisters generated in this study were produced during the solution heat treatment – no blisters were observed on the castings in the as-cast condition.

To provide a quantitative estimate of the impact of process parameters on the level of blistering, the size of the largest blister in each of five separate areas of the step casting was measured as shown in Figure 2, and the scoring system shown in Table 2 was used. The score from each of the five separate areas was averaged, providing an overall score for each processing condition evaluated. The morphology and chemical composition of the inner surfaces of blisters were analyzed by SEM and EDS.

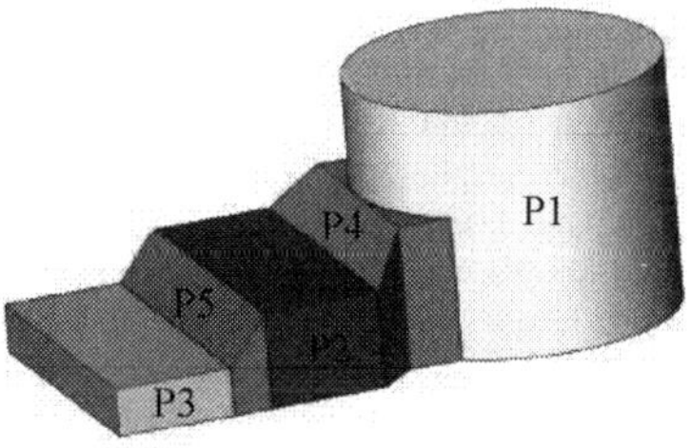

Figure 2 Five separate areas of step casting used to measure the blisters

Table 2 Method used to characterize the size of blisters

Blister Size	Score
< 0.1 mm	0
0.1 to 0.5 mm	1
0.5 to 1.0 mm	2
>1.0 mm	3

Results and Discussion

The first item to mention is that there was a large variation in the level of observed blistering, even for a series of castings produced under the same condition. For example, for castings produced using the same processing parameters, some castings had large (>1 mm) blisters after heat treatment, while other castings had only small blisters (<0.5 mm). Therefore, the results discussed in this paper are the average values for a number of castings produced under the same conditions (generally the results for between 5 and 10 separate castings have been averaged).

Table 3 summarizes the impact upon blistering of changing the intensification pressure for castings produced using a slug temperature of 580°C and plunger velocity of 0.50 m/s. The observation that the average size of the blisters became larger with increasing intensification pressure suggests that high intensification pressure is indeed causing the blisters. However, high intensification pressure maximizes the quality of the castings by helping to minimize shrinkage porosity, and it is normal to use an intensification pressure of close to 1000 bar for the production of high integrity semi-solid castings.

Table 3 Impact of the intensification pressure on the average blister size

Intensification Pressure (bar)	Plunger velocity (m/s)	Slug temperature (°C)	Blister score
413	0.50	580	0.20
900	0.50	580	1.13
1120	0.50	580	1.15

As the high intensification pressure of 1120 bar can improve the quality of the semi-solid castings, this pressure was used for the remainder of castings produced in this study. Table 4 shows the impact of plunger velocity on the average size of blisters for castings produced using a slug temperature of 580°C (which corresponds to a liquid fraction of about 46% [6]). The data in Table 4 indicates that the average blister size increased considerably with higher plunger velocities. Comparable data is shown in Table 5 for castings produced with a higher slug temperature of 590°C (about 58% liquid [6]), and in this case the blisters are smaller and show no relation to plunger velocity. This result is surprising, as it was expected that the die filling behavior would be more turbulent with the lower viscosity semi-solid metal produced at the higher slug temperature, resulting in a higher level of entrapped air and larger blisters. The reason for the larger blisters formed on the castings with the lower slug temperature is not fully understood at this stage.

Table 4 Impact of the plunger velocity on the average blister size at slug temperature of 580°C

Intensification Pressure (bar)	Plunger velocity (m/s)	Slug temperature (°C)	Blister score
1120	0.12	580	0.67
1120	0.25	580	0.83
1120	0.50	580	1.15

Table 5 Impact of the plunger velocity on the average blister size at slug temperature of 590°C

Intensification Pressure (bar)	Plunger velocity (m/s)	Slug temperature (°C)	Blister score
1120	0.12	590	0.30
1120	0.25	590	0.28
1120	0.50	590	0.36

Figure 4 shows SEM photographs of the inner surfaces of blisters from castings produced in this study. Visually the inner surfaces of the blisters appeared clean and shiny, and there were local smooth surfaces and dimple fracture morphology, caused by the expansion of air or gas entrapped in the defect during the solution heat treatment.

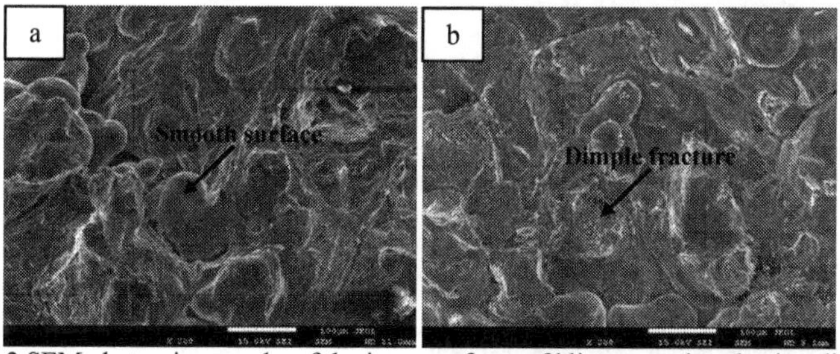

Figure 3 SEM photomicrographs of the inner surfaces of blisters produced using different plunger velocities (slug temperature 580°C and intensification pressure 1120 bar)
a) 0.12m/s; b) 0.25m/s

In comparison, Figure 5 shows an optical photomicrograph of the inner surface of a blister reported in Reference [3], from a casting produced using high levels of plunger lubricant. The inner surface of this blister was very rough and contained a dark-colored material, most likely produced from the plunger lubricant.

Figure 4 Inner surface of a blister produced using high levels of an oil-based plunger lubricant

Figure 5 shows an EDS analysis of the two types of blisters shown in Figure 3 and Figure 4. The blister produced using the high level of plunger lubricant contained relatively high levels of carbon and oxygen, while the blister produced in this study had essentially no carbon and only a low level of oxygen. This suggests that the blisters in this study were produced by entrapment of air, and not from lubricants applied to the die.

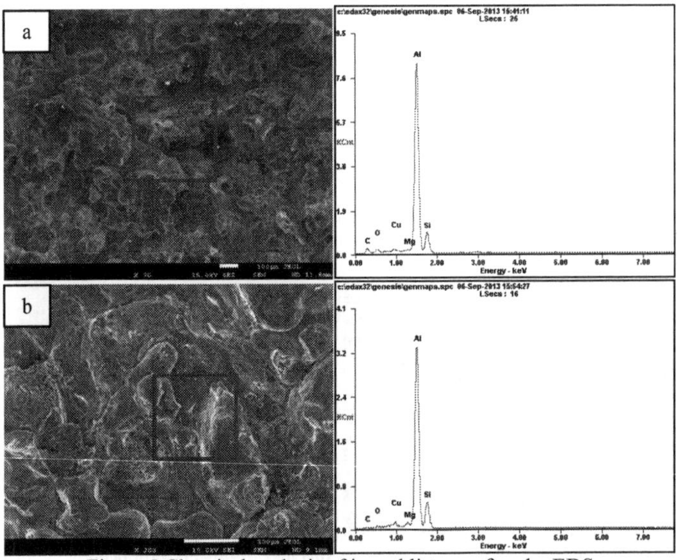

Figure 5 Chemical analysis of inner blister surface by EDS
a) Blister produced by heavy lubricant; b) Blister produced in this study

Conclusion

The results presented here confirm that blistering in semi-solid castings can originate from the entrapment of air, as well as from lubricants applied to the faces of the dies. The results from this study indicated:

1) With increasing of intensification pressure, the average size of the blisters became larger.
2) Plunger velocity has great influence on blister when the slug temperature was 580°C, however, there was little change of blister size when the slug temperature was 590°C

Acknowledgments

The authors would like to acknowledge financial support by the International Science and technology cooperation projects of China (2012DFA50300), by the National Science and Technology Support Program of China (2012BAE14G01) and by the Innovation Foundation of the General Research Institute for Nonferrous Metals (52611).

References

[1] Steve Midson, "Minimizing Blistering During T6 Heat Treating of Semi-Solid Castings," *Die Casting Engineer*, 2011
[2] Adam Kopper, "Die Casting Plunger Lubricant Success Story: T6 Heat Treatable Die Castings", 113th NADCA Metal Casting Congress, NV, Las Vegas, NV, 7-10 April, 2009, Paper number T09-023.
[3] Y.F. He et al., "Impact of Die and Plunger Lubricants on Blistering during T6 Heat Treatment of Semi-Solid Castings," 2013 NADCA Die Casting Congress, Louisville, KY, 16-18 September, 2013, paper number T13-012
[4] Q. Zhu, S.P. Midson, "Semi-solid moulding: a competition to cast and machine from forging in making automotive complex components," *Trans. Nonferrous Met. Soc. China*, 20 (2010), s1042-s1047.
[5] G. Wallace et al., "High-quality aluminum turbocharger impellers produced by thixo-casting," *Trans. Nonferrous Met. Soc. China*, 20 (2010), s1786-s1791.
[6] Daquan Li et al., "Evolution of Microstructure and Mechanical Properties of the Thixo-diecast 319s Alloy during Heat Treatment," *Materials Science Forum*, 765 (2013), 511-515.

Shape Casting: 5th International Symposium 2014
Edited by: Murat Tiryakioğlu, John Campbell, and Glenn Byczynski
TMS (The Minerals, Metals & Materials Society), 2014

INTEGRATED CASTING-HEAT TREATMENT PROCESS ROUTE FOR NEAR NET SHAPE ADI CASTING PRODUCTION

A. Meena[1], M. El Mansori[1]

[1]Arts et Métiers ParisTech – MSMP Laboratory ; Rue Saint Dominique, BP 508 ; Châlons-en-Champagne, 51006, France

Keywords: Austempered ductile iron (ADI); Integrated casting-heat treatment; Ausferrite; Near net shape.

Abstract

This paper focuses on developing an innovative process route for Austempered Ductile Iron (ADI) casting production. The innovative process route introduces an integrated approach towards casting and heat treatment practices for the production of near-net shape light-weight ADI casting in a permanent mould. It is based on the fundamental correlation between the production parameters and its combined influences on the microstructure to get the desired mechanical properties and performance in ADI castings. Casting and heat treatment practices are implemented efficiently in a control manner using thermal analysis adaptive system (melt quality) and fluidized bed heat treatment facility (controlled and uniform heat treatment) respectively to optimize the foundry practices for ADI production. The influence of austempering time on the microstructural characteristics, mechanical properties, and strain hardening behaviour of ADI was studied. Optical microscopy, scanning electron microscopy (SEM), and X-ray diffraction (XRD) analyses were performed to correlate the mechanical properties with microstructural characteristics. It was observed that the mechanical properties of resulting ADI samples were influenced by the microstructural transformations and varied retained austenite volume fractions obtained due to different austempering time. The results indicate that the strain-hardening behaviour of the ADI material is influenced by the carbon content of retained austenite.

Introduction

Austempered Ductile Iron (ADI) is an alloyed, and heat treated ductile iron. The attractive properties offered by ADI are accredited to a unique "ausferrite" microstructure that is induced by austempering heat treatment process [1-2]. During the heat treatment of ADI, either the ferritic or pearlitic ductile iron is first austenitized and then austempered [2]. Depending upon the austempering time two important stage reactions occurred during austempering process [3-5]. At the initial stage, the primary austenite decomposes into ferrite ( ) and high-carbon austenite ( ). If the austempering time is too long, then second stage reaction sets in, where high-carbon austenite further decomposes into ferrite and carbide. The stage II reaction is undesirable since it causes the embrittlement of structure and degrades the mechanical properties of ADI due to presence of carbides. During the stage I reaction, individual plates of ferrite separated from each other by thin layers of carbon saturated austenite nucleate at austenite grain boundaries and grow. As the reaction proceeds, the carbon diffusion ahead of the ferritic needle becomes more difficult and the growth of the ferrite plate ceases, resulting in an ausferrite matrix [6-7]. At this stage, the matrix is likely to have optimized ausferrite microstructure with no carbides.

During the last few years, industrial applications of ADI have grown enormously. This fact can be accredited to the improvement achieved in casting technology as well as in the heat treatment technology. Many researchers [8-11] in the past try to optimize the ausferrite microstructure by applying various processing technologies. While the unique ausferrite microstructure developed by the heat treatment of ductile iron provides many remarkable engineering properties to ADI, certain characteristics impose limitations on its widespread use. These include (a) requirement of defects free ductile iron casting with a good dimensional accuracy and surface finish for ADI production, (b) optimization of heat treatment parameters, (c) saving energy and time during its large-scale industrial production, and (d) achieve better combinations of strength and ductility properties. To solve these problems a new processing technology was developed for the direct manufacturing of ADI using the continuous casting-heat treatment process using a permanent mould. In this technology, the ductile iron samples obtained using the permanent moulds are first austenized during the post solidification stage followed by austempering heat treatment in the fluidized bed and then air cooled at room temperature to get ADI material. The influence of austempering time on the microstructural characteristics, mechanical properties and strain hardening behavior of ADI samples was studied. Optical microscopy, scanning electron microscopy (SEM) and X-ray diffraction (XRD) were performed to correlate the mechanical properties with microstructural characteristics.

Experimental procedure

Material

The melting was conducted in an induction furnace with a holding capacity of 100 kg, power of 120 kW and a frequency of 50 Hz. The charge materials were steel scraps, iron, carbon additive, and ferrosilicon. The melt was initially superheated up to 1773 K (1500 °C) and alloying elements are added to the charge materials. At 1723 K (1450 °C) the melt was treated with FeSi-Mg (Ferro Silicon Magnesium) alloy using tundish cover treatment ladle for the spherodization of the melt followed by inoculation process using the ferrosilicon based alloyed to increase the nodule count [12-13]. The melt then followed into the permanent mould, and the casting process started. The permanent mould used in the present study was made up of gray cast iron (high thermal fatigue resistance) [12]. The permanent mold was used to meet the challenges presented by the modern manufacturing industry: higher production rate of near net shape ADI castings and the in-situ heat treatment for austempering [12]. Each specimen is in the shape of rectangular block with size of 182mm ×29mm ×16mm. The mould was pre-heated in the temperature range of 448-498 K (175-225 °C) before the melt was poured into the cavity. The permanent mould was preheated to avoid the gas absorption through dislocations of steams and contact surface [14]. The mould cavity was coated with thick silica coating to provide an insulating barrier between the liquid metal and permanent mould and to avoid sticking of the molten metal to mould [15]. The production process was performed by maintaining a regular time interval of 150-240 s for each production cycle (one cycle produce two specimens) including casting and ejection processes. The temperature of the mould and ejected casting was measured by using K-type thermocouples located at selected locations. The chemical composition of the obtained samples for each austempering time is determined by using the X-Ray Fluorescence (XRF) method. The mould temperature was maintained by using flame heaters, and the temperature is controlled by using forced-air cooling.

Heat treatment

Specimens were produced by using a continuous casting-heat treatment process [12-13, 16-17]. In this process, the ductile iron samples obtained by using a permanent mould are first austenitized during the post solidification stage followed by austempering heat treatment in fluidized bed and then air cooled at room temperature to get ADI material. The ductile iron samples were then taken out of casting in the temperature range of 1273-1373 K (1000-1100 °C) and put in a muffle furnace for austenitization. The austenitization treatment was then carried out at 1203 K (930 °C) for 90 min [12]. After austenitization, the specimens were quenched down to the temperature range of 673-773 K (400-500 °C) in the first fluidized bed furnace (at room temperature) to get a controlled and average quenching rate of 7 °C/s. The quenching was done in a fluidized bed at room temperature to avoid the pearlite region [17]. The specimens were then austempered in second fluidized bed for different austempering time of 60, 90, 120 and 150 min at 653 K (380 °C) and then air cooled. For each austempering time four specimens were produced. Fluidized bed technology is being increasingly used in the heat treatment industry, due to the increased energy efficiency, more uniform temperature distribution and reduced treatment time [18]. The average quenching rate (7 °C/s) was obtained by performing quenching rate experiments on the cylindrical ductile iron sample (diameter: 32 mm) under the same experimental conditions using the K-type thermocouple.

Characterization methods

The microstructural characterization of the samples was carried out at three different sections (bottom, middle and top) of the samples. The obtained samples were sectioned using wet abrasive cutting method at different sections and mounted in resin for wet grinding and polishing to prepare the samples for metallographic analysis. Grit reductions of 180, 400, 600, 800, and 1200 silicon carbide are used. A final surface polish of 3 μm and 1 μm grain size was also performed. The polished surfaces are then etched with 4% Nital (96-98 mL ethanol and 2-4 mL nitric acid (HNO3)) for 2-5 seconds at room temperature. Microstructures were observed by optical microscope (OM) and scanning electron microscope (SEM). The graphite nodule counts at different zones were determined on the un-etched sample surface by taking the average of three different regions (100X) at each section (bottom, middle and top) of the samples using an optical microscope. The retained austenite volume fractions and its carbon content was measured by X-ray diffractometry method. XRD analysis was done using a monochromatic Co Kα radiation at 30 kV and 30 mA. An Inel CPS-120 X-ray diffractometer was used for the analysis.

Results and discussion

Microstructure

Based on average graphite nodule counts and average graphite nodule size, two different zones exhibiting slightly different microstructures were observed. First is the outer zone at the outer surface which is approximately 0.5-1 mm thick and second is the centre zone which is at the interior of the samples. Image analysis of optical micrographs of un-etched samples at different sections (bottom, middle and top) gave an average graphite nodule density in the range of 900 ± 150 nodules/mm^2 at centre zone with a mean graphite nodule size in the range of 13-17 μm whereas average graphite nodule density at the outer zone is in the range of 1300 ± 160 nodules/mm^2 with a mean graphite nodule size in the range of 5-10 μm. The variations in the average graphite nodule counts and its average size in the resultant microstructure can be explained due to the distinctly different cooling rates during solidification at the centre and outer zones. The outer zone is in contact with the metallic mould, solidified rapidly resulting in higher nodule counts with smaller size compared with those in the interior zone of the casting.

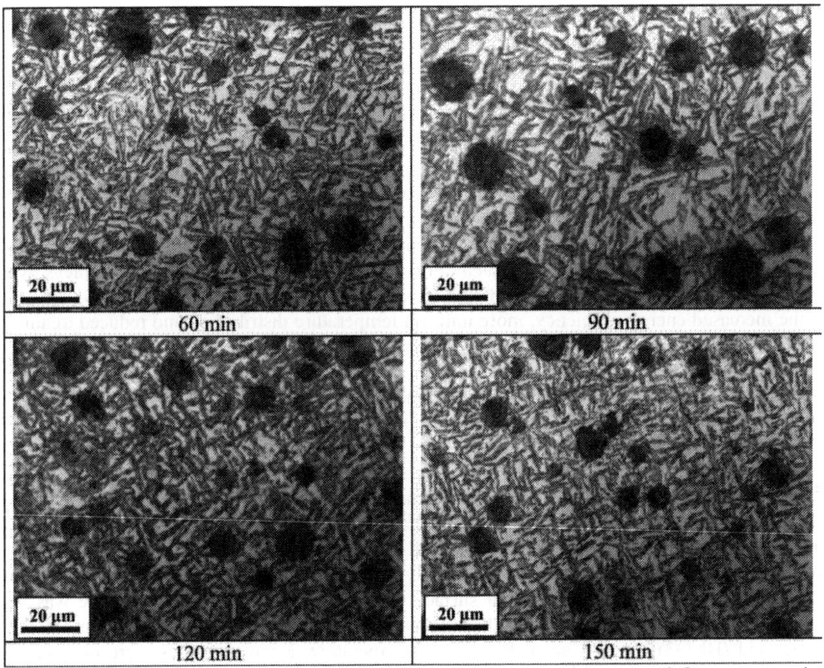

Figure 1. Optical micrographs of resultant ADI microstructure austempered for austempering time.

Due to the in-situ heat treatment of ductile iron samples, the obtained microstructure consists of a matrix having dark needle-shaped ferrite and bright austenite with graphite nodules embedded in it. The dark needle-shaped ferrite and bright austenite between the needles in the micrograph constitute an ausferrite matrix, and the bright bulky region is untransformed austenite volume [2, 19]. Different morphologies of austenite, ferrite and graphite nodule were observed based on the zones (centre and outer) and austempering time. The austempering time during the heat treatment process of ductile irons play a critical role in the development of microstructural characteristics of ADI and its mechanical properties. The combined effect of austempering temperature (653 K (380 °C)) and times (60, 90, 120, and 150 min) on the obtained ADI microstructures are shown in Fig. 1. The coarse and feathery characteristic of ferrite was found on increasing the austempering time.

Austempering kinetics

The austempering time have a marked influence on the hardness, retained austenite volume fraction and its carbon content as shown in Fig.2a. All these microstructural characteristics can be further used to gain information regarding the kinetics of stage I and stage II reactions during austempering process. The change in measured properties (retained austenite volume content, carbon content of retained austenite, ferrite cell size, etc.) during the initial and later stages of austempering reflects the progress of stage I and II reactions [20].

The austempering kinetics was also followed by observing the hardness alteration of the obtained ADI samples as a function of austempering time. The influence of austempering time on the hardness values of resulting ADI samples are shown in Fig. 2b.

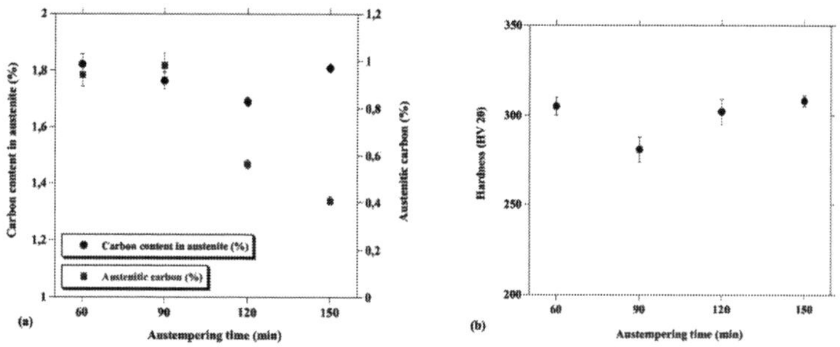

Figure 2. Influence of austempering time on the (a) Carbon content of the retained austenite (wt%) and (b) Hardness (HV20).

It can be observed that the hardness values show decreasing behavior for shorter treated time (60-120 min) and after that they show increasing behavior with austempering time (120-150 min). It was observed that as the austempering time increases, more of the softer phase of high-carbon ausferrite formed and the average matrix hardness decreases (**Fig. 2b**). This continues until the onset of stage II of the reaction, after which the hardness is expected to rise slightly due to the formation of carbides during stage II reaction. In the present study, the austempering kinetics (stage I) is further accompanied by an increase in the retained austenite volume fraction in the range of 60-90 min. The stage II reaction of the austempering kinetics is explained by a decrease in the retained austenite volume fraction in the range of 120-150 min.

Conclusions

A new processing technology was investigated for the direct manufacturing of light weight ADI casting using a continuous casting-heat treatment process. The microstructural alterations and its correlation with mechanical properties and strain hardening behavior of the obtained ADI samples with respect to different austempering time were studied. Based on the results obtained, the following conclusions can be drawn:

- The continuous casting-heat treatment process leads to the formation of ausferrite matrix (needle-shaped ferrite and high-carbon austenite with graphite nodules embedded in it) in the obtained ADI samples.
- The variations in measured properties (retained austenite volume content, carbon content of retained austenite, and ferrite cell size) along with hardness alteration indicate the occurrence of stage I reaction in the austempering time range of 60-90 min whereas stage II reaction occur in the austempering time range of 120-150 min.
- Higher retained austenite volume fractions at lower austempering time leads to the lower strength values but higher elongation %. On the other hand, lower retained austenite volume fractions at higher austempering time leads to the higher strength values and lower elongations %.

References

1. L.C. Chang, "Carbon content of austenite in austempered ductile iron," *Scripta Materialia*, 39 (1) (1998), 35-38.
2. F. Klocke, M. Arft, and D. Lung, "Material-related aspects of the machinability of Austempered Ductile Iron," *Production Engineering*, 4 (5) (2010), 433-441.
3. J. Achary, and D. Venugopalan, "Microstructural development and austempering kinetics of ductile iron during thermomechanical processing," *Metallurgical and Materials Transactions A: Physical Metallurgy and Materials Science*, 31 (10) (2000), 2575-2585.
4. U. Batra, S. Ray, and S.R. Prabhakar, "Effect of Austenitising Temperature on Microstructure and Wear Properties of Low Carbon Equivalent Austempered Ductile Iron," *Journal of Materials Engineering and Performance*, 12 (5) (2003), 597-601.
5. Y. Amran, A. Katsman, P. Schaaf, and M. Bamberger, "Influence of Copper Addition and Temperature on the Kinetics of Austempering in Ductile Iron," *Metallurgical and Materials Transactions B*, 41 (5) (2012), 1052-1058.
6. A. Trudel, and M. Gagne, "Effect of composition and heat treatment parameters on the characteristics of austempered ductile irons," *Canadian Metallurgical Quarterly*, 36 (5) (1997), 289-298.
7. P.A. Blackmore and R.A. Harding, "The effects of metallurgical process variables on the properties of austempered ductile irons," *Journal of Heat Treating*, 3 (4) (1984), 310-325.
8. A.A. Nofal, and L. Jekova, "Novel processing techniques and applications of austempered ductile iron," *Journal of the University of Chemical Technology and Metallurgy*, 44 (3) (2009), 213-228.
9. J. Achary, "Tensile properties of austempered ductile iron under thermomechanical treatment," *JMEPEG*, 9 (2000), 56-61.
10. J.F. Janowak, and R.B. Gundlach, "Approaching austempered ductile iron properties by controlled cooling in the foundry," *Journal of Heat Treating*, 4 (1) (1985), 25-31.

11. J. Yang, and S.K. Putatunda, Improvement in Strength and Toughness of Austempered Ductile Cast Iron by a Novel Two-Step Austempering Process," *Materials and Design*, 25 (3) (2004), 219-230.
12. A. Meena, and M. El Mansori, "Drilling performance of green Austempered Ductile Iron (ADI) grade produced by novel manufacturing technology," *International Journal of Advanced Manufacturing Technology*, 59 (1) (2012), 9-19.
13. A. Meena, and M. El Mansori, "Study of dry and minimum quantity lubrication drilling of novel austempered ductile iron (ADI) for automotive applications," *Wear*, 271 (9-10) (2011), 2412-2416.
14. P. Beeley, "*Foundry technology, 2nd edition*," Butterworth-Heinemann, Oxford, 2001, pp. 443-622.
15. S.V. Shepel, and S. Paolucci, "Numerical simulation of filling and solidification of permanent mold castings," *Applied Thermal Engineering*, 22 (2) (2002), 229-248.
16. A. Meena, M. El Mansori, and P. Ghidossi, "Machinability of austempered ductile iron (ADI) produced by integrated green technology of continuous casting and heat treatment processes," *AIP Conference Proceedings*, 1315 (2010), 1521-1526.
17. A. Meena, M. El Mansori, P. Ghidossi, and A. Mkaddem, "Anti-friction coating for drilling of green Austempered Ductile Iron (ADI) grade," *AIP Conference Proceedings*, 1353 (2011), 1800-1805.
18. R.E. Haimbaugh, "Practical Induction Heat Treating", *ASM International*, 2001, 83-119.
19. S. Zhou, K. Zhang, Y. Wang, J.F. Gu, and Y.H. Rong, "The Mechanism of High Strength-Ductility Steel Produced by a Novel Quenching-Partitioning-Tempering Process and the Mechanical Stability of Retained Austenite at Elevated Temperatures," *Metallurgical and Materials Transactions A*, 43 (3) (2012), 1026-1034.
20. B. Bosnjak, B. Radulovic, K. Pop-Tonev, and V. Asanovic, "Influence of microalloying and heat treatment on the kinetics of bainitic reaction in austempered ductile iron," *JMEPEG*, 10 (2001), 203-211.

Shape Casting: 5th International Symposium 2014
Edited by: Murat Tiryakioğlu, John Campbell, and Glenn Byczynski
TMS (The Minerals, Metals & Materials Society), 2014

SWAGE CASTING OF A356 (AlSi7Mg0.3) ALLOY

Huseyin Murat Lus[1]

[1]Yildiz Technical University, Department of Metallurgical and Materials Engineering, Davutpasa, Istanbul 34210, Turkey

Keywords: Swage Casting, Squeeze Casting, Semi-solid Casting, A356 Alloy

Abstract

Swage casting, a new semisolid metal processing technology, has been developed in Balkan Center for Advanced Casting Technologies (BCACT) for manufacturing near-net shape components from leightweight metals. Components with one rotating axis such as a cylinder can be produced on a swage casting machine from molten metal in a one-step operation. The present study was undertaken to investigate the effect of intensive shearing action on the morphological evolution of as swage-cast A356 (AlSi7Mg0.3) alloy. Compared with conventional squeeze casting, the experimental results show that intensive melt stirring during swage casting, promotes the formation of net-globular and rosette-shaped grains.

Introduction

Cast Aluminum alloys and especially Aluminum-Silicon alloys are widely used in engineering stuructures and components for many decades. Today large percentage of Al castings part are produced using High-pressure Die-casting Process (HPDC) due to low production costs, near-net shape production and good surface finish [1]. However, there are some casting defects such as gas porosity and micro-voids in HPDC parts which prevents from the application of high-safety applications [2]. To compensate these drawbacks, new casting techniques have been developed such as semi-solid casting, squeeze casting, etc. Swage casting is a general term to specify a fabrication technique which combines advantages of squeeze, centrifugal and semi-solid casting Methods. In this new casting method, at the beginning liquid alloy is poured into rotational already heated lower die. As a result of the high-speed centrifugal action, the melt rises and covers the inner surface of lower die. Due to chill effect the melt starts to solidify partly on the surface of lower die. Then, immediately the upper die is lowered and partly solidified alloy is squeezed. During both squeezing and centrifugal actions partly solidified alloy is sheared and also the static upper die acts as chill. As a result of this, dendrites in the middle of the thin section between upper and lower dies break into small particles [3]. This work presents results of the study on microstructures of a A356 (AlSi7Mg0.3) alloy fabricated by squeeze and swage casting

Experimental Procedures

A356 alloy family is the most popular alloy used in squeeze casting and semi solid metal processing and used in automative components such as cast wheels, pump bodies, cylinder blocks, etc [4]. The raw material used in this study, was a commercially purchased A356 alloy and had the composition shown in Table 1. Because of relatively broad freezing range, this alloy exhibits some degree of porosity which affects ductility and fatigue resistance of cast parts.

Table 1. Chemical Composition of A356 Alloy used in this study and nominal solidification range.

Element	Si	Fe	Cu	Mg	Zn	Al
Content (wt%)	7.35	0.413	0.22	0.32	0.06	Rest.
Solidification Range [5]	555-615 °C					

The mould used in swage casting is a bomb body like shape and is shown in the figure 1. The inner and outer diameters at the middle of this mould are 48 mm and 67 mm respectively whereas the height of it is 200 mm. In both cases, first of all, the lower and upper dies were heated up to 180 °C to increase the fluidity of alloy and for swage casting while the lower die was rotating around the central axis of the casting machine with 500 rpm, the molten aluminum was poured into the die cavity. Then immediately liquid alloy was squeezed under 10 MPa approximately for 10 s.

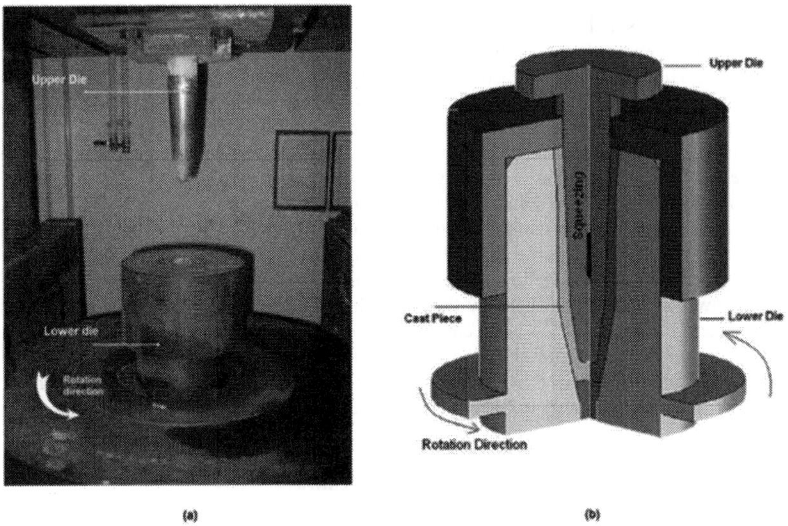

Figure 1. (a) Swage casting machine and die assembly (b) Schematic illustration of swage casting process and solid shape of cast piece.

Same amount of liquid metal was cast with under 10 MPa pressure without rotating lower die. Casting conditions used for each method are given in the Table 2.

Table 2. Process parameters

Casting Method	Squeezing Pressure (MPa)	Squeezing Time (s)	Mould Temperature (°C)	Pouring Temperature (°C)	Lower Mould Speed (rpm)
squeeze	10	10	180	735	-
swage	10	10	180	745	500

The specimens for optical microscope were prepared mainly sampling, grinding, polishing and etching with Keller's reagent.

The microstructure investigation and Shape factor measurements were carried out using LEICA image analyzer. The shape factor used in this study was the inverse of sphericity and calculated as;

$$\text{Shape Factor} = \frac{p^2{}_{\alpha}}{4\pi A_{\alpha} 1.064} \tag{1}$$

Where Aα and Pα represent the area and perimeter of the primary phase of the microstructure and 1.064 is the correction factor. The value of shape factor is 1.0 in the case of a perfect spheroid while the higher the shape factor, the higher the dendritic morphology [6]. The measured globule was the size of the primary phase in the microstructure.

Results

For comparison, corresponding microstructure of the squeeze-cast (a) and swage-cast (b) samples can be seen in Figure 2. In the squeeze casting case, the structure presents homogeneous grain size and very refined dendrite morphology surrounded by the eutectic. This is attributed to higher heat transfer coefficient as a result of the better contact between the metal and the die surface during solidification improved heat transfer across the metal/die interface [7]. It is also worth mentioning that the effect of pressure reduced the shrinkage and gas porosity as can be seen in this figure. The average secondary dendrite arm spacing in squeeze casting is 17μm at the center of cast piece. The similar results were reported by Lima *et al* [8] and Lynch *et al* [9].

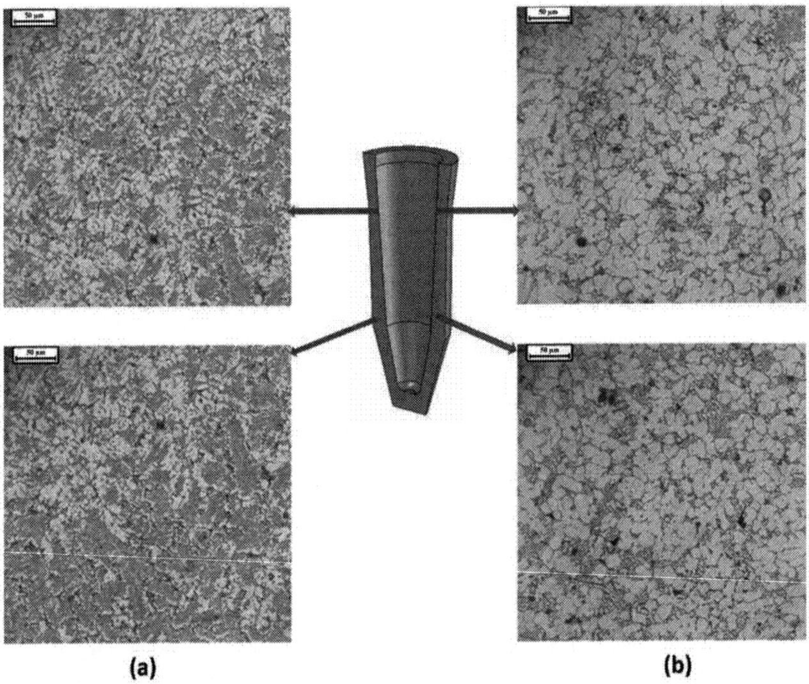

Figure 2. Optical micrographs of the middle section of cast pieces taken from bottom and upper region, (a) squeeze cast and (b) swage cast samples.

On the other hand, in the swage casting, final microstructure is fine dendritic only at both surfaces but spherical and rosette shapes between them (Figure 3). It is observed that the α-Al phases have spheroidized and coarsened to some extent, as compared with squeeze-cast alloy. This can be explained with shear action occurred during rotation. Firstly liquid metal solidifies as dendritic at the both surfaces due to chill effect. Then growing dendrites from both surfaces break into small particles in the middle of thin section because of high rotational speed of lower mould. Some of them remain as spherical particles whereas some of them transform to rosette shapes depending on the shear rate and solidification time [10]. The shearing occurs in the semi-solid region affects the efficiency of the globularisation, It may be said that the higher the shear rate occured at the center, the more spherical the grain became.

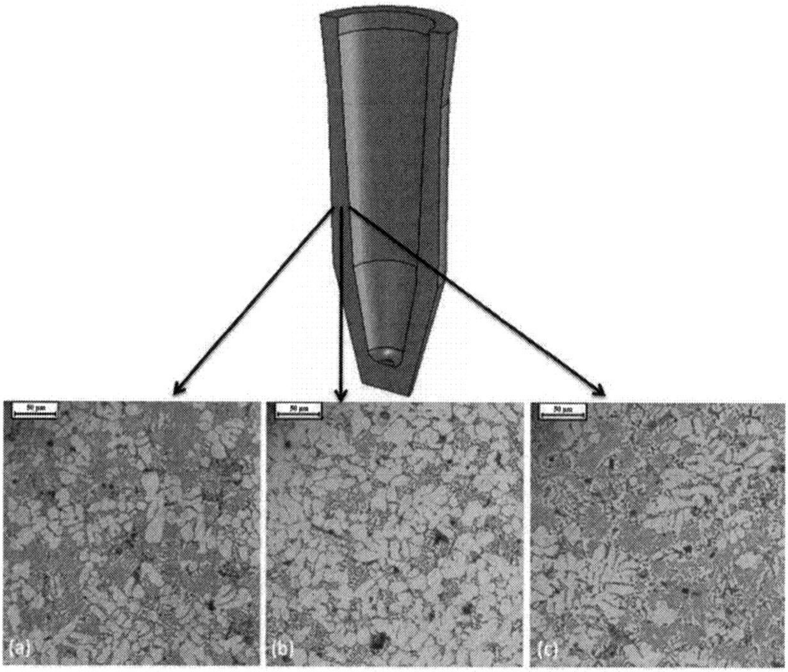

Figure 3. Typical microstructure variation across the thickness of the swage cast specimen from the (a) outer surface to the (b) center and to the (c) inner surface.

Figure 4 shows the calculated average diameter of α-Al particles and shape factors. Swage cast part reveals that the shape factor of the samples was decreased with the increase of distance from outer surface toward the center of cast part. The minimal shape factor value presented in this work is 1.8 at along with the center of swage-cast piece. It is also worth to note that solidification is a function of the solid/liquid interfacial energy, the system tries to reduce excess surface area to the minimum possible [11,12]. This phenomenon may helped the coarsening and the globularisation mechanisms to some extent.

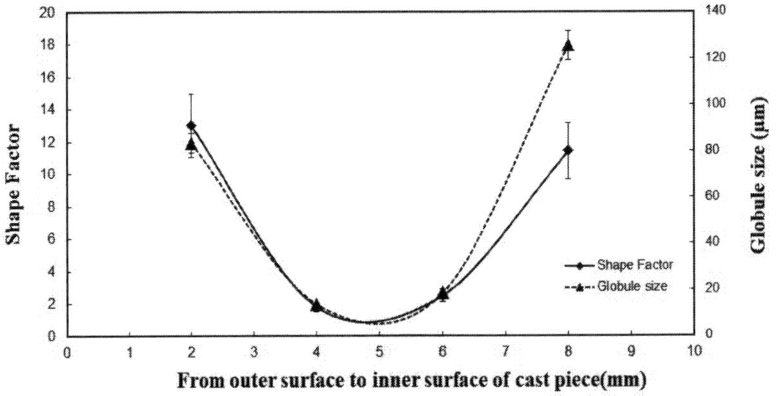

Figure 4. Variations of shape factor and globule (rosette) size with thickness of cast part from outer surface to inner surface. Perfect globule = 1. (Thickness of cast part is ≈ 10 mm)

Summary

In this study, A356 die casting alloy is cast both squeeze and swage casting techniques. The satisfying non-dendrite microstructure can be obtained by swage casting method and globularisation mechanism appears to produce spherical grains with 1.8 average values of shape factor at the center of cast piece. In the swage-casting process, the shearing effect is the most probable cause of the promotion of the formation of spherical α-Al phases throughout center of cast piece.

Acknowledgement

The author wishes to thank Yildiz Technical University, BCACT (Balkan Center for Advanced Casting Technologies) for their help in experiments.

References

1. Z. Fan, X. Fang and S. Ji, "Microstructure and mechanical properties of rheo-diecast (RDC) aluminium alloys," *Materials Science and Engineering* A 412 (2005), 298–306

2. M. R. Ghomasci and A. Vikhrov, "Squeeze casting: an overview," *J. Mater. Process. Technol.*, 101 (2000), 1–9.

3. H.M. Lus, A. Turkeli, N.G. Kinikoglu, "Swage Casting of A380 Alloy,"*Materials and Design*, 32 (2011), 3570–3577.

4. Diran Apelian, *Aluminum Cast Alloys: Enabling Tools for Improved Performance*, (World Wide Report, published by NADCA, 2009)

5. J.R. Davis, *Metals Handbook: Properties and Selection: Nonferrous Alloys and Special-Purpose Material. Vol: 2* (Ohio, ASM International, 1990)

6. M. Paes, E.J. Zoqui, "Semi-solid behavior of new Al–Si–Mg alloys for thixoforming," *Materials Science and Engineering A*, 406 (2005), 63–73

7. A. Maleki, B. Niroumand, A. Shafyei, "Effects of squeeze casting parameters on density, macrostructure and hardness of LM13 alloy," *J. Mater Sci Eng A*, A428 (2006), 135–140.

8. C.S. Lima, A.J. Clegg, N.L. Loh, "The reduction of dendrite ARM spacing using a novel pressure-assisted investment casting approach", *Journal of Materials Processing Technology* 70 (1997), 99-102

9. R.F Lynch, R.P. Olley, and P.C.J Gallagher, "Squeeze Casting of aluminum" *AFS Transactions*, 83 (1975), 755-60.

10. M.C. Flemings, "Behavior of metal alloys in the semi-solid state," *Metall Transactions*, 22A (1991), 957–981.

11. D. Apelian, S. Shivkumar and G. Sigworth, "Fundamental Aspects of Heat Treatment of Cast Al-Si-Mg Alloys" *AFS Transactions*, 97 (1989), 727-742.

12. M.T. Abou El-khair, "Microstructure characterization and tensile properties of squeeze-cast AlSiMg alloys" *Materials Letters*, 59 (2005) 894– 900.

Shape Casting: 5th International Symposium 2014
Edited by: Murat Tiryakioğlu, John Campbell, and Glenn Byczynski
TMS (The Minerals, Metals & Materials Society), 2014

Additive Manufacturing Supports the Production of Complex Castings

A. Druschitz[1], C. Williams[2], D. Snelling[2], M. Seals[3]

[1]Virginia Tech, Department of Materials Science and Engineering, 213 Holden Hall, 445 Old Turner Street, Blacksburg, VA 24061, USA
[2]Virginia Tech, Department of Mechanical Engineering, 114F Randolph Hall, 460 Old Turner Street, Blacksburg, VA 24061, USA
[3]ATI-Allvac, 2020 Ashcraft Ave., Monroe, NC 28110 USA

Keywords: additive manufacturing, binder jetting, vat photopolymerization, fused filament fabrication, sand casting, investment casting

Abstract

Additive manufacturing is being used in a variety of ways to support the production of complex castings. Some of the common additive manufacturing processes include fused filament fabrication, vat photopolymerization, powder bed fusion, binder jetting, and material jetting. In this paper, the authors discuss the use of (i) binder jetting technology to fabricate sand molds for casting complex, cellular structures and (ii) fused filament fabrication & vat photopolymerization to produce complex investment casting patterns. Binder jetting of foundry sand molds allows the realization of cast structures that are impossible to mold using conventional methods. The structures are lightweight, multi-functional and may provide exceptional blast protection. With regards to investment casting, wax is currently the primary material used for producing expendable patterns due to a desirable combination of thermal expansion, thermal conductivity and melting point. However, wax is not a typical printed material. A variety of polymers are available for additive manufacturing and, as would be expected, only a few are suitable for use as expendable patterns for investment casting. The best polymers for use as expendable patterns for investment castings are PMMA, epoxy resin containing a reactive diluent and ABS.

Introduction

In today's fast paced, rapidly changing, economically sensitive market place, the ability to reduce product development time, and thus, rapidly introduce new products, at low cost is critical. Additive manufacturing is meeting this challenge. Additive manufacturing has been used in the casting industry for many years to quickly produce full-size models that engineers can examine and determine if their math models are correct and accurate. The casting industry quickly realized that these models could also be used as patterns to rapidly produce prototype castings, thus the phrase "rapid prototyping" was used in the foundry industry rather than "additive manufacturing". To produce a "rapid prototype", a computer-aided design (CAD) model is required, which was not the norm in pre-1990's manufacturing. But, times have changed, and most everything is designed (modeled) on computers today. With the change in the availability of CAD packages, the use of rapid prototyping has grown quickly and a wide variety of additive manufacturing techniques have been developed. The purpose of this paper is to describe how the shaped casting industry can use this "disruptive" technology.

Castings are used in 90% of all manufactured goods [1]. Castings can be produced from most any metal and may have simple or complex shapes. The molds that molten metals are poured into to produce a casting may be reusable, such as the metal molds used for gravity or high pressure die casting, or expendable, such as sand molds or ceramic investment casting molds. This paper focusses on expendable mold casting processes. Specifically, this paper describes how additive manufacture of sand molds and polymers can be used as "disruptive processes" to rapidly produce new, revolutionary casting concepts and complex shapes, which, here-to-fore were impossible or impractical to produce.

Molds for sand casting are typically produced by forming a combination of sand plus binder over a pattern. Since the sand mold must be removed from the pattern, the pattern must have draft and no undercuts; this greatly limits the shapes that can be produced. The molds also must have a parting line, i.e., the interface between the top and bottom molds. Dimensional tolerances are very good within a mold but the tolerance is much larger across a parting line since how well the two halves fit together is somewhat variable. To form internal passageways, bonded sand cores can be added. Sand molding is a versatile process and castings made using sand molding processes may weigh a few ounces to tens of tons.

Investment casting or the "lost wax" process has been used for the production of castings for centuries. In this casting process, an expendable pattern is coated with a ceramic, the expendable pattern is melted or burned out, the ceramic fired to attain sufficient strength, molten metal is poured into the ceramic mold and, after cooling, the ceramic is removed from the casting. Advantages of this process are high dimensional accuracy, excellent surface finish, ability to fill very thin sections (<1 mm is easily obtainable) and nearly unlimited shape capability. For high volume production, wax is typically injected into a metal die, which somewhat limits the shape capability. However, for the artist, any shape that can be formed from wax can be cast. An alternative to wax is polymers.

The characteristics of a desirable material for expendable patterns for investment casting are 1) low coefficient of thermal expansion, 2) low thermal conductivity, 3) low compressive strength, 4) low melting or softening point and 5) complete combustion to gaseous products during burn-out. Some of these characteristics minimize the stresses that cause cracking in the ceramic coating during pattern burn-out. During pattern burn-out, the ceramic shell has very low tensile strength and is easily deformed and/or cracked. One printing technique that helps meet these requirements is printing "sparse", which means that the printed object has a solid surface skin but is hollow inside. Various honeycomb type internal structures can be printed to provide sufficient strength for handling but maximize the collapsibility of the structure during burn-out, thus allowing the structure to collapse onto itself rather than creating tensile stresses in the ceramic shell.

The additive manufacturing processes described in this paper are 1) binder jetting technology to fabricate (i) sand molds and (ii) expendable patterns for investment casting and 2) fused filament fabrication and vat photopolymerization to produce expendable patterns for investment casting. All of these processes allow the economical manufacture of cast structures that are impossible to mold using conventional methods.

Production of Sand Molds and Expendable Patterns for Investment Casting Using Binder Jetting Technology

One form of 3D printing spreads a layer of uniform thickness of particulate material on a moveable platen, binder is jetted where desired to cause bonding of the particulate material within the layer and to the underlying layer, a new layer is then spread, and the process repeated until the desired shape is complete [2-5]. The particulate material can be a polymer coated refractory material, such as gypsum or silica sand, or a polymer, such as PMMA (polymethylmethacrylate). When the "binder" polymerizes, the particles are bonded together. Examples of this method for producing molds for casting are the ZCast™ direct metal casting process [6] and the ExOne™ process [7]. An example for producing expendable patterns for investment casting is the Voxeljet process [8]. In these processes, molds, cores or expendable patterns are produced on a 3D printer from CAD models. For the ZCast process, the molds and cores require a subsequent curing cycle [9]. ExOne molds and cores do not require subsequent processing.

These technologies allow the casting engineer the flexibility to produce castings that were previously impossible to produce using conventional molding processes that require patterns with draft and parting lines. With 3D printing, neither draft nor parting lines are required. As a matter of fact, backdraft is permissible and features that previously required separate cores can be integrated into the molds. Simply put, almost anything that can be designed mathematically can now be produced as a casting. The primary limitation on these processes is the removal of unbonded sand from the molds and cores. Also, the number of binder/sand systems is limited, but this limitation is being addressed.

Producing a Turner's cube was often a test for new machinists. The machinist had to 1) determine how to produce the desired shape and 2) operate the equipment at the level of expertise necessary to produce the desired shape. A Turner's cube casting was created in CAD and a sand mold was produced by binder jetting. There was an issue with removal of unbonded sand as evidenced by the incomplete fill in several sections. This was a very challenging casting for a group of new casting engineers. The cast Turner's cube is shown in Figure 1.

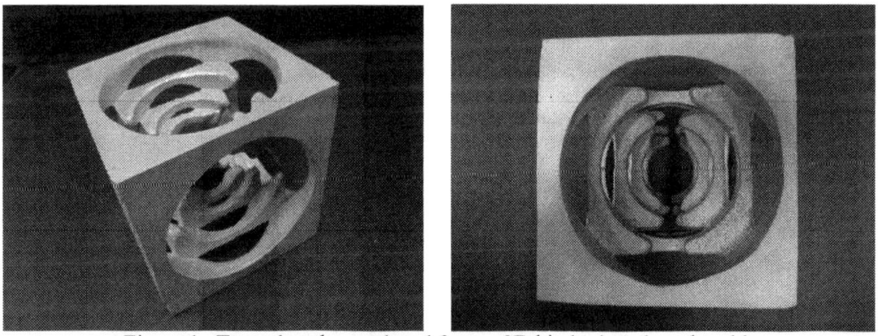

Figure 1. Turner's cube produced from a 3D binder jetted sand mold.

Complex, cellular structure castings cannot be produced by any process other than 3D mold printing. These casting are currently being produced at Virginia Tech using a combination of ordinary chemically bond sand and 3D printed cores. The 3D printed core forms the complex, cellular structure and the ordinary chemically bond sand forms the remainder of the mold (at low cost). The details of producing the complex, cellular structure castings has been discussed in detail elsewhere [9-11]. These castings have been produced in ferrous (iron & steel) and non-ferrous alloys. Complex, cellular structure castings are shown in Figure 2.

Figure 2. Complex, cellular structure castings produced from 3D binder jetted sand molds.

The mechanical properties of the complex, cellular structure castings shown in Figure 2 are currently under investigation. Modeling has indicated that the energy absorbing/dissipating properties of these structures under blast conditions should be exceptional. These castings are multi-functional structures and their use is only limited by the imagination of the designer. Examples of potential uses for cellular structure castings are 1) blast resistant panels for armored vehicles, 2) blast & fire resistant doors, bulkheads or walls on ships, 3) lightweight elevator floors for high load applications, 4) heat resistant surfaces for vertical take-off and landing aircraft if combined with water or air cooling or 5) a network structure to enhance the structural integrity of protective or ablative coatings.

PMMA is an excellent material for producing expendable patterns for investment casting. Bonded PPMA has very low pattern expansion and produces very low residual ash during burn-out [8]. The only negative for PMMA is that it is somewhat brittle so the expendable patterns need to be handled carefully. Due to the desirable physical properties of PMMA, normal investment casting shell thicknesses can be used successfully. An example of a binder jetted PMMA expendable pattern and the resulting investment casting is shown in Figure 3.

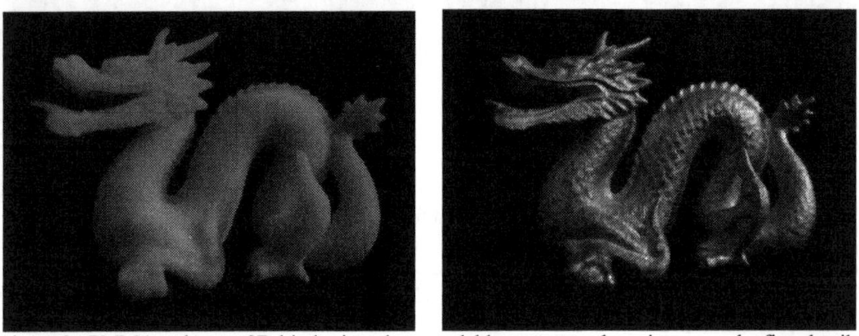

Figure 3. PMMA polymer, 3D binder jetted, expendable pattern and casting; note the fine details (scales) on the casting.

Fused Filament Fabrication and Vat Photopolymerization for Expendable Patterns

A variety of polymers (nylon, ABS, epoxy resin, etc.) are used in additive manufacturing and some are better than others if the end use is an expendable pattern for investment casting. Fused filament fabrication machines heat a polymer to slightly above their melting point and then force the polymer through a die (extrude the polymer) onto a moveable platen. The extruded polymer is deposited layer by layer until the complete part has been built. An alternative to polymer extrusion is vat polymerization. In this process, a vat of liquid polymer is locally polymerized (turned to solid) using UV light or laser energy. A digital file with surface geometry is sent to the printer which directs the polymerizing energy to the desired locations and layer upon layer of polymer is built on a moveable platen. Scaffolding is built alongside the part to provide support for overhanging structures and internal voids, therefore permitting the ability to produce most any shape. These support structures need to be removed along with any entrapped liquid polymer before use. When printing polymers for expendable patterns, there is also an advantage to making the pattern with the least amount of polymer possible since 1) this reduces cost (less polymer consumed) and 2) this reduces the amount of polymer that must be removed from the mold during burn-out. Printing a shape with a solid skin and a honeycomb internal structure is termed "sparse" printing and is very effective in improving the performance of printed polymer expendable patterns [12].

A truss structure and a turbine wheel with airfoil shaped vanes were investment cast using expendable patterns produced from ABS polymer via fused filament fabrication [13, 14]. These structures cannot be easily or economically produced using conventional molding techniques. These castings could be produced using a two-piece mold plus complex core assemblies, but this would produce parting line flash that would need to be removed and the dimensional accuracy would not be nearly as good due to the fit-up tolerances that would be needed for the molds and cores. Printed ABS polymer expendable patterns were successfully used for these castings but the investment casting shell had to be made 50% thicker to prevent cracking during polymer burnout. A truss structure investment casting and a turbine wheel investment casting produced using polymer extrusion printed expendable patterns are shown in Figures 4 and 5.

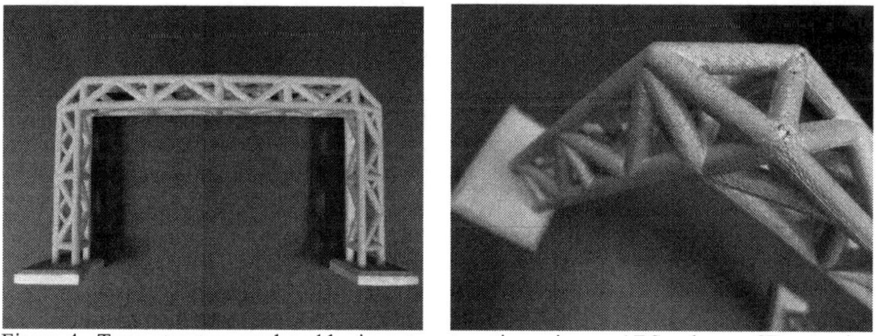

Figure 4. Truss structure produced by investment casting using an ABS polymer, 3D extrusion printed, expendable pattern.

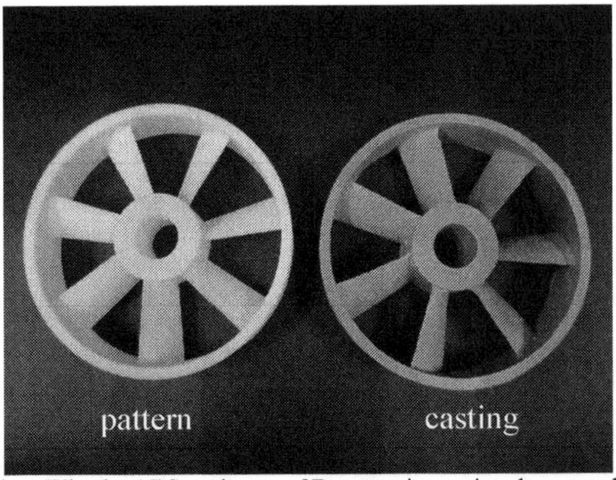

Figure 5. Turbine Wheel: ABS polymer, 3D extrusion printed, expendable pattern and investment casting.

Vat polymerization and sparse printing was used to produce expendable patterns for testing from an epoxy resin containing reactive diluents [15]. Expendable patterns for investment casting have been produced using this process since 1998 [11]. A sphere was investment cast using an expendable pattern produced from vat polymerization. The pattern was "sparse" printed. The internal honeycomb structure was visible in the pattern but only the external steps were apparent on the casting. The pattern and casting are shown in Figure 6.

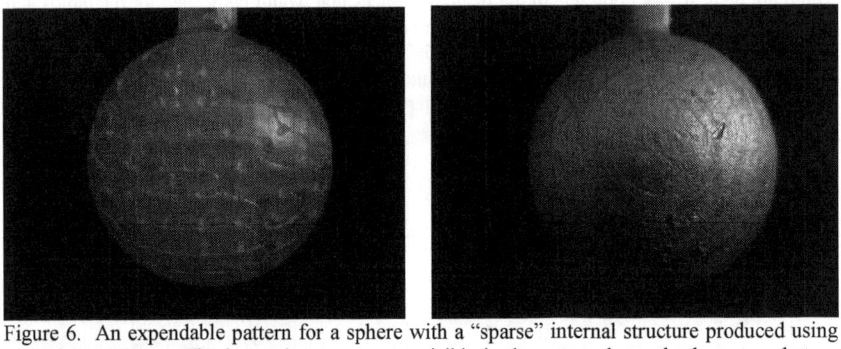

Figure 6. An expendable pattern for a sphere with a "sparse" internal structure produced using vat polymerization. The internal structure was visible in the pattern but only the external steps were apparent on the casting.

Concluding Remarks

Additive manufacturing has demonstrated the ability to transform the casting industry. Numerous additive manufacturing techniques and materials are currently available to make shaped castings available by the next day or within a few days; giving new meaning to fast-to-market.

Acknowledgements

The authors would like to acknowledge ExOne for printing the cellular structure sand cores, Voxeljet for binder jet printing the PMMA dragon pattern and test patterns, the Virginia Tech DREAMS Labs for printing the ABS polymer truss structure, turbine wheel and test specimens via fused filament fabrication, and 3DSystems for producing test specimens using vat photopolymerization. The authors would also like to acknowledge the Virginia Tech alumni and students (Shawn McKinney, Patrick Stockhausen, Ruthie Kay and Shawn Bottoms) who designed, molded, poured and finished the shaped castings highlighted in this paper.

References

1. "Metalcasting: Who and What Are We?" American Foundry Society presentation, American Foundry Society, Schaumburg, IL.
2. D.T. Pham and R.S. Gault, "A Comparison of Rapid Prototyping Technologies," *International Journal of Machine Tools & Manufacture*, 38 (1998) pp.1257-1287.
3. T. Grimm, *User's Guide to Rapid Prototyping* (Society of Manufacturing Engineers, 2004).
4. R.I. Noorani, *Rapid Prototyping: Principles and Applications* (John Wiley & Sons, Inc., 2005).
5. K.V. Wong and A. Hernandez, "A Review of Additive Manufacturing," *ISRN Mechanical Engineering*, Article ID 208760 (2012).
6. http://www.zcorp.com/en/Solutions/Castings--Patterns--Molds/spage.aspx, 3DSystems, 333 Three D Systems Circle, Rock Hill, SC 29730.
7. http://www.exone.com/, ExOne™, 2341 Alger Drive, Troy, MI 48083.
8. http://www.voxeljet.de/en/, Voxeljet.
9. D.A. Snelling et al, "Mitigating Gas Defects in Castings Produced from 3D Printed Molds," *117th Metalcasting Congress Conference Proceedings*, St. Louis, MO (2013).
10. N.A. Meisel, C.B. Williams, and A. Druschitz, "Lightweight Metal Cellular Structures via Indirect 3D Printing and Casting," *International Solid Freeform Fabrication Symposium*, Austin, TX (2012).
11. D. Snelling et al, "The Effects of 3D Printed Molds on Metal Castings," *International Solid Freeform Fabrication Symposium*, Austin, TX (2013).
12. T. Mueller, "Guide to Casting Using QuickCast™ Patterns," 3DSystems, 333 Three D Systems Circle, Rock Hill, SC 29730.
13. M. Seals, S. McKinney, and P. Walsh, "Comparison of Wax and 3D Printed Investment Casting Pattern Materials," *Materials Science & Technology 2012 Conference & Exhibition Conference Proceedings*, Pittsburg, PA (2012).
14. M.E. Seals et al, "Evaluation of 3D Printed Polymers for Investment Casting Expendable Patterns," accepted for publication in *118th Metalcasting Congress Conference Proceedings*, Schaumburg, IL (2014).
15. VisiJet™ SL Clear MSDS, http://www.3dsystems.com/sites/www.3dsystems.com/files/24672-S02-00-C-MSDS-US-English-VisiJet-SL-Clear.pdf

Shape Casting: 5th International Symposium 2014
Edited by: Murat Tiryakioğlu, John Campbell, and Glenn Byczynski
TMS (The Minerals, Metals & Materials Society), 2014

EVOLUTION OF FILLING SYSTEM DESIGN FOR AN A356-T6 ALUMINUM HOUSING CASTING

Joseph Chvala, Murat Tiryakioğlu, Nick Hudyma, Paul Eason

University of North Florida, School of Engineering, 1 UNF Drive, Jacksonville, FL 32224

Keywords: Campbell's 10 Rules; vortex gate; porosity

Abstract

The results of a student project to produce two housing halves for an engine dynamometer are presented in this paper. The filling systems were designed to minimize turbulence and damage to the metal. The results of small changes in the filling system design are discussed. A new method of using different parts of the filling system (rigging) to evaluate the results of quality improvement efforts is introduced.

Introduction

The Osprey Racing team at the University of North Florida designed a water brake engine dynamometer. The design involves two identical housings, as presented in Figure 1. A decision was made to produce the housings by the sand casting process. A356 cast aluminum alloy was chosen based on its castability and mechanical properties. Patterns for the housing were prepared by the fused deposition modeling rapid prototyping process. Although the initial intent was to produce a pair of housings, the project scope was later expanded to examine how changes in the filling and feeding system would affect the quality of the casting. There were three trials with slightly different filling system designs. This paper summarizes the methods used in the project to assess the quality of the casting nondestructively and final results and recommendations.

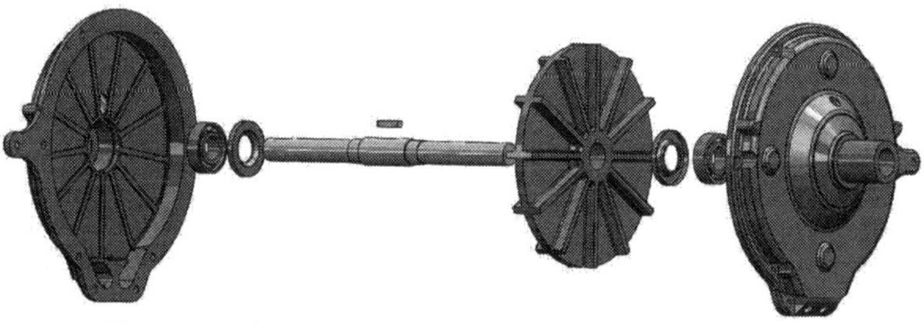

Figure 1. Exploded view of the engine dynamometer assembly.

Experimental Details

The dimensions for the filling system of the castings were calculated based on the practice recommended by Campbell [1]. The dimensions of the filling system, presented in Figure 2, were calculated based on a flow rate of 0.5 kg/s and a velocity of the melt entering the casting of 0.25 m/s. This velocity is well below the critical velocity of 0.5 m/s for aluminum [2]. The dimensions of the feeder were calculated based on (i) expected volumetric shrinkage of the casting, (ii) feeder (riser) efficiency (~14% [1]) and (iii) an expected solidification time that is 20% longer than that of the casting. Due to the shape of the part, the feeder was located below the casting during filling. After the completion of filling, a lid was placed over the mold which was then rotated 180° to relocate the feeder above the casting. The dimensions of the sprue and the feeder were not changed between the three filling system iterations. The runner design incorporated a flow-off volume by extending the runner past the feeder. The end of this extension was tapered to a point to help prevent a back wave. A no-bake binder (Lino-Cure™) was used for all sand molds. In all three trials, virgin A356 ingots were used and castings were poured at 720°C (100°C superheat). In Trials 1 and 2, a powder flux was added to the melt for degassing and the melt was subsequently stirred. No flux was used in Trial 3.

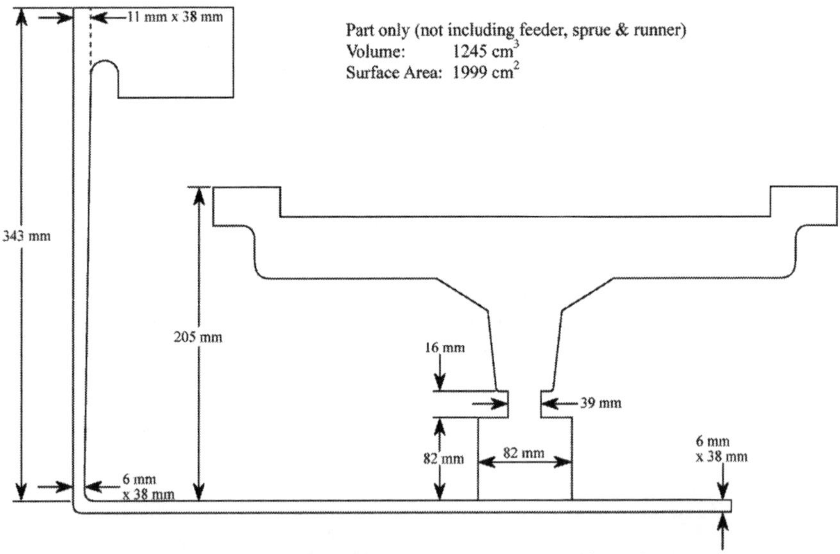

Figure 2. Details of the filling system used in this study.

Trial 1:
The pouring basin in Trial 1 was approximately 38 mm deep and included a radiused weir to control the flow into the sprue, Figure 3.a. The filling system did not include a vent and there was no mechanism to determine whether the mold was completely filled. The runner was arranged to pass under the center of the feeder.

Trial 2:
In Trial 2, the pouring basin kept the same basic dimensions and radiused weir – only the depth was increased to 89 mm, as shown in Figure 3.b. In addition, a vent was added to the mold, both to eliminate any possible back pressure and to allow a visual indication of when filling is completed.

Trial 3:
The only change between Trials 2 and 3 was the runner design; the runner joined the feeder tangentially (vortex gate), Figure 3.c. Moreover the runner was not extended past the feeder.

 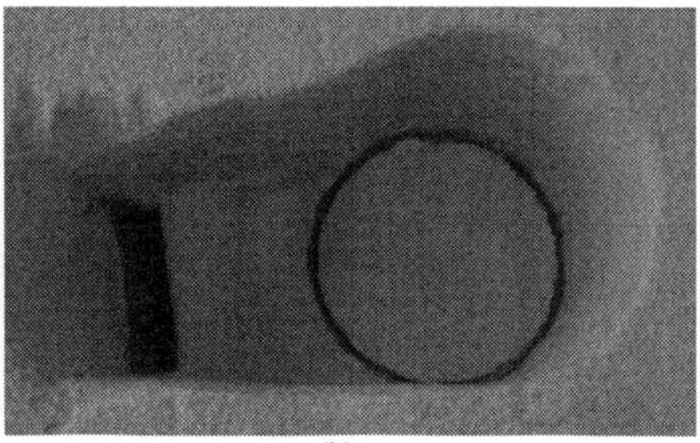

(a) (b)

(c)

Figure 3. The pouring basins in Trial 1 (a) and Trial 2 (b) and the vortex gate in Trial 3 (c).

The sprues and feeders from the three trials were heat treated to T6 condition along with the castings for improved machinability. Parts of the sprues, approximately 20 mm downstream from the pouring basin and the cross section of the feeders were machined by using a 76.2 mm (3 in.) face mill tool with six inserts. The machined surfaces were ground and finally polished to expose all internal pores. Subsequently, all surfaces were scanned with 600 dpi resolution and images were analyzed by using the ImageJ software [3] from the National Institutes of Health.

Results and Discussion

Initial Observations:

The castings after they were taken out of the sand mold are shown in Figure 4. In Trial 1, it was determined that the shallow depth of the pouring basin did not allow sufficient flow into the sprue to keep it completely full. Subsequently, a minor misrun was detected in Trial 1, as indicated by the arrow in Figure 4.a. In Trials 2 and 3, the deeper pouring basin allowed for a much faster filling rate and the sprue was completely full during the pour. Initial observation of surface finish indicated that the casting in Trial 1 had the roughest surface of the three and castings in the other two trials had almost the same surface roughness.

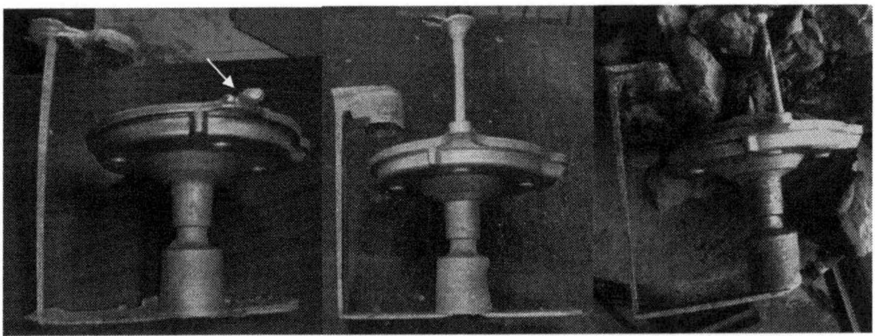

Figure 4. Castings with filling systems attached: (a) Trial 1 (b) Trial 2, and (c) Trial 3.

Nondestructive Evaluation of Internal Defects

The sections of the sprues are presented in Figure 5. Note that sprues for Trials 1 and 2 contain many pores, whereas that for Trial 3 has only a few pores compared to the first two. The same observation can be made for feeders, presented in Figure 6. Note that there is significant external shrinkage in the feeder

(a) (b) (c)

Figure 5. Sections of the sprues: (a) Trial 1, (b) Trial 2, and (c) Trial 3.

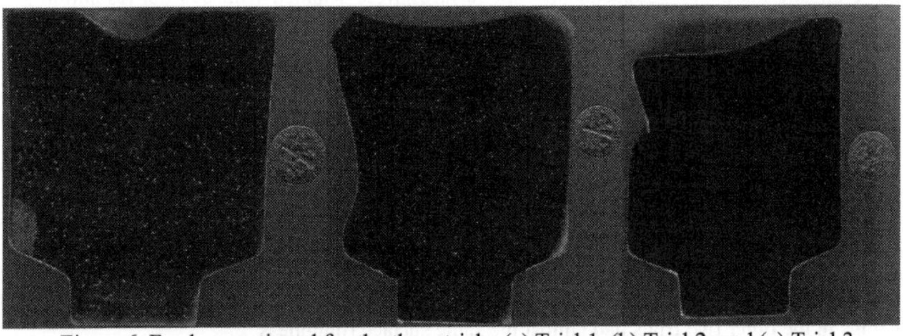

Figure 6. Feeders sectioned for the three trials: (a) Trial 1, (b) Trial 2, and (c) Trial 3.

Areas of pores shown in Figures 5 and 6 were measured by the ImageJ software. Equivalent diameters, d_{eq}, were calculated ($=\sqrt{4A/\pi}$) after excluding any pore with an area of less than 0.1 mm^2, as recommended by Dispinar and Campbell [4]. Analysis of d_{eq} results showed that three-parameter lognormal distribution provided the best fit to all datasets, consistent with results on pore sizes in Mg die castings [5]. The probability density functions, f, for pore sizes in sprues and feeders in all trials are presented in Figure 7.

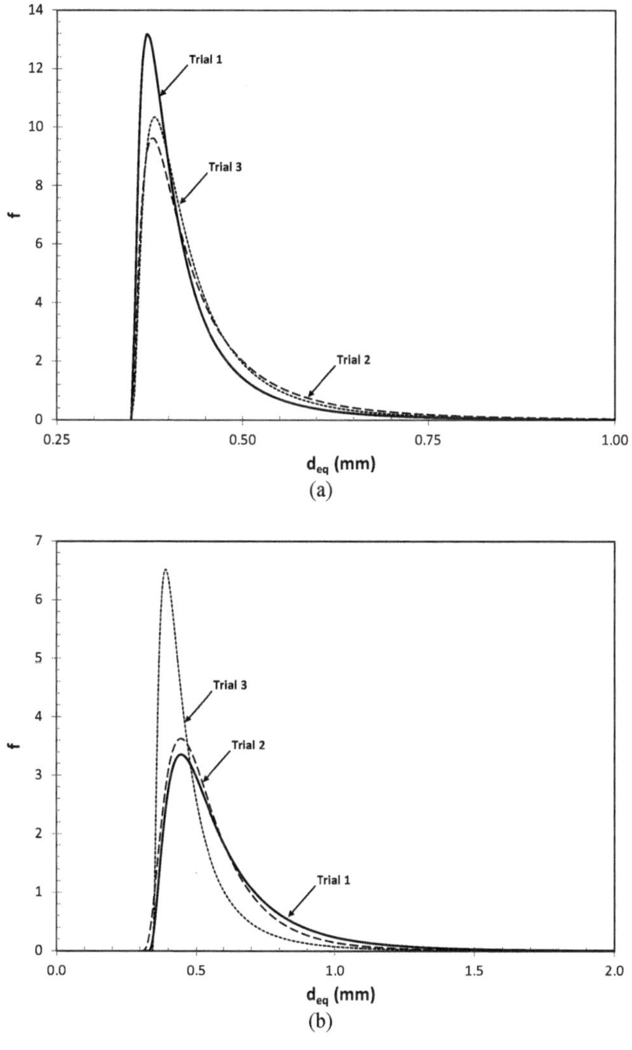

Figure 7. Lognormal distributions for equivalent diameters of pores in (a) sprues and (b) feeders.

It is noteworthy that the size distribution of pores in the sprues of three trials is very similar with those for Trials 2 and 3 being almost identical. Because the depth of the pouring basin was the same in Trials 2 and 3, it can be expected that sprues from Trials 2 and 3 have similar pore size distributions. The main difference between the results from these two trials is the number of defects per unit area, as can be seen in Table 1, in which numbers of pores per $100mm^2$ area are presented. Although the size distribution of pores in sprues in Trials 2 and 3 are similar, there are almost 30 times more pores in Trial 2 than in Trial 3. This result is a clear indication that the powder flux added for degassing severely degraded the melt quality in Trials 1 and 2. The powder particles entrained surface oxides which led to the unintended consequence of higher number of defects than what would be possible without the flux addition as in Trial 3.

Table 1. The number of defects per 100 mm^2 in sprues and feeders for all trials.

	Sprue	Feeder
Trial 1	10.39	19.08
Trial 2	11.38	18.76
Trial 3	0.35	4.53

The number of pores in feeders given in Table 1 is consistent with the results for pores in sprues. The size distribution of pores in Figure 7.b., however, clearly shows that the tangential runner design in Trail 3 has resulted in smaller pores, as evidenced by the pronounced peak at lower d_{eq} values. In contrast, Trials 1 and 2 produce almost identical pore size distributions.

It is noteworthy that Trials 1 and 2 gave similar results in which the numbers of defects arriving in the feeder were approximately doubled when compared to those at the sprue. It can be expected that the melt suffered a modest amount of damage during its passage through these filling systems because all gravity filling systems are ultimately merely damage limitation systems [6]. For Trial 3, Table 3 indicates that the number of defects was increased by over 10 times between the sprue and the feeder. However taking the ratios of the number of defects at the sprue and the feeder may be misleading. The damage given to the metal while flowing through a gravity filling system is expected to be additive. Therefore the three trials need to be evaluated based on the increase in the number of defects between the sprue and the feeder, which was 8.69, 7.38 and 4.18 for Trials 1, 2 and 3, respectively. These numbers indicate the beneficial effects of the deeper pouring basin (Trial 2) and the vortex gate (Trial 3).

The method of sectioning parts of the filling and feeding system presented in this study is of potential value to foundries. The 'rigging', which would be normally be discarded and remelted, should be heat treated and the macrostructure should be analyzed. The results of any quality improvement effort, such as changes in the filling system and/or melt treatments, can be easily quantified and documented. Although the effects of filling system design can be isolated to some extent from the erratic and large variability of melt quality from melt to melt, there seems no substitute for the accuracy in comparison which can be made by pouring all three castings at the same time from the same melt.

Conclusions

1. Using powder flux for degassing degraded the melt quality significantly by entraining surface oxides. The melt quality was significantly better when no flux was used.
2. The increase in the number of defects between the sprue and feeder was smallest when a vortex gate was incorporated.
3. A new method of analyzing the defects in different parts of the 'rigging', which is normally discarded and remelted, is introduced. The new method can be used to quantify and document the results of any improvement effort, such as changes in the filling system and/or melt treatments.

Acknowledgements

The authors would like to thank Alcoa for donating the ingots used in this study, and Braddock Metallurgical, Jacksonville, FL, for heat treating the parts.

References

1. J. Campbell: Complete Casting Handbook: *Metal Casting Processes, Techniques and Design*, Elsevier Science, 2011.
2. J.J. Runyoro, S.M.A. Boutorabi, J. Campbell: *AFS Transactions*, 100 (1992) 225–234.
3. http://rsbweb.nih.gov/ij/
4. D. Dispinar, J. Campbell, in this volume.
5. M. Tiryakioğlu: *Mater. Sci. Eng. A*, 465 (2007) 287–289.
6. J. Campbell, personal communication, October 29, 2013.

Shape Casting: 5th International Symposium 2014
Edited by: Murat Tiryakioğlu, John Campbell, and Glenn Byczynski
TMS (The Minerals, Metals & Materials Society), 2014

Recent Advances on the Solidification Processing of Cast Energetic Materials

Ruslan Mudryy[1], Shian Jia[2], and Laurentiu Nastac[2]

1. U.S. ARMY, RDECOM-ARDEC, Armaments Engineering and Technology Center, Picatinny, NJ 07806, USA, email: ruslan.s.mudryy@us.army.mil

2. THE UNIVERSITY OF ALABAMA, Department of Metallurgical and Materials Engineering, P. O. Box 870202, Tuscaloosa, Al, 35487, USA, email: lnastac@eng.ua.edu

Abstract

This paper investigates the solidification of highly viscous energetic materials cast into a projectile. Active cooling and heating (ACH) control solidification technology as well as mechanical vibration (MV) are applied to achieve unidirectional solidification and to reduce cracks, gas pores, and shrinkage defects and to decrease the detrimental gap size between the projectile and the solidified energetic material. A comprehensive numerical model was developed to simulate the solidification processes during casting of energetic materials, as well as the resulting induced thermal stresses. The optimized design parameters of the proposed technologies are developed based on numerical modeling and experiment work.

A detailed comparison between the latest experiments performed at the University of Alabama, Solidification Laboratory, obtained with electrical heating and water cooling and with and without mechanical vibration is provided in this paper. In these experiments, a special wax material (*e.g.*, Chlorez 700S) that has similar thermo-physical properties with the IMX-104 explosive material was used. Experiments performed at the USARMY ARDEC using the IMX-104 explosive material with steam heating and water cooling are also presented in this paper. These experiments are being used to further validate the numerical model.

Introduction

Solidification and Casting processes are widely employed in the manufacture of energetic materials. The cooling conditions applied in the casting process can affect the quality of the final cast in terms of void formation, residual stress distributions, and mold separation. Substantial shrinkage is also observed due to the density change upon solidification [1-3]. Residual stresses are known to be closely related to the formation of cold cracks and hot tears during casting. The formation of a gap between the mold and the cast material is of critical importance due to its deleterious effect on heat removal and on crack formation. In the casting of energetic materials all these defects can significantly impair the detonation velocity, Gurney energy, and insensitive munitions characteristics of the formulation, and lead to catastrophic accidents in explosives handling [4-5]. Imposition of carefully controlled cooling condition is thus critical in optimizing the cast quality that could help avoid such destructive effects.

Figure 1 presents the typical mold filling and solidification defects in a 120 mm IMX-104 mortar. The air bubble (entrapped during the mold filling) and the separation and shrinkage formed during solidification are severe. These defects are detrimental to the quality of the 120 mm mortar. Optimization of the process design as well as the design of improved rigging systems is required to minimize/eliminate these defects.

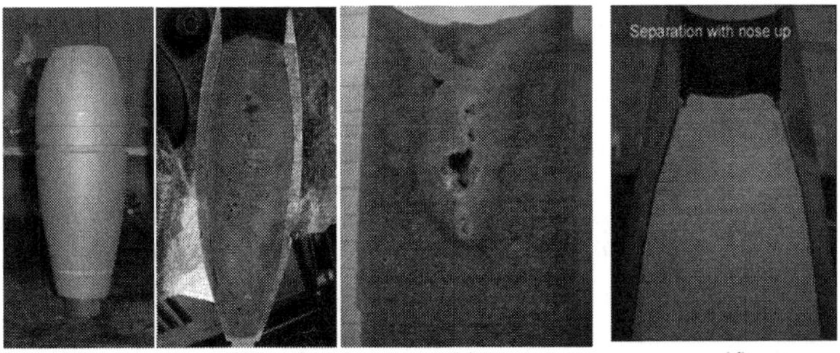

(a) (b) (c) (d)
Figure 1. Casting position (a) and pouring and solidification defects (air bubble and shrinkage cavities ((b) and (c)) and separation (gap) (d)) in the 120 mm mortar.

In the current work, a comprehensive numerical model was developed using ANSYS FLUENT and ANSYS MECHANICAL [6] software to accurately simulate the transport phenomena as well as induced thermal stresses encountered in the casting process of recently developed explosive, consisting of RDX-binder mixtures. An enthalpy method, successfully exploited by many authors [7-8], was used to simulate the solid/liquid phase change process. Instead of working entirely in terms of the temperature of explosive material, an enthalpy function is defined which represents the total heat content per unit mass of the material. The advantage of such a reformation of the problem is that the necessity to track the position of the solid/liquid interface is eliminated. Shrinkage effects and the resulting velocities induced in the melt have been largely neglected in the literature due to the difficulties involved in multiphase pressure-velocity coupling, and the interaction between free surface dynamics and solidification volume change [9]. In order to track the shrinkage shape that is critical in simulation of explosive casting process, NOVAFLOW&SOLID software [10] was used. Effective shrinkage was calculated at each time step and the volume of the solidified material was then subtracted from the liquid phase in the control volumes that contain the interface.

Experimental measurements of physical properties such as thermal conductivity, specific heat, thermal expansion coefficient, and liquid viscosity were conducted using Diamond TMA (www.tmadiamond.com), Hot Disk Instrument, and Brookfield viscometer. The stress-strain relationship was measured using simple compression technique developed in-house [11].

A new technology named "Active cooling and heating" (ACH) that can be used for melt cast process of energetic materials in order to improve product quality is described in this paper. The purpose of the ACH is to use controlled (active) heating and cooling of the mold to try to achieve unidirectional solidification with minimum solidification related defects including gap separation and solidification shrinkage. Figure 3 illustrates the experimental mold setup and cast wax material that shows no shrinkage and gap separation. Optimal cooling parameters predicted by numerical simulations can be easier controlled using the ACH technology. Maintaining higher temperature along a riser and gradually decreasing temperature at the bottom part will keep solid front flat propagating upward. This helps to reduce the excessive thermal stress formed due to large temperature gradient and provides control of shrinkage shape. New design of the riser geometry with electrical or steam heating was proposed after calculation of the thermal modulus distribution inside the projectile.

Numerical Model

The studied energetic material IMX-104 is assumed to be isotropic, incompressible and Newtonian fluid with a very high viscosity due to the high volume fraction of solid particles. The constituents of IMX-104 are DNAN (2, 4-dinitroanisole), NTO (3-Nitro-1, 2, 4-triazol-5-one) and RDX. Of these three constituents only DNAN undergoes solid/liquid phase change while NTO and RDX remain in solid, crystalline form during the entire process. The solidifying melt was modeled as a single material with temperature-dependent density, thermal expansion coefficient and Young's modulus, with all other properties remaining the same in both solid and liquid phases. The variation of density with temperature was described by Boussinesq approximation. The numerical model was based on solving the system of governing equations for conservation of mass, momentum, and energy:

$$\frac{\partial \rho}{\partial t} + \nabla \cdot (\rho \vec{u}) = 0 \tag{1}$$

$$\frac{\partial \rho \vec{u}}{\partial t} + \nabla \cdot (\rho \vec{u} \vec{u}) = -\nabla p + \nabla \cdot \left[\mu \left(\nabla \vec{u} + \nabla \vec{u}^T\right)\right] + \rho_\infty \vec{g} \beta_l (T - T_\infty) + \frac{(1 - f_l)}{\left(f_l^3 + \varepsilon\right)} A_{mush} \vec{u} \tag{2}$$

$$\frac{\partial \rho h}{\partial t} + \nabla \cdot (\rho \vec{u} h) = \nabla \cdot (k \nabla T) \tag{3}$$

where A_{mush} is the mushy zone constant, f_l the liquid fraction, ε =0.001 is used to prevent division by zero, and $h = c_p T + f_l \Delta H$, the specific enthalpy of the melt.

In the absence of detailed visco-elastic-plastic material properties, the material was treated as an isotropic thermo-elastic material:

$$\frac{\partial}{\partial t}\left(\rho \frac{\partial \vec{w}}{\partial t}\right) = \nabla \cdot \left[\mu \left(\nabla \vec{w} + \nabla \vec{w}^T\right)\right] + \nabla \left(\lambda \nabla \cdot \vec{w} + (3\lambda + 2\mu)\beta_s T\right) - \left\{\nabla \cdot \left[\frac{\sigma_{ij}^d \sigma_{kl}^d}{\sigma^2}\left(\frac{9G^2}{H' + 3G}\right)\nabla \vec{w}\right]\right\} + \vec{b} \tag{4}$$

where $\lambda = \frac{\upsilon E}{(1+\upsilon)(1-2\upsilon)}$, μ are Lame's coefficients, H' is the plastic modulus, G is the shear modulus and b is the body force.

In order to analyze the shrinkage and void formation caused by the density change during solidification, the algorithm employed must be capable of tracking a moving free surface. The volume of fluid (VOF) method is employed in this work since it can handle free surface movement and has been previously applied to study solidification shrinkage. For the casting problem considered, only two phases, i.e., IMX-104 and air is present in the system. When the density change upon solidification is taken into consideration, the governing equations for conservation of mass, momentum, and energy will have the additional source terms due to the density difference between the solid and liquid phases of IMX-104. The resulting system of equations is solved by NOVAFLOW&SOLID software.

Experimental Results

The experimental setup with ACH at the University of Alabama (UA) is shown in Figure 2. Figure 3 illustrates the experimental mold setup and the cast wax material that shows no shrinkage and gap separation. A special wax material (*e.g.*, Chlorez 700S) was used in these cast experiments. The wax material has similar thermo-physical properties with the IMX-104 explosive material and it can be safely used in the UA lab for process optimization. Electrical heating (60 W and 150 W heaters) and water cooling systems were applied in the experiments to control the solidification process in such a way to minimize both the gap separation and the shrinkage defects in the top portion of the mortar.

(a) (b) (c)

Figure 2. Experimental setup for the active cooling and heating (ACH) of the 120 mm mortar: (a) top view of the solidified mortar in the water tank and (b) 60W electrical heating system and (c) 150W electrical heating system.

(a) (b)

Figure 3. Experimental mold setup (a) and (b) Cast wax (Chlorez 700S) material without shrinkage and gap separation (preheated mold, 60W electrical heating, without water cooling).

Figure 4 show the experimental setup of the mortars with MV and with and without water cooling. The process conditions are explained in the following paragraph. The wax material was melt in a furnace at 423K for 9 hours. The mold (mortar case) was preheated to 423K. The heating cable was set at 373K for 2 hours after pouring the wax material. Water cooling was done in three steps (similar heights) (i) water flow rate at 1.26 l/min for 20 min (ii) 0.95 l/min for 30 min; (iii) 0.63 l/min for 40 min.

Figures 5 and 6 present the temperature experiments for the setup with MV system as illustrated in Fig. 4, without and with water cooling, respectively. In Figures 5 and 6, T1 (or channel 1) is the thermocouple located inside of the cast material (neck area), T2 (or channel 2) is the thermocouple inside the thermal isolation (neck area), and T3 (or channel 3) is the thermocouple located in the middle height of the mortar case beneath the thermal isolation area. It can be seen from Figs. 5 and 6 that the addition of both heating and vibration had positive effects on the temperature evolution profiles of the 120 mm mortar. Fluidity of the wax material (that helps feeding the solidification shrinkage) was significantly increased due to the use of mechanical vibration.

(a) (b) (c)

Figure 4. Experimental setup with MV: (a) mortar assembled on the MV system (without water cooling) (b) water cooling tank connected to the MV system and (c) mortar assembled (with 150W electrical heating) inside the water cooling tank.

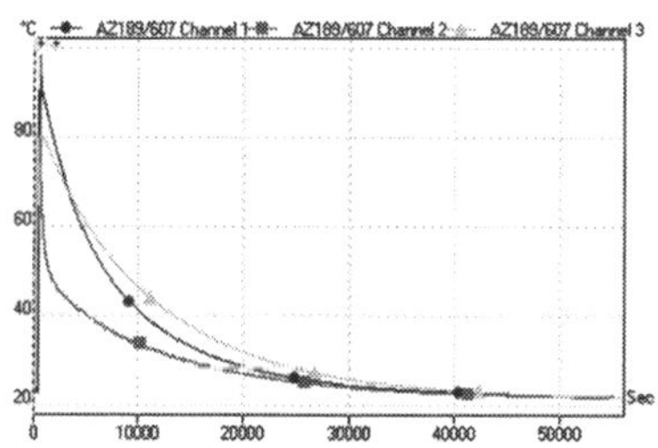

(a) without MV without heating

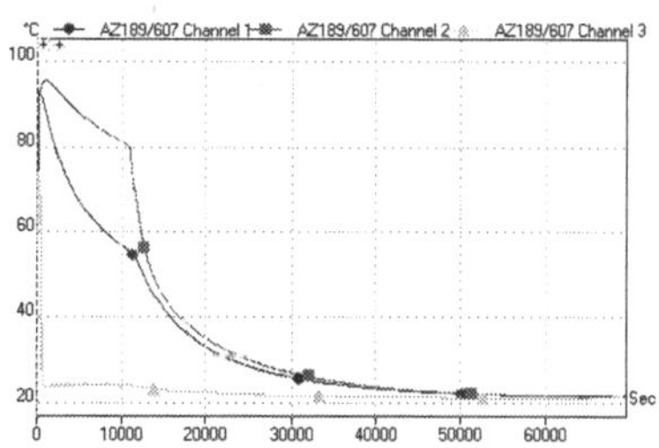

(b) without MV with heating

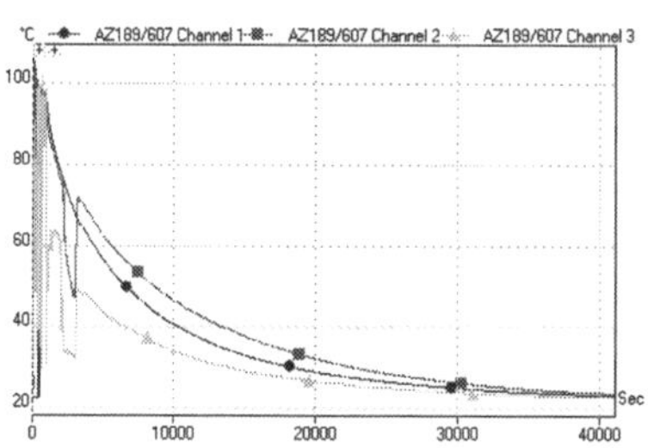

(c) with MV without heating

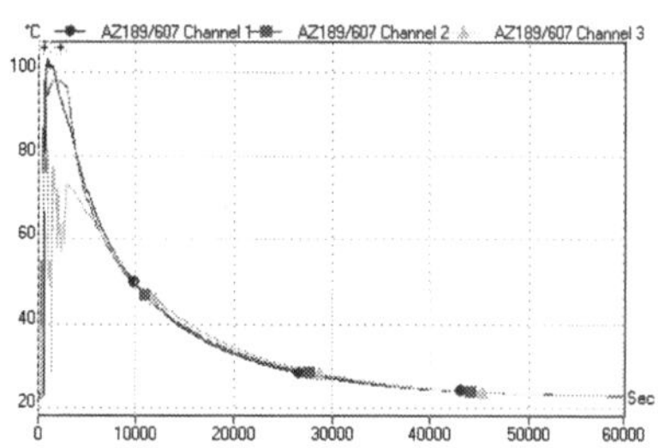

(d) with MV with heating

Figure 5. Temperature measurements in the wax experiments (without water cooling).

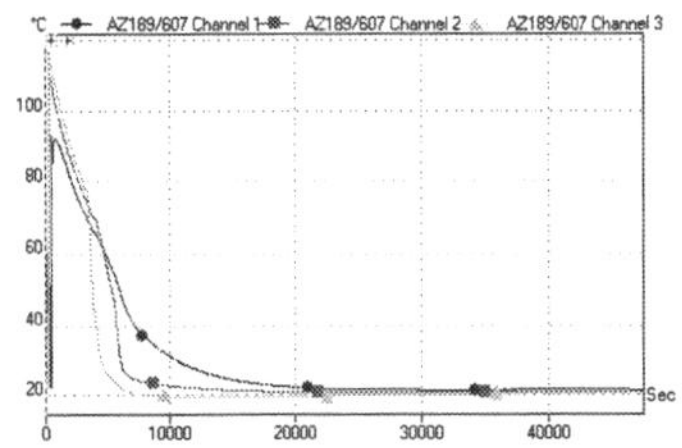

(a) without MV without heating

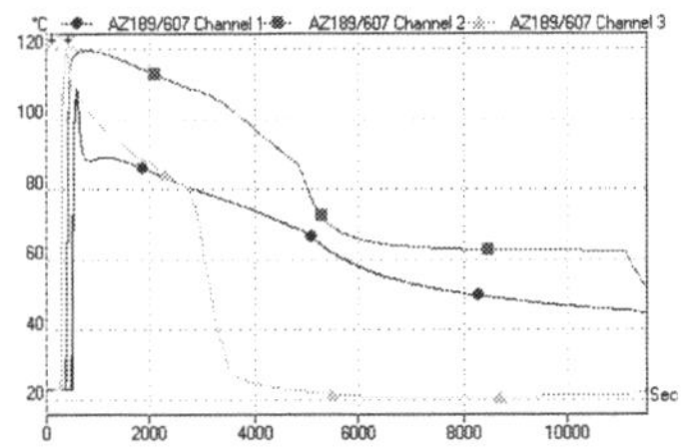

(b) without MV with heating

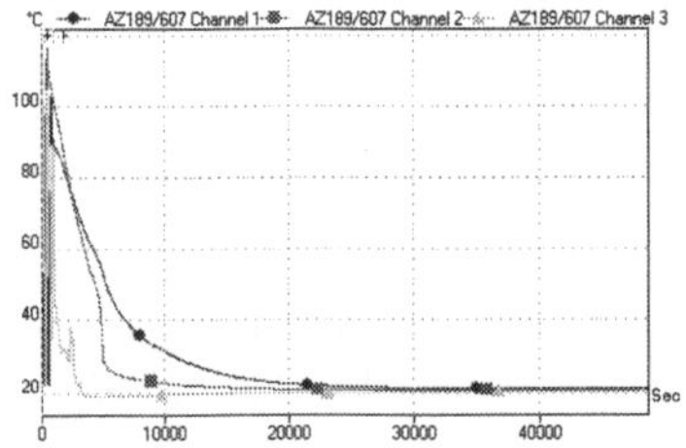

(c) with MV without heating

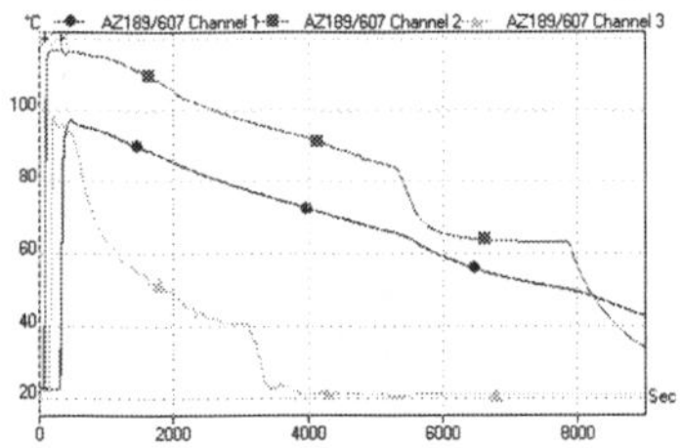

(d) with MV with heating

Figure 6. Temperature measurements in the wax experiments (with water cooling).

Controlled directional solidification was achieved in the 120 mm mortar (Figs. 5 and 6) by using the electrical heating and water cooling technology (ACH). The solidification shrinkage in the top portion of the mortar was completely eliminated. Also, no gap separation was observed in these mortars.

Figure 7 showed some of the experiments performed at USARMY ARDEC facility using IMX-104 material cast into 120 mm mortars. The experiments were performed under different solidification conditions. Improved process conditions (Figs. 7 c and 7d) eliminated the major solidification related defects shown in Figs. 7a and 7d.

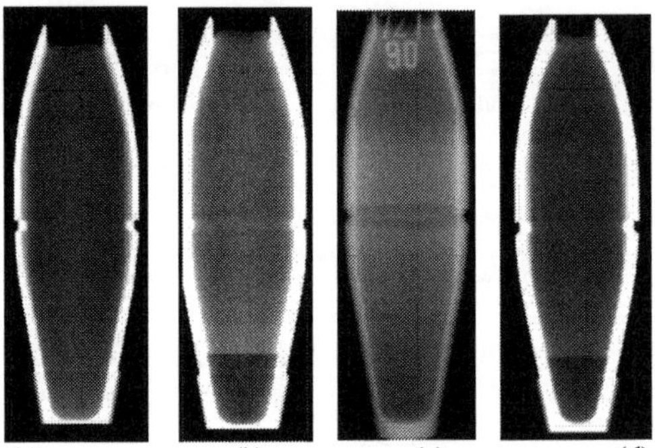

(a) (b) (c) (d)

Figure 7. USARMY ARDEC Experiments: (a) macro-shrinkage (top portion) (b) longitudinal macro-cracks (c) and (d) no macro-defects.

Numerical Simulation Results: Design Improvements

NOVAFLOW&SOLID [10] and ANSYS'S FLUENT [6] were used in this study to perform all casting simulations needed to improve the mold design. The results of the original design are shown in Refs. [12-14]. The new design consists of several improvements including the geometry modification of the top riser, mold preheating and application of mechanical vibration as well as application of electrical or steam heating techniques in the neck area between the top riser and the mortar. The results of the ACH with electrical heating are shown in Figures 6-8 in Ref. [14]. It was demonstrated in Ref. [14] that the active heating can be applied to improve feeding and therefore can eliminate the macro-shrinkage in the neck area.

Figures 8 and 9 present the results with the without steam heating. The steam heating technology needs to be used for safety reasons. Based on the results in Figure 9, it can be concluded that steam heating can successfully replace the electrical heating.

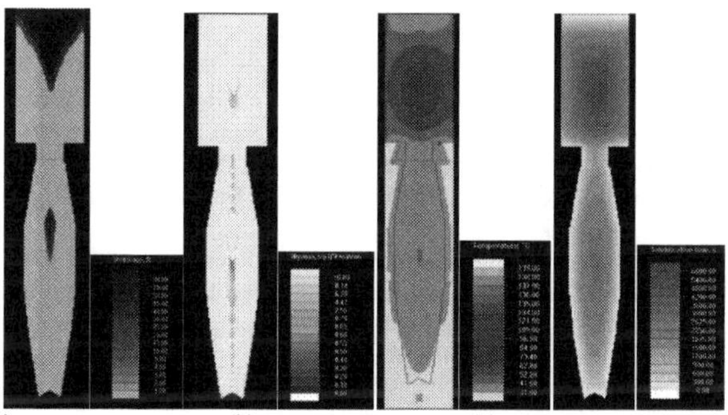

(a) (b) (c) (d)

Figure 8. Predicted micro-shrinkage (a), macroshrinkage (b), temperature (c) and solidification time (d) in the 120 mm mortar cooled in water (no steam heating, Al riser preheated at 110 C).

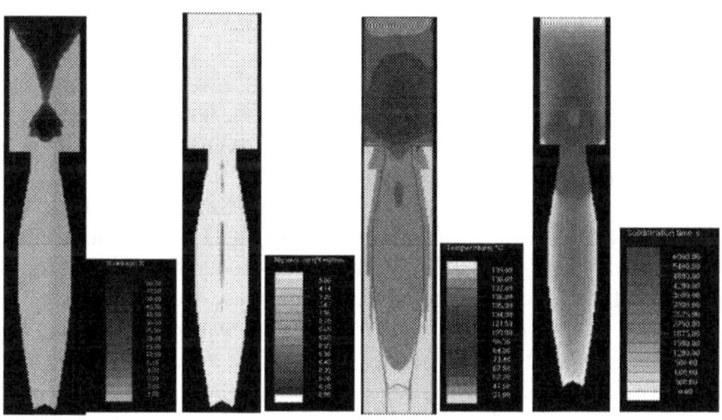

(a) (b) (c) (d)

Figure 9. Predicted micro-shrinkage (a), macroshrinkage (b), temperature (c) and solidification time (d) in the 120 mm mortar cooled in water (steam heating, Al riser preheated at 110 C).

Concluding Remarks and Future Work

Experimental and modeling studies have been performed to improve the design and process conditions of energetic materials cast into a projectile. The proposed improvements led to significant decrease in casting and solidification shrinkage defects in the cast projectiles. The improved process design consists of using ACH technologies with proper modifications in the top riser geometry, pouring rates, mold preheat temperature, and water cooling conditions. MV was also studied. It was shown that: (i) ACH with MV can improve the directional solidification conditions of the mortar that can further decrease the solidification shrinkage, cracking tendency and gap separation and (ii) steam heating can replace successfully the electrical heating.

Future work will include a detailed study regarding the mechanical vibration (MV) and steam heating effects on the solidification of energetic materials cast into projectiles.

References

1. D. Sun, S. Annapraganda, S. Garimella, and S. Singh, "Analysis of Gap Formation in the Casting of Energetic Materials", *Numerical Heat Transfer A*, 51:415-444, 2007.
2. D. Sun and S. Garimella, "Numerical and Experimental Investigation of Solidification Shrinkage", *Numerical Heat Transfer A*, 52:145-162, 2007.
3. R. Mudryy, "Thermo-Mechanical Modeling of Melt Casting of Explosives", *Proc. of AIChe Annual Meeting*, Salt Lake City, Utah 2010.
4. U. Teipel, Energetic Materials, Wiley-VCH, Berlin, Germany, 2005.
5. R. Mudryy, "Solidification Modeling of DNAN Based Explosive Compositions", *Proc. of 42nd International ICT Conference*, Karlsruhe, Germany, June 27- July 01, 2011.
6. ANSYS FLUENT and ANSYS MECHANICAL: www.ansys.com
7. A. Wood, S. Ritchie, and G. Bell, "An Efficient Implementation of the Enthalpy Method", *International Journal for Numerical Methods in Engineering*, vol.17, pp. 301-305, 1981.
8. V. Voller and C. Swaminathan, "Generalized source-based method for solidification phase change", *Numerical Heat Transfer B*, vol. 19, pp. 175-189, 1991.
9. D. Sun, S. Garimella, S. Singh, and N. Naik, "Numerical and Experimental Investigation of the Melt Casting of Explosives", *Propellants, Explosives, Pyrotechnics*, vol. 30, pp. 369-80, 2005.
10. NOVAFLOW&SOLID software, NOVACAST AB, Sweden: www.novacast.se.
11. D. Wiegand, "Mechanical failure of composite plastic bonded explosives and other energetic materials", *Proc. 12th Int. Detonation Symposium*, San Diego, CA, 11-16 Aug 2002, Office of Naval Research, Department of Navy, pp. 744-750.
12. R. Mudryy and L. Nastac, "Numerical Modeling and Simulation Analysis of Casting of Energetic Material in a Projectile Controlled by the ACH Solidification and Mechanical Vibration Technologies," *AICHE2012*, October 28-November 2, Pittsburgh, PA, 2012.
13. R. Mudryy and L. Nastac, "Computational Multi-Phase Modeling of Cast Energetic Materials", *Proc. of the Modeling of multi-scale Phenomena in Materials Processing – III,* Eds: A. Sabau, L. Nastac and A. Rollett, TMS2013, San Antonio, Tx, March 3-7, 2013, *EPD Congress 2013,TMS, pp. 31-38, 2013.*
14. R. Mudryy and L. Nastac, "Solidification Processing of Cast Energetic Materials", *Proc. of the Symposium: Solidification, Deformation and Related Processing,* The 8th Pacific Rim Intern. Conference on Advanced Materials and Processing, *TMS*, Hawaii, August 4-9, 2013.

SHAPE CASTING:
5th International Symposium 2014

Solidification and Microstructure I

Shape Casting: 5th International Symposium 2014
Edited by: Murat Tiryakioğlu, John Campbell, and Glenn Byczynski
TMS (The Minerals, Metals & Materials Society), 2014

THERMODYNAMICS-BASED COMPUTATIONAL APPROACH TO Al-Cu ALLOYS: GRAIN REFINEMENT

J.H. Li[1*], C. Promer[2], A. Jahn[2], B. Oberdorfer[2], S. Wurster[3], F. Martin[4], P. Schumacher[1, 2]

[1*] Chair of Casting Research, Department of Metallurgy, University of Leoben, A-8700, Leoben, Austria (jie-hua.li@hotmail.com)
[2] Austrian Foundry Research Institute, Parkstrasse 21, Leoben, Styria, A-8700, Austria
[3] Department of Materials Physics, University of Leoben, A-8700, Leoben, Austria
[4] Department of Physical Metallurgy and Materials Testing, University of Leoben, A-8700, Leoben, Austria

Keywords: Al-Cu alloy; ThermoCalc software; Solute segregation; Growth restriction factors.

Abstract

To describe the solute effect on grain refinement, the growth restriction factor (Q) in multicomponent multiphase Al alloys has been often evaluated using a simple summation of the Q values of the individual constituents taken from the binary alloy diagram. Such kind of evaluation can lead to mistakes, or completely fail when an intermetallic phase, even in a trace amount, solidifies prior to the primary α-Al. A more accurate method to evaluate growth restriction factor (Q_{true}) from thermodynamic descriptions is to calculate the initial slope in the development of constitutional supercooling (ΔT) with the phase fraction of the growing solid phase (f_s). In this contribution, ThermoCalc software (with TTAl5 database) was used to evaluate the Q_{true} in a series of Al-Cu based alloys with Ti, Zr and Sc additions. This investigation demonstrates that thermodynamic-based alloy design can provide a significant tool to develop novel Al alloys.

Introduction

Al-Cu based alloys have been widely used in the automotive and aerospace industry due to their high yield strength and good fatigue resistance. However, their large freezing ranges lead to a high hot tearing tendency when compared to Al-Si based alloys. Grain refinement of Al-Cu based alloys can improve their castability, in particular improving hot tearing. However, there is still a debate on the nucleation and subsequent mechanisms for grain refinement.

Grain refinement of Al alloy has been extensively investigated for several decades both in industry and in academic, not only for developing efficient grain refiners, but also for achieving a better understanding of the grain refinement mechanism[1-12]. During grain refining of Al alloys, Al-Ti-B grain refiners have been widely investigated, due to their higher nucleation potency and wider industrial application. Various theories regarding the grain refinement mechanisms of Al-Ti-B refiners, such as the particle theory, the phase diagram theory, the duplet nucleation theory, the peritectic bulk theory, have been proposed and reviewed in [1]. Recently, the free growth theory[1-5], and modified free growth theory[6] have also been proposed. Despite of the difference between these theories, it is generally accepted that Ti has a multiple role within the melt. Firstly, Ti provides a substrate in the form of TiB_2. Secondly, excess Ti provides an enriched Ti region leading to the formation of an Al_3Ti monolayer necessary for the nucleation of Al on the stable boride substrates (TiB_2)[9-12]. Thirdly, excess Ti provides an effective growth restriction factor (Q)[3, 4, 6, 7]. In other words, the presence of excess Ti affects both the heterogeneous nucleation and growth of Al alloys. The combined effects of enhanced copious potent nuclei and growth restriction result in the formation of desirable, small, uniform, equiaxed Al grains.

The growth restriction factor (Q) is proportional to the constitutional undercooling at the dendrite tip and can directly be used as a criterion for the grain refinement in Al alloys with strong potential nucleation particles. To describe the solute effect on grain refinement, the growth restriction factor (Q) in multicomponent multiphase alloys has often been evaluated using a simple summation of the Q values of the individual constituents taken from the binary alloy diagram. The evaluated Q is denoted as Q_{sum}. Such kind of evaluation can lead to mistakes, or completely fail when an intermetallic phase (e.g. Al_3Zr, Al_3Ti), even in trace amounts, solidify prior to the primary Al matrix[13,14]. Another method to evaluate growth restriction factor (Q) is a quasi-binary equivalent method[15]. The evaluated Q is denoted as Q_{Eq}. Although both methods have been validated, a more accurate method to evaluate growth restriction factor (Q) is to calculate the initial slope in the development of constitutional supercooling (ΔT) with the phase fraction of the growing solid phase (f_s) from thermodynamic descriptions[14, 16]. The evaluated Q is denoted as Q_{true}.

In this contribution, ThermoCalc software (with TTAl5 database) was used to evaluate the Q_{true} in a series of Al-Cu based alloys with Ti, Zr and Sc additions. The evaluated Q_{true} was compared with the evaluated Q_{sum} and Q_{Eq} using other two different methods. This investigation demonstrates that thermodynamic-based alloy design can provide a significant tool to develop novel Al alloys.

Experimental material and procedures

A series of Al-4.0Cu based alloys (wt. %, used throughout the paper, in case not specified otherwise) with Ti, Zr and Sc additions were prepared using commercial purity Al ingots (99.7), an Al-25 Cu master alloy pre-prepared in induction melting using high purity Al (99.998) and high purity Cu (99.999), and Al-10Ti, Al-10Zr, Al-2.2Sc master alloys, respectively. The nominal compositions are listed in Table I. It should be noted here that some trace elements (i.e. Fe, Mn and Si) are also present in Al-4.0Cu based alloys. However, these elements are not included for ThermoCalc calculations. ThermoCalc calculations (non-equilibrium (Scheil)) were performed to evaluate Q_{true}.

Each batch, weighting about 6 kg, was melted in a resistance furnace at 720 °C. A reference sample was taken from the melt in order to identify the grain size before inoculation. The nucleant particles (TiB_2) were added using commercial grain refiner rod (Al-5.0Ti-1.0B). The concentration of the nucleant particles (TiB_2) is about 0.01 wt % (100 ppm).

The melt was stirred with a graphite rod for 20 s after inoculation. The samples were taken from the melt at 5 min after the grain refiner addition and tested using a standard TP-1 method. The samples were sectioned 38 mm from the bottom surface. Standard metallographic procedures were performed to prepare these sections for grain size measurements. The samples were etched using a mixture of 13 g boric acid, 35 g HF, 800 ml H_2O at a voltage of 20 V for 45 seconds. All images used for grain size measurement were taken from the centre of the samples using optical microscopy in a polarized mode at the same magnification. The reported grain sizes were measured from at least 20 images using line-intersect method.

Table I: The nominal composition of a series of Al-4.0Cu based alloys with Ti, Zr, and Sc additions. (*: free Ti; ♯: total Zr)

Alloy No.	Cu	Si	Fe	Mn	Ti*	Zr♯	Sc	Al
1	4.0	-	-	-	-	-	-	Bal.
2	4.0	-	-	-	0.1	-	-	Bal.
3	4.0	-	-	-	0.2	-	-	Bal.
4	4.0	-	-	-	-	0.25	-	Bal.
5	4.0	-	-	-	-	0.5	-	Bal.
6	4.0	-	-	-	-	-	0.25	Bal.
7	4.0	-	-	-	-	-	0.5	Bal.
8	4.0	-	-	-	-	0.25	0.25	Bal.
9	4.0	-	-	-	0.1	-	0.25	Bal.
10	4.0	-	-	-	0.1	0.25	-	Bal.

Results

ThermoCalc calculation Al-4.0Cu based alloys with Ti, Zr and Sc additions

Growth restriction factor (Q) can be evaluated using a simple summation of the Q values of the individual constituents taken from the binary alloy diagram, as listed in equation 1.

$$Q_{sum} = \Sigma_i m_i c_0^i (k_i - 1) \qquad (1)$$

where, for each element i, m is the liquidus gradient, c_0 is the composition, and k is the binary partition coefficient. The relative data for determining Q is listed in Table II.

Table II: Phase diagram data for determining Q in binary Al alloys[15].

Element	k	m	Max. conc. (wt.%)	m(k-1)
Cu	0.17	-3.4	33.2	2.8
Ti	7-8	33.3	0.15	~220
Zr	2.5	4.5	0.11	6.8
Si	0.11	-6.6	~12.6	5.9
Cr	2.0	3.5	~0.4	3.5
Ni	0.007	-3.3	~6	3.3
Mg	0.51	-6.2	~3.4	3.0
Fe	0.02	-3.0	~1.8	2.9
Mn	0.94	-1.6	1.9	0.1
Sc	0.64	-9.1	0.55	3.3

Growth restriction factor (Q) can be also evaluated using a quasi-binary equivalent method [15], as listed in equation 2, for Al-Si based alloys.

$$Si_{Eq} = Si + \Sigma Si_{Eq}^{x_i}[wt\%] \qquad (2)$$

where, equivalent is calculated as the sum of the contribution of the individual elements. For additional element (X_i, e.g. Cu, Ti), $Si_{Eq}^{X_i}$ can be determined using a mathematical model (equation 3).

$$X_{Eq}^{i} = a_0^{X_i} + b_0^{X_i} X_i + c_0^{X_i} X_i^2 \quad (3)$$

where, $a_0^{X_i}$, $b_0^{X_i}$, $c_0^{X_i}$ are the polynomial coefficients, respectively, as listed in Table III, X_i is the concentration of the elements (wt.%). However, it should be noted that the coefficients listed in Table III were validated in Al-Si based alloys. For Al-Cu based alloys investigated here, the coefficients may not be valid, and should be determined in an Al-Cu system, thus questioning the validation of this approach.

Inserting values calculated from equation 2 into equation 1, the Q_{Eq} values for Alloys 1-3 are determined. Due to a lack of the polynomial coefficients of Zr and Sc, no attempt was taken here to evaluate the Q_{Eq} values for Alloys 4-10.

Table III: Polynomial coefficients for various binary Al-X_i alloys representing the most common major and minor element of the Al-Si based alloys. $a_0 = 0$ for the elements presented in this table[15].

Element	b_0	c_0
Ti	-0.8159	0.009927
Zr	-	-
Ni	0.5644	-0.0285
Mg	0.0258	-0.0088
Fe	0.6495	0.0003
Cu	0.529	-0.027
Mn	0.8221	-0.0349

A more accurate method to evaluate growth restriction factor (Q) is to calculate the initial slope in the development of constitutional supercooling (ΔT) with the phase fraction of the growing solid phase (f_s) using equation 4.

$$Q_{true} = \frac{d\Delta T_c}{df_s}|_{f_s \to 0} \quad (4)$$

where, ΔT_c is the rate of development of constitutional supercooling, and f_s is the fraction solid. For clarity, Figure 1 shows the evaluation of Q_{true} in Alloy 1 (Al-4Cu based alloy).

The determined Q values for Alloys 1-10 are listed in Table IV. When comparing the evaluated Q values using three different methods, it was found that (i) the Q value evaluated using equation 1 is very close to that evaluated using equation 4, if the primary phase is α-Al matrix, rather than Al_3Ti (Alloy 3) or Al_3Zr (Alloy 5). This indicates that equation 1 is valid, and can be used to evaluate the Q value in dilute Al alloys. (ii) equation 2 is not valid when Ti is present, indicating that a re-evaluation of the coefficients is required in Al-Cu based alloys. (iii) In the case of Ti and Zr addition, the Q value evaluated using equation 4 (24.2) is much less than that evaluated using equation 1 (32.9). (iv) In the case of the addition of Sc (Alloys 6, 7) and Zr (Alloys 4, 5), the evaluated Q_{true} value is very close despite of higher concentrations of each element, indicating that an asymmetric Q values is obtained once an primary intermetallic is formed. This is not the case for Q_{sum}.

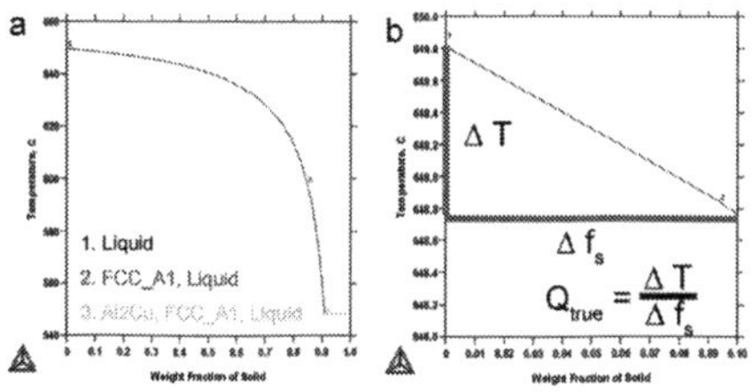

Figure 1 ThermoCalc calculation (a), and the evaluation of Q_{true} (b) in Alloy 1 (Al-4Cu based alloy). The primary phase is FCC_A1 (α-Al) solidified from liquid. Eutectic Al_2Cu forms subsequently.

Table IV: Comparison of Q values determined using equation 1, equation 2 and equation 4.

	Equation 1 (Q_{sum})	Equation 2 (Q_{Eq})	Equation 4 (Q_{true})
Alloy 1	11.2	9.9356	10
Alloy 2	33.2	17.93	34
Alloy 3	44.2*	7.09	35
Alloy 4	12.9	-	11.3
Alloy 5	14.6	-	11.8
Alloy 6	12.03	-	11
Alloy 7	12.85	-	11.1
Alloy 8	13.73	-	11.6
Alloy 9	34.03	-	32.6
Alloy 10	32.9	-	24.2

*Note: For Alloy 3, Q_{sum} value (44.2) was evaluated using 0.15 Ti, although 0.2 Ti was added.

Experimental investigation on Al-4.0Cu based alloys with Ti, Zr and Sc additions

Figure 2 shows a typical as-cast microstructure of Alloy 1 (Al-4Cu based alloy). The grain size is very large, about 705 ± 68 μm. The addition of TiB_2 (100 ppm) and excess Ti (0.1, Alloy 2) greatly decreases the grain size to 70 ± 5 μm, as shown in Figure 3. However, further increasing excess Ti (0.2, Alloy 3) results in a slight increase of grain size (80 ± 4 μm) due to the formation of primary Al_3Ti phase prior to α-Al phase (not shown here).
The addition of Zr (0.25, Alloy 4) does not greatly decrease the grain size. The grain size is about 142 ± 13 μm (Figure 4). Further increasing the Zr content (0.5, Alloy 5) does not result in a decrease in grain size. The grain size is about 131 ± 10 μm (Figure 5).
The addition of hypoeutectic Sc (0.25, Alloy 6) does not greatly decrease the grain size. The grain size is about 282 ± 44 μm (Figure 6), much higher than that (142 ± 13 μm) with 0.25 Zr addition (Figure 5). This can be attributed to the higher solute solubility of Sc compared with Zr in Al (Table II). At the same addition level (e.g. 0.25), no significant fraction of primary Al_3Sc phase form as nucleation sites for α-Al. Further increasing Sc (0.5, Alloy 7) decrease the grain

size by additional Q but remains hypoeutectic (less than 0.55, Table II). The grain size is about 140 ± 8 μm. However, the combined addition of Zr and Sc (Alloy 8) greatly decrease the grain size. The grain size is about 110 ± 10 μm (Figure 7).

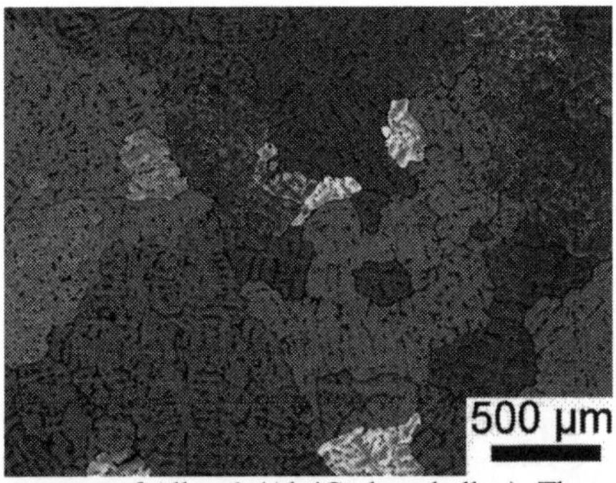

Figure 2 As-cast microstructure of Alloy 1 (Al-4Cu based alloy). The grain size is about 705 ± 68 μm.

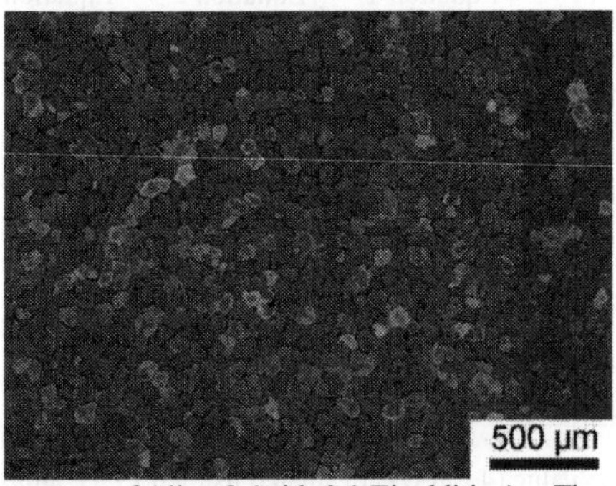

Figure 3 As-cast microstructure of Alloy 2 (with 0.1 Ti addition). The grain size is about 70 ± 5 μm.

Figure 4 As-cast microstructure of Alloy 4 (with 0.25 Zr). The grain size is about 142 ± 13 μm.

Figure 5 As-cast microstructure of Alloy 5 (with 0.5 Zr addition). The grain size is about 131 ± 10 μm.

Figure 6 As-cast microstructure of Alloy 6 (with 0.25 Sc addition). The grain size is about 282 ± 44 μm.

Figure 7 As-cast microstructure of Alloy 8 (with 0.25 Zr and 0.25 Sc). The grain size is about 110 ± 10 μm.

Discussions

The Q values (Table IV) evaluated using equation 1 and equation 4 can be used to interpret the change of grain size (Figures 2-7) observed using TP-1 test. For example, in the case of Ti addition, the presence of excess Ti (0.1, Alloy 2) increases growth restriction factor (Q) sharply (Table IV) from 10 (Alloy 1) to 34 (Alloy 2). A higher growth restriction factor (Q) leads to a significant decrease of the grain size from 705 ± 68 μm (Figure 1) to 70 ± 5 μm (Figure 3, as shown in Figure 8, which is similar to cases of TiB_2 addition in Al alloys. This strongly indicates that the evaluation of growth restriction factor using equation 1 (Q_{sum}) and equation 4 (Q_{true}) is reliable.

Comparing the Q_{true} and Q_{sum} values for Alloy 2 and Alloy 3, it is apparent that only Q_{true} predicts for the identical cooling condition. Similar grain size is obtained for Alloy 2 (70 ± 5 μm) and Alloy 3 (80 ± 4 μm). Similarly, Q_{true} also predicts similar grain size for Alloy 4 (142 ± 13 μm) and Alloy 5 (131 ± 10 μm). Overall, Q_{true} gives a better evaluation of growth restriction factor.

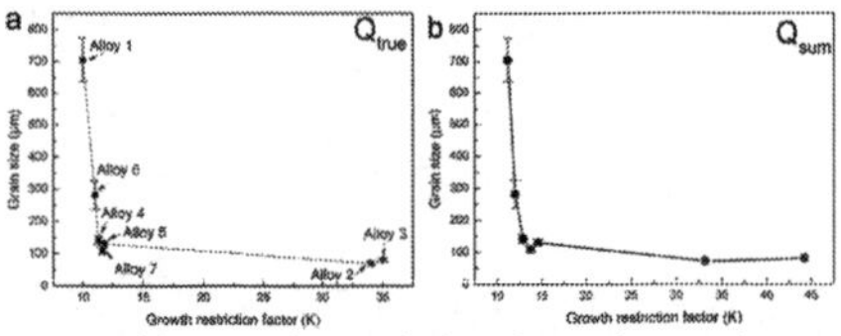

Figure 8 Grain size decreases with growth restriction factor increasing.

In the case of Zr and / or Sc addition, the evaluated growth restriction factor (Q_{sum} and Q_{true}) is nearly the same (about 11, Table IV, Alloys 6-8). However, the grain size changes greatly, especially for a combined addition of Zr and Sc (Figure 7). One important question arises, why the grain size changes greatly even with the same growth restriction factor. Is growth restriction factor a dominant factor affecting the grain refinement, or should any other mechanisms (i.e. heterogeneous nucleation and interactions between solutes) also be taken into consideration ?

A strong interaction between solutes, i.e. forming the τ phase during solidification, has been reported that Al-Si-Ti ternary alloys[13]. This interaction affects the evaluation of growth restriction factor. However, unlike Al-Si-Ti systems, Al-4.0Cu based alloys do not exhibit a strong interaction between solutes (i.e. Cu, Ti), because no Cu-rich phase forms during solidification.

The reported Zr-poisoning between TiB_2 and Zr[17] suggests that an interaction between TiB_2 and Zr to forming ZrB_2, which does not act as a good nucleation site. However, here no TiB_2 is present, and above the composition for the onset of the peritectic reaction (0.11, Table II), Al_3Zr is formed acting as nucleation site and the remaining Zr as a poor growth restrictor. This hypothesis can be further supported by the lower evaluated growth restriction factor (Q) (Table IV, Alloy 10).

The interaction between solutes definitely affects the heterogeneous nucleation of α-Al, i.e. enhancing or deactivating the nucleation potency. In the case of Ti addition, the presence of excess Ti and thus the possible formation of Ti-rich layers will reduce the growth velocity of the nucleated crystals and increase the maximum undercooling achievable before recalescence[7]. This allows more particles to be active in nucleation and, consequently, increases the number density of the active particles, giving rise to a finer grain size. In the case of combined addition of Zr and Sc, three aspects should be taken into consideration. Firstly, the formation of $Al_3(Zr, Sc)$ reduces the lattice parameter of both the stable DO_{23} and metastable $L1_2$ Al_3Zr, thus decreases the mismatch between the nucleation sites particles and α-Al matrix (Table V), and finally promotes the nucleation of α-Al[18,19]. Secondly, the combined addition of Zr and Sc reduces the solute solubility of each other[18]. More nucleating sites are available for the nucleation of α-Al. Thirdly, in the case of single Sc addition, most Al_3Sc (if any, once Sc content is above 0.55) formed through an eutectic reaction will be pushed to the grain boundaries, while in the case of the combined Zr and Sc addition, a so-called shell-core $Al_3(Zr, Sc)$ formed through a eutectic reaction, will be surrounded by α-Al, as a good nucleation site. For clarity, a schematic diagram is shown in Figure 9. Some typical SEM images taken from Alloy 5 (0.5 Zr) and Alloy 8 (0.25 Zr and 0.25 Sc) are shown in Figure 10. Clearly, in the case of a single Zr addition (0.5), an Al_3Zr particle was located at the center of the grain, and acted as a nucleation site (Figure 10a). In the case of combined additions of Sc and Zr, $Al_3(Zr, Sc)$ forms during solidification. Al_3Zr solidified as a core and was covered by an Al_3Sc shell (Figure 10c).

Table V: Lattice parameters and mismatch between the particles and Al matrix[18].

Particles	Lattice parameters (nm)	Mismatch	Orientation ship
Al_3Ti	DO22 a = 0.3851 c = 0.8608	0.23	$(221)_{Al}$ // $(011)_{Al3Ti}$, $[110]_{Al}$ // $[210]_{Al3Ti}$
Al_3Zr	DO23, $L1_2$ a = 0.4048	0.8	Any plane and directions
Al_3Sc	$L1_2$ a = 0.4105	< 1.5 (1.2)	Any plane and directions

The diffusion of solute elements (i.e. Ti, Sc, Zr and Cu) should also be taken into consideration when discussing the grain refinement of Al alloys. The partition behaviour of Ti (k = 7-8 in pure Al) is much stronger than that of Cu (k = 0.17), Zr (k = 2.5) and Sc (k = 0.64). It can be expected that Ti partitions more easily into the nucleants than Cu, thus forming a Ti-rich region. The possible formation of a Ti-rich region will affect the interface structure (i.e. ordering or disordering)[20], reducing the interface energy, and thus reducing the required undercooling for nucleation[1]. Then, smaller nucleants can be activated. This suggestion is fully consistent with a modified free growth model[6], which suggests that the nucleation potency of inoculation particles is reduced by the solute field (Ti-rich region) that develops close to existing, growing equiaxed grains under near isothermal conditions. Solute suppressed nucleation leads to much lower nucleated grain densities, higher nucleation undercooling and longer times to recalescence when further nucleation events are halted. Thus, it can be concluded that activating more nucleants, rather than growth restriction factor (Q), is the key factor for grain refinement. In other words, a high growth restriction factor is necessary for nucleants to be activated, but the enhanced heterogeneous nucleation is the dominant factor for grain refinement of Al alloys. In foundry practices, attempts should be made to enhance the heterogeneous nucleation of α-Al, and thus to achieve a desirable, small, uniform, equiaxed Al grains.

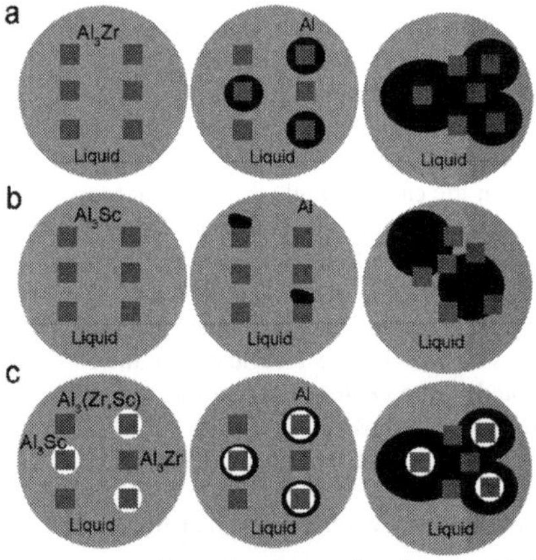

Figure 9 Schematic diagram showing (a) peritectic reaction for Zr addition (higher than 0.11), (b) eutectic reaction for Sc addition (higher than 0.55), and (c) eutectic reaction for Zr and Sc addition.

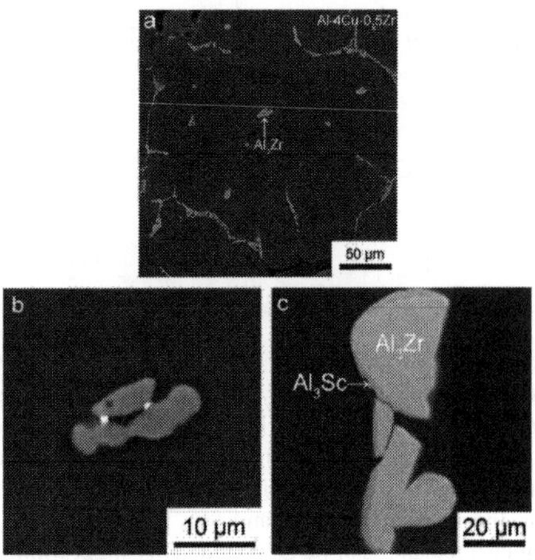

Figure 10 SEM images taken from Alloy 5 (0.5 Zr) and Alloy 8 (0.25 Zr and 0.25 Sc), showing an Al_3Zr particle was located at the center of the grain (a,b) in the case of Zr (0.5) addition, and another Al_3Zr was covered by an Al_3Sc shell (c) in the case of Zr (0.25) and Sc (0.25) addition.

Conclusion

1. In Al-Cu based alloy, the evaluated Q_{sum} and Q_{true} values are nearly identical due to low interaction between the alloying constituents.

2. A more accurate method to evaluate growth restriction factor is to calculate the Q_{true} using equation 4 at complex alloy systems. Under identical cooling condition in the TP-1 test, Q_{true} predicts grain size more accurately than Q_{sum}.

3. For dilute Al alloy, the evaluated Q values can be used to interpret the change of grain size observed using TP-1 test.

4. More attention should be paid when the summation of Q values of the individual constituents taken from the binary alloy diagram is used to evaluate growth restriction factor (Q) because a mistake or a complete fail may occur when α-Al is not the primary phase.

5. The presence of excess Ti results in an enhanced grain refinement due to increased growth restriction factor and enhanced heterogeneous nucleation.

6. The combined Zr and Sc additions enhanced the heterogeneous nucleation, and thus resulting in an enhanced grain refinement, although the growth restriction factor remained unchanged.

7. Heterogeneous nucleation is a dominant factor for grain refinement of Al alloys.

References

[1] A.L. Greer, A.M. Bunn, A. Tronche, P.V. Evans, D.J. Bristow. Acta Mater., 48 (2000) 2823-2835.

[2] T.E. Quested, A.L. Greer. Acta Mater., 52 (2004) 3859-3868.

[3] T.E. Quested, A.L. Greer. Acta Mater., 53 (2005) 4643-4653.

[4] T.E. Quested, A.T. Dinsdale, A.L. Greer. Acta Mater., 53 (2005) 1323-1334.

[5] T.E. Quested, A.L. Greer. Acta Mater., 53 (2005) 2683-2692.

[6] D. Shu, B.D. Sun, J. Mi, P.S. Grant. Acta Mater., 59 (2011) 2135-2144.

[7] H. Men, Z. Fan. Acta Mater. 59 (2011): 2704-2712.

[8] H.T. Li, M. Xia, Ph. Jarry, G.M. Scamans, Z. Fan. Journal of Crystal Growth. 314 (2011) 285-292.

[9] P. Schumacher, A.L. Greer. Mater. Sci. Eng. A, 181-182 (1994): 1335-1339.

[10] N. Iqbal, N.H. van Dijk, T. Hansen, L. Katgerman, G.J. Kearley. Mater. Sci. Eng. A, 386 (2004): 20-26.

[11] N. Iqbal, N.H. van Dijk, S.E. Offerman, N. Geerlofs, M.P. Moret, L. Katgerman, G.J. Kearley. Mater. Sci. Eng. A, 416 (2006): 18-32.

[12] N. Iqbal, N.H. van Dijk, S.E. Offerman, M.P. Moret, L. Katgerman, G.J. Kearley. Acta Mater., 53 (2005): 2875-2880.

[13] J. Gröbner, D. Mirković, R. Schmid-Fetzer. Mater. Sci. Eng. A, 395 (2005): 10-21.

[14] R. Schmid-Fetzer, A. Kozlov. Acta Mater. 59 (2011): 6133-6144.

[15] A.M. Mitrašinović, F.C. Robles Hernández. Mater. Sci. Eng. A, 540 (2012): 63-69.

[16] G.N. Haidemenopoulos, A.I. Kastsamas, H. Kamoutsi. Metall. Mater. Trans. A., 41 (2010) 888-899.

[17] A.M. Bunn, P. Schumacher, M.A. Kearns, C.B. Boothroyd, A.L. Greer. Mater. Sci. Tech., 15 (1999): 1115-1123.

[18] C. Booth-Morrison, Z. Mao, M. Diaz, D.C. Dunand, C. Wolverton, D.N. Seidman. Acta Mater. 60 (2012): 4740-4752.

[19] O. Beeri, D.C. Dunand, D.N. Seidman. Mater. Sci. Eng. A, 527 (2010): 3501-3509.

[20] J.S. Wang, A. Horsfield, P.D. Lee, P. Brommer. Phys. Rev. B, 82 (2010) 144203.

Shape Casting: 5th International Symposium 2014
Edited by: Murat Tiryakioğlu, John Campbell, and Glenn Byczynski
TMS (The Minerals, Metals & Materials Society), 2014

Control Diffusion Solidification (CDS): An Overview of Mechanism and Application

Reza Ghiaasiaan[1], Sumanth Shankar[1],and Diran Apelian[2]

[1]Light Metal Casting Research Centre (LMCRC), McMaster University, Hamilton, ON, Canada L8S 4L7
[2]Metal Processing Institute (MPI), Worcester Polytechnic Institute (WPI), Worcester, MA, USA 01609.

Keywords: control diffusion solidification, CDS, near net shaped casting, Aluminum alloys, Al wrought alloys, Al-Zn-Mg-Cu, Al 7050

Abstract

The Controlled Diffusion Solidification (CDS) is an innovative casting technology that is being successfully employed in the near net shaped casting of Al based wrought alloys; specifically the 2xxx and 7xxx series. The novelty of the CDS technology is that it could mitigate the notable casting defect in conventional solidification of the Al wrought alloys, namely, hot tearing and enables the alloy to be cast into near net shaped components. This publication presents an overview of the phenomenological mechanism that lead to the non-dendritic morphology of the primary Al phase in the as-cast microstructure that facilitates a viable opportunity to cast the wrought alloys into high integrity and structural near net shaped castings. The salient mechanical properties of the cast component with AA7050 alloy in various heat treatment tempers will be presented along with microstructural characterization. Additionally, potential application for the technology to be adapted to various Al shaped casting processes will be explored and presented.

Introduction

Amongst all the families of aluminum alloys, the 7xxx series alloys (Al-Zn-Mg-Cu) and 2xxx series alloys (Al-Cu-Mg) lend themselves to be used in the structural automotive and aerospace castings due to their high strength to weight ratio [1] and good ductility. Typically, the 7xxx series Al wrought alloys in their as-cast condition do not offer their best mechanical properties and that is because of their microstructural deficiencies such as the intermetallic phases, significant solute segregation in the large primary Al grains, solid solubility limitations and most importantly, macro-segregation and hot tearing [2, 3, 4]. The hot tearing could be alleviated or mitigated during solidification by controlling the process parameters to either significantly refine the Secondary Dendrite Arm Spacing (SDAS) of the primary Al phase [5, 6] or alter the morphology of this phase to non-dendritic. Presently, the Semi-Solid Metal (SSM) processing techniques in thixoforming and rheocasting are popular routes to modify the dendritic morphology of the solidifying Al phase by applying physically or thermally induced convection to enhance the copious nucleation event and uniform distribution of the Al nuclei [7, 8, 9] within the solidifying alloy melt. However, these SSM processes are not commercially favored by industry due to the high capital investment and complexity (lack of reliability) in these processes [10, 11]. Further, recent publications [12, 13] have attributed the further difficulty in the feeding of the liquid in the inter-dendritic regions during solidification as accentuated by the non-Newtonian flow behavior of the liquid alloys to further complicate the SSM process methodologies. Therefore, there is a real and urgent need for the Al casting industry, especially, the automotive and aerospace sectors, to develop alloys and processes to enable high integrity and structural near net shaped castings from high performance alloys such as the 7xxx series of wrought alloys.

The Controlled Diffusion Solidification (CDS) is a viable processing technology that enables near net shaped castings of Al wrought alloys [14, 15, 16, 17] by mitigating the inherent hot tearing defect during solidification by altering the morphology of the primary Al phase to non-dendritic. The CDS process involves the mixing of two precursor-melts at specific melt temperatures and solute compositions, respectively to obtain a slurry of the resultant desired alloy at around its liquidus temperature; subsequently (almost immediately), this resultant alloy is introduced (pressure or pressure-less) into a mould cavity to enable near net shaped casting of the component [18].

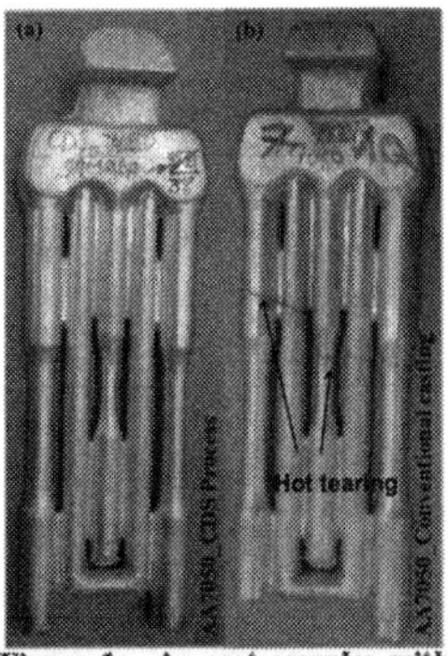

Figure 1: As cast samples with AA7050 in tilt pour gravity casting process, (a) CDS and (b) conventional with hot tears.

Figure 1 (a) and (b) compare the microstructure of the cast part obtained by conventional solidification of AA7050 alloy (Al-Zn-Mg-Cu) and the CDS process technology, respectively; wherein, the former presents significant problems while the latter presents a good cast microstructure devoid of such defects. In Figure 1 (a) and (b), shows micrographs for conventionally cast 7050 and that obtained by the CDS process, respectively; wherein, significant hot cracking is repeatedly observed in the former part and this was repeatedly absent in the latter presenting a good casting of the AA7050 alloy.

Background

Prior to addressing the working mechanisms of the CDS technology, it would benefit to address our knowledge of conventional casting processes in light of the fundamentals of the solidification involved in the same. It is common knowledge [19] that the three legs on which solidification stands on are heat, mass and fluid flow within the various phases. The cellular/dendritic morphology of the primary solidifying phase is a result of the instability of the growing solid/liquid interface in a direction opposite to the heat extraction [19, 20, 21]; it stems from the pile-up of the solute atoms ahead of the solid/liquid interface resulting in a solute gradient in the liquid coupled with the temperature gradient in the liquid ahead of the interface as effected by the solidification process conditions. Figure 1 (a) to (d) presents the schematic images of the solute and thermal conditions in the liquid ahead of the solidifying interface, that create the criteria of constitutional undercooling and result in the instability of the growing solid/liquid interface to give rise to cellular/dendritic primary phase front.

An enabling solidification processing technology for near net shaped casting of single phase Al wrought alloys will arise from avoiding the instability of the growing solid/liquid (S/L) interface during solidification; in other words, the conditions that lead to the constitutional undercooling will have to be mitigated. The high value of the velocity of the S/L interface, R, such as in rapid solidification processes, would provide a stable and unperturbed interface growth resulting in a planar interface; however, such high values of R is rarely feasible to reproduce in a commercial casting operation for near net shaped manufacturing. The alternate route to mitigate constitutional undercooling would be to maintain a nearly homogeneous solute concentration in the liquid ahead of the S/L interface (no solute gradient) which would result in a zero gradient trace of the liquidus temperature in the liquid, ahead of the S/L interface, therefore maintaining a stable S/L interface growth. This condition would be represented by a horizontal line for the solute gradient in the liquid at C_L^* in Figure 2 (b) to obtain a horizontal (zero gradient) trace of the liquid temperature in the liquid at T^* in Figure 2 (d). In a commercial casting process this condition to favor non-dendritic solidification could be attained by inducing significantly rapid solute transport in the liquid ahead of the S/L interface by forced convection in the same liquid. Once, the alloy liquid in the early stages (low fraction solid) of the two solidification regime is introduced into the mould cavity, the growth of the S/L interface within the mould would renew the solute pile-up at the interface and the constitutional undercooling condition will begin to

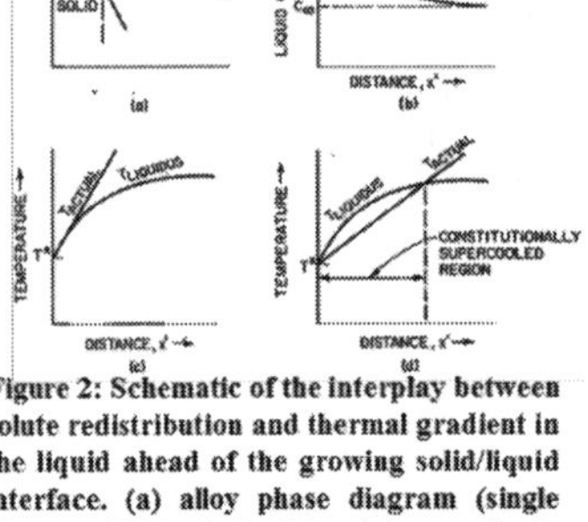

Figure 2: Schematic of the interplay between solute redistribution and thermal gradient in the liquid ahead of the growing solid/liquid interface. (a) alloy phase diagram (single phase solidification), (b) solute profile in liquid, (c) thermal profile in liquid and (d) constitutional undercooling [19].

appear and strengthen; because, the forced convection of the liquid will be removed within the mould. Hence, to continue maintaining a stable S/L interface with the mould, the favorable condition would be to enable innumerable and evenly distributed growing nuclei of the primary phase with the alloy so that the proximity of two growing S/L interface is such that the solute fields in the liquid ahead of each interface would impinge before a pronounced renewal of the solute pile-up and gradient ahead of them, respectively. This would result in maintaining a nearly stable S/L interface until the completion of the solidification and present favourable non-dendritic primary phase morphology.

In summary, a nearly stable S/L interface growth during solidification resulting in a nearly non-dendritic morphology of the primary Al phase in casting Al alloys would be attained by creating the following environment during the solidification processing:

- Significantly large nucleation event of the primary Al phase in the alloy melt during controlled cooling of the liquid from a superheated temperature above liquidus to that of temperature in the two phase (solid + liquid) slurry with low fraction solid ($f_s < 0.3$).
- Simultaneous re-distribution of the Al nuclei in the alloy slurry to achieve a nearly uniform distribution prior to injecting the semi-solid slurry into the mould cavity (usually pressure assisted).

Presently, the notable rheocasting processing routes that attempt to achieve the above mentioned conditions in a commercial process are the Semi Solid Rheocasting (SSR), Sub-Liquidus Casting (SLC), New Rheocasting Process (NRP) and Continuous Rheoconversion Process (CRP) [22, 23] route of the SSM processing attempts to achieve a stable S/L interface growth during solidification by mechanically induced forced convection in the liquid coupled with controlled cooling of the liquid to the semi-solid slurry state (f_s < 0.3) prior to casting the component with a pressure assisted process such as low / high pressure die casting in a meal mould cavity. Each of these rheocasting processing has respective proprietary technologies to achieve the two required conditions for the nearly non-dendritic solidification. The main drawbacks of these technologies in enabling such solidification are the high capital cost of infrastructure, introduction of unwanted artifacts such as oxides from the rapid break down of the Al alloy melt surface and create high scrap and unpredictable flow of the semi-solid slurry into the mould cavity due to lack of fluid flow knowledge. The time required to bring a large volume of alloy melt from a superheated liquid state to the favorable semi-solid state while inducing the desired nucleation and distribution phenomena is significantly high (in minutes) compared to the desired productivity of the cast component from such alloys (in seconds), resulting in the requirement of multiple casting stations being set up to append to the high capital investment in the casting process infrastructure. Hence, none of the rheocasting technologies have found prevalent acceptance and use in the commercial Al casting industry.

Further, the present day rheocasting processing technologies do not lend themselves to casting Al wrought alloy into near net shaped casting; in other words, these technologies do not perform well when casting single phase Al alloys and hence, only used in two-phase casting alloys. Figure 3 (a) and (b) present the temperature drop below the liquidus temperatures of AA7050 Al wrought alloy and Al-7wt%Si casting alloy to attain a semi-solid slurry with a solid fraction, fs = 0.3, respectively. Typically, the wrought alloys have a very narrow range of operating temperature (<7 degrees for AA7050 in Figure 3 (a)) within which the liquid alloy would have to be brought down to from a superheated state while creating the copious nucleation and distribution processes in the rheocasting technology. Whereas, in the Al casting alloys, such as Al-7Si in Figure 3 (b), this operating temperature range is quite significant; thus, presenting a favorable operation window for the casting process to be successful, repeatable and reproducible. The trace of the fs as a function of temperature in the wrought alloy makes them a very difficult candidate for rheocasting process.

The CDS casting technology was innovated to circumvent the above-mentioned problems in the present day rheocasting processing routes:

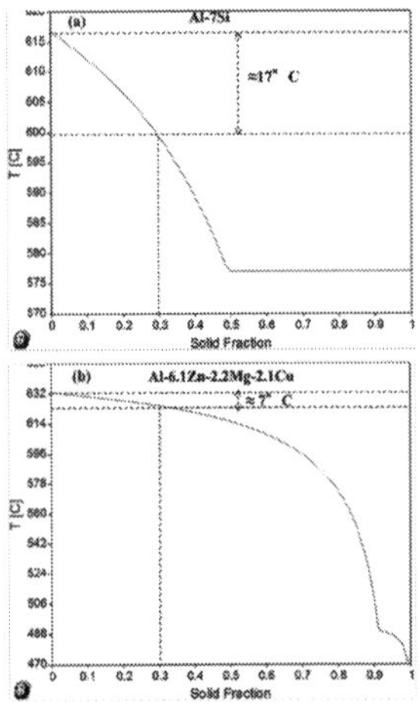

Figure 3: Trace of solid fraction (f_s) as a function of temperature for (a) AA7050 wrought alloy and (b) Al-7Si casting alloy to show the temperature drop to obtain a semi-solid slurry of 0.3 solid fraction

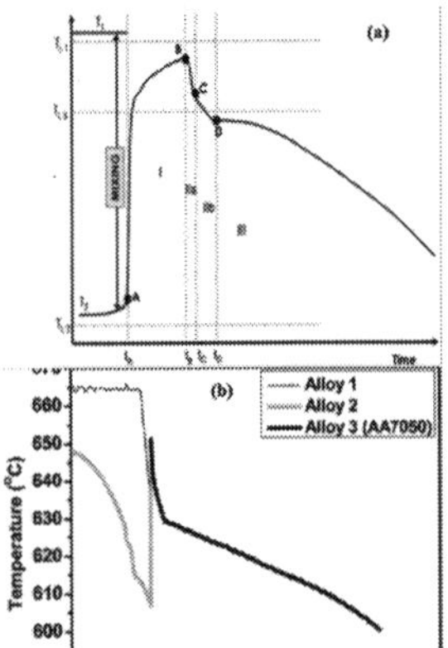

Figure 4: Thermal data obtained during the CDS technology. (a) schematic presenting the salient features in the mechanism [14] and (b) thermal data obtained while casting AA7050 alloy with the CDS technology

- Mixing of the two precursor alloys such that the alloy with the higher thermal mass (temperature and mass fraction) is mixed into the alloy with the lower thermal mass to instantaneously (in seconds) create the desired copious nucleation event of the primary Al while being well distributed in the resultant alloy of the desired composition.
- Homogenization of the temperature field in the resultant alloy (in seconds) as a result of the forced convection from the mixing process to create a nearly uniform temperature distribution in the resultant alloy that is around its liquidus temperature (±5 degrees); thus, creating the favorable liquid slurry with a very low fraction solid uniformly distributed in the entire volume of the large alloy melt.
- Casting of the near liquidus resultant alloy with the copious nuclei of Al phase into a mould cavity with or without the assistance of pressure during metal injection. The laminar flow of the liquid along with the innumerable nuclei of Al in the mould cavity has shown to fill thin wall sections (~2mm) of sand and metal mould cavities without a pressure assisted injection process.
- Solidification of the alloy slurry with the well distributed Al nuclei within, while maintaining the stability of the growing S/L interface to result in a non-dendritic morphology of the primary Al phase.
- The precision of obtaining the alloy melt innumerable and well distributed nuclei of Al phase within as dictated by the nature of mixing the two precursor alloys with a relatively different thermal mass ensures high degree of repeatability and reproducibility in the process and further, enables the adaptation of this technology to cast Al wrought alloys into near net shaped components.

In the 1980s, Apelian et al [24] initially dabbled with the idea of diffusional solidification to enable the rapid-cycle casting of steel, and laid the foundation for the isothermal diffusional solidification principles that purely depended upon the solute redistribution in an isothermal environment to enable solidification. The CDS technology ventures a step further into this philosophy by introducing a combined variance of thermal and solute fields as the initial condition for solidification. Recently, Symeonidis [18] laid the fundamental framework of the mechanism of the CDS technology followed by Khalaf [14], who presented an in-depth analysis of the mechanism of the same. Figure 4 (a) and (b) presents a schematic of the thermal data obtained during the solidification with the CDS technology and the thermal curve obtained during the casting of the AA7050 wrought alloy, respectively; various significant regions of interests are demarcated in the schematic by notations A, B, C and D.

The following critical stages of the CDS mechanism are described with reference to the typical thermal data shown in Figure 4 (a) [14].

Segment AB (Stage I): At this stage of *"mechanical mixing"*, the Alloy1 continuously entering the Alloy2 will break down into small masses (at temperature T1) in the resultant mixture (at a temperature less than T1) to form the resultant Alloy3 (AA7050). The exposure of hot liquid of Alloy1 to the low temperature liquid mixture in this stage of mixing results in the nucleation of broken liquid pockets of Alloy1 resulting in a well distributed and innumerable nucleii of primary Al phase form Alloy 1 with a low solute content.

Segment BCD (Stage II: (a) and (b)): At this stage of *"re-distribution of the thermal and solute fields in the resultant mixture"*, convective cells (which are akin to Bernard cells [25]) form in the mixture where the nucleation and growth of the primary Al phase takes place in a nearly isothermal melt with significant gradients of solute elements in the micro-scale environment. The significantly slow rate of solute homogenization in the resultant mixture of Alloy 3 presents a unique solute distribution in the liquid ahead of the growing S/L interface. Figure 5 (a) presents the unique solute and temperature distribution ahead of the S/L interface in the CDS process; the solute field has a positive gradient ahead of the interface and triggers the diffusion of the solute atoms towards the interface which is opposite to that observed in conventional solidification as shown in Figure 5 (b). Thereby, the trace of the liquidus temperature ahead of the growing S/L interface in the CDS process is negative and opposite to that of conventional solidification. Hence, any gradient in the liquid temperature caused by the extraction of heat in the casting mould will only result in negligible to no constitutional undercooling ahead of the S/L interface during growth and hence maintain the stability of this interface resulting in a non-dendritic morphology of the primary Al phase in the solidified microstructure.

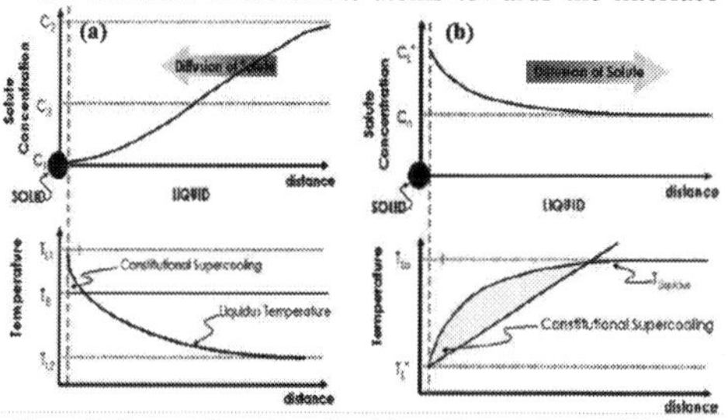

Figure 5: Comparative schematic solute and temperature redistribution regimes ahead of the solidifying solid-liquid interface are presented for (a) the CDS process and (b) conventional solidification [28].

Point D nucleation (stage III): At this stage the final nucleation events in CDS process occur. At this stage, as the temperature field and solute concentration field of the residual liquid (small volume fraction) have reached a nearly homogenized state, where they can nucleate and grow in a constrained volume to present a small / negligible pockets of rosette shaped morphology of the primary Al phase [14,15].

Khalaf et al [14] succeeded in obtaining non-dendritic Al morphology of various aluminum wrought alloys series such as 2024, 6082, 7005 and 7075 using the CDS process in laboratory scale experiments. Moreover, he proposed a set of conditions for a successful CDS process [14,28,26]: (a) the difference between the liquidus temperature of the precursor alloys before mixing should be more than 50-80K, (b) the maximum temperature attained during mixing of the two alloys should preferably be marginally above the liquidus temperature of the resultant alloy to facilitate complete filling of the casting mould, (c) the mass ratio between the two precursor alloys should be at least 3.

The CDS technology has been adapted to several casting processes with successful outcomes: sand, gravity permanent mould, tilt pour gravity and high pressure die casting. It is notable that the mechanism of the technology prefers a slower casting process in order to obtain a nearly globular morphology of the primary Al phase. When the resultant alloy slurry is injected into the mould cavity, the heat is extracted by the mould wall and away from the casting; in Figure 5 (a), there is a competition between the rate of change of the gradients of the actual temperature $\left(\frac{dG_{actual}}{dt}\right)$ affected by the casting process and the trace of the liquidus temperature $\left(\frac{dG_{Liqidus}}{dt}\right)$ affected by the diffusion of the solute atoms towards the growing S/L interface. Typically the diffusion of the solute atoms towards the S/L interface and hence, $\left(\frac{dG_{Liqidus}}{dt}\right)$ is predominantly controlled by the diffusion process. The limiting conditions in the various casting processes incorporating the CDS technology to cast Al wrought alloys, along with the resultant effect on the stability of the growing S/L interface is presented in Table 1

Table 1: Limiting conditions of the re-distribution of both the solute and temperature fields in the liquid ahead of the growing S/L interface in casting Al wrought alloys in various casting processes using the CDS technology.

Casting Process with CDS Technology	Solute and Thermal Condition	Interface Stability
Sand Mould	$\left(\frac{dG_{actual}}{dt}\right) \approx \left(\frac{dG_{Liquidus}}{dt}\right)$	Stable interface and nearly equiaxed primary phase morphology
Metal Mould Gravity	$\left(\frac{dG_{actual}}{dt}\right) \geq \left(\frac{dG_{Liquidus}}{dt}\right)$	Stable interface and non-dendritic morphology. The interface may perturb slightly depending on the difference between the two rates of change of temperature gradients.
Chilled Metal Mould (pressure assisted)	$\left(\frac{dG_{actual}}{dt}\right) > \left(\frac{dG_{Liquidus}}{dt}\right)$	Tends to cause instability in the S/L interface due to imposed constitutional undercooling, however, stable interface prevails because of the high velocity of the S/L interface imposed by the process.

Table 1 shows that in a range of casting processes incorporating CDS technology, the resultant morphology of the primary Al phase is predominantly non-dendritic and thereby mitigates the incidence of hot tearing to present a viable route to near net shape cast with Al wrought alloys. Notably, the CDS technology is easily incorporated with any conventional casting process (Table 1) with an additional furnace to hold the two required precursor alloys (minimal capital investment).

In the subsequent sections of his publication, an example of shaped casting AA7050 in a tilt pour gravity casting process is presented along with microstructure and mechanical property characterization. The nomenclature used in the experiments are: *alloy1* and *alloy2* are the two precursor alloys with higher and lower thermal masses, respectively; *alloy3* is the final desired AA7050 alloy; T_{L1}, T_{L2} and T_{L3} are the liquidus temperatures of alloy1, alloy2 and alloy3, respectively; T_1, T_2 and T_3 are the melt superheated temperatures for alloy1, alloy2 and alloy3, respectively; and m_1 and m_2 are the masses of alloy1 and alloy2, respectively with mr being the ratio m_1:m_2.

Materials and Experiments

Several isopleth of multi-component phase diagrams of Al-Zn-Mg-Cu are critically investigated to choose the appropriate compositions and initial temperatures of Alloy1 and Alloy2, respectively, such that the difference between T_{L1} and T_{L2} is greater than 55 °C [14,18] and the mixing of Alloy1 into Alloy2 yields the desired composition of Alloy3 at a temperature T3 around the value of T_{L3}. The initial temperatures of Alloy1 and Alloy2 are typically about 5~10 °C above the respective liquidus temperatures of T_{L1} and T_{L2}. To obtain the nominal composition, two precursor alloys, Alloy1 and Alloy2, were developed from review of three multi-component thermodynamic phase diagrams isopleth for precursor alloys of Alloy1 and Alloy2 with a predetermined mass ratio of 3 (mr = 3) to obtain the resultant casting of Al 7050 alloy.

Based on the design of Alloy1 and Alloy2 from thermodynamic phase diagrams, the process parameters for all the AA7050 alloy was defined: T_{L1} = 650 °C, T_{L2} = 589 °C, $T_{L3,\,experiment}$ = 636 °C and $T_{L3,simulation}$= 632.7 °C. The Alloy1 and Alloy2 were freshly made from commercial purity raw/master alloy materials and held in individual electric resistance holding furnaces. For industrial scale CDS experiments with Tilt Pour Gravity Casting (TPGC) machine, the melts of Alloy1 and Alloy2 were degassed using a rotary degasser with high purity Ar gas purged at 6 L.min-1 and 120 RPM for 30 min. The tilt pour casting equipment along with the mould for casting tensile test bars and the laboratory scale CDS test stet ups which were specially designed and built for the CDS process [27]. The tensile bar was designed in accordance to the design specified in the ASTM B557-06 standard document [28]. The die mold was preheated to 375°C and the pouring cup in the tilt machine was preheated with about ten shots of some scrap Al alloy melt. The specific required amount of Alloy2 was poured into the pouring cup and subsequently, Alloy1, in the specific amount, was poured into the Alloy2 to form the mixture in the pouring cup. Without much delay (~2s) after completion of pouring the required amount of Alloy1, the tilt machine was activated to enable the filling of the mould and subsequent solidification of the cast part. The temperature of the die mould was continuously monitored by a K type thermocouple to enable consistent die opening and closing to maintain a standard and repeatable casting cycle.

The castings from the TPGC machine were subjected to the uniaxial tensile test using an Instron 8800 machine with an MTS Frame equipped with a 250 kN Model 312 MTS Load cell coupled with an on-line Extensometer of 50mm gauge and connected to an on-line data acquisition system; all uniaxial tensile tests were carried out a load speed of 1mm/min. Microstructure evaluation of the sample sections were carried out using Nikon light optical microscope and Scanning Electron Microscope Model JEOL 7000. The samples for microstructural evaluation were sectioned from the gauge length of the tensile test bar castings. The heat treatment of the cast components was carried out as: F (as-cast) and T4 with solution treatment at 750 K for 24 h followed by quenching in room temperature water bath and subsequently underwent natural ageing at room temperature for more than 96 h.

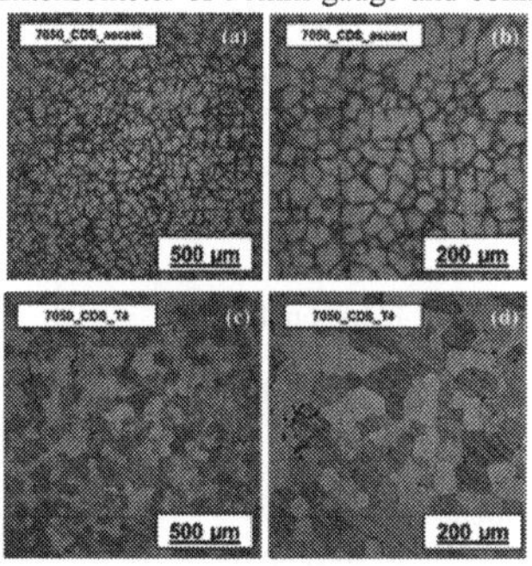

Figure 6: Typical microstructure in optical microscope of AA7050 with CDS and TPGC process. (a) and (b) as-cast structure, and (c) and (d) T4 heat treated structure.

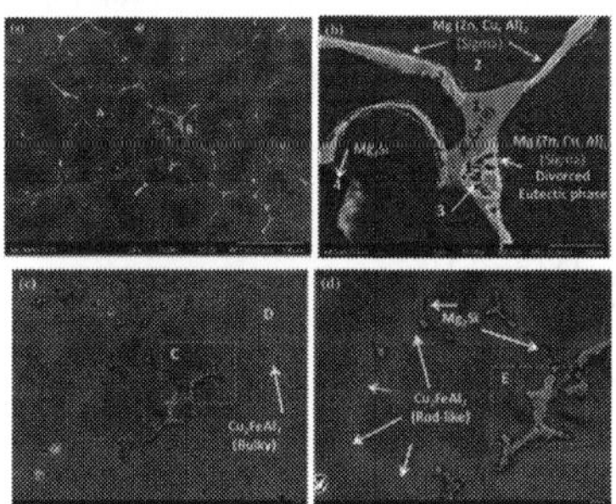

Figure 7: Typical microstructure in scanning electron microscope of AA7050 cast using the tilt pour gravity casting process coupled with CDS technology. (a) and (b) as-cast structure, and (c) and (d) T4 heat treated structure

Results

Figure 6 presents typical microstructure of the AA7050 alloy casting in a TPGC process coupled with the CDS technology. The photograph of the casting was shown in Figure 1 where in no discernable evidence of any hot tearing was observed in any casting. In Figure 6, the morphology of the primary Al phase was repeatable and non-dendritic in all castings. The inter-dendritic secondary phases obtained during solidification in the as-cast condition are shown in the SEM image of Figure 7 (a) and (b); wherein, large bulky and complex multi-component phases evolve in the final stages of solidification. Figure 7 (c) and (d) present the SEM images of the T4 heat treated samples showing significant alteration in the type and morphology of the secondary phases leading to a more uniform phase composition across the sample microstructure.

Table 2 presents the typical uniaxial properties of the AA7050 alloy cast with the TPGC coupled with CDS technology; these properties represent the feasibility and proof-co-concept of the technology in producing sound near net shaped components with the CDS technology.

Table 2: Tensile properties of AA7050 cast samples by CDS.

Temper	UTS Mpa	YS (0.2%) Mpa	%El
T4	447.5 (±14.3)	315.9 (±9.8)	7.3 (±1.7)

Further research is currently underway to significantly improve the mechanical properties to high structural application by re-defining the heat treatment cycles for the AA7050 alloy produced by net shaped casting process. This is a new area of research as these alloys have seldom been cast as such in the past.

Summary

This study has demonstrated the feasibility of casting the AA7050 Al wrought alloy into near-net shaped casting parts using the novel technology of control diffusion solidification (CDS) coupled with the tilt pour gravity casting machine process. Typical non-dendritic morphology of the primary Al phase in the cast microstructure and the uniaxial mechanical properties both show that the castings of the AA7050 alloy are sound and lend themselves to structural applications. The heat treatment development is currently underway and preliminary studies have shown yield strengths in excess of 550 MPa after suitable T6 temper in these castings. This publication merely intends to present a valid and economically viable option to enable Al shaped castings with a significantly improved (>200 %) yield strength without compromising on the ductility of the castings. Significant light weighting measures could now be designed and adapted in automotive components by exploring the use of this technology in manufacturing both Al and Mg parts using the respective wrought alloys.

References

[1] M.M. Sharma, M.F. Amateau, T.J. Eden: *Mater. Sci. Eng. A*, 2006, vol. 424A, pp. 87-96.

[2] S. Shankar, D. Apelian: *Metall. Mater. Trans. B*, 2002, vol. 33B, pp. 465-76.

[3] D.W. Suh, S.Y. Lee, K.H. Lee, S.K. Lim, and K.H. Oh: *J. Mater. Process. Technol.*, 2004, vol. 155, pp. 1330-36.

[4] C.C. Chang, J.G. Yang, C. Ling, C.P. Chou: *IACSIT Int. J. Eng. Technol.*, 2010, vol. 3, pp. 269-73.

[5] S. Lin: Masters Thesis, Université du Québec à Chlcoutimi, Montreal, Canada, 1999.

[6] S. Lin, C. Aliravci, M.O. Pekguleryuz: *Metall. Mater. Trans. A*, 2007, vol. 38A, pp. 1056-68.

[7] M.C. Flemings: *Metall. Trans. B*, 1991, vol. 22B, pp. 269-93.

[8] D.H. Kirkwood: *Mater. Des.*, 2000, vol. 21, pp. 387-94.

[9] T. Marlaud, A. Deschamps, F. Bley, W. Lefebvre, and B. Baroux: *Acta Mater.*, 2010, vol. 58, pp. 248-60.

[10] X. Fan, D. Jiang, Q. Meng, B.Y. Zhang, and T. Wang: *Trans. Nonferr. Met. Soc. China*, 2006, vol. 16, pp. 577-81.

[11] A. de Figueredo: *Sci. Technol. Semi-Solid Met. Process.*, North American Die Casting Association, 2001.

[12] S.Shankar, Y.W.Riddle , M.M. Makhlouf: *Metall. Mater. Trans. A*, 2004, vol. 35A, pp. 3038-43.

[13] M. Jeyakumar, S. Shankar: *Mater. Sci. Forum*, 2011, vol. 690, pp. 226-29.

[14] A.A. Khalaf: PhD Thesis, McMaster University, Hamilton, Canada, 2009.

[15] A.A. Khalaf, S. Shankar: *Metall. Mater. Trans. A*, 2011, vol. 42A, pp. 2456-65.

[16] D. Saha, S. Shankar, D. Apelian, M.M. Makhlouf: *Metall. Mater. Trans. A*, 2004, vol. 35A, pp. 2174-80.

[17] D. Apelian ,M.M. Makhlouf, D. Saha: *Mater. Sci. Forum*, 2006, vol. 519-21, pp. 1771-76.

[18] K. Symeonidis: PhD Thesis, Worcester Polytechnic Institute (WPI), Massachusetts, USA, 2009

[19] M.C. Flemings: Solidification Processing, Mc Graw Hill, New York, USA 1974.

[20] J. W. Rutter and B. Chalmers, Can. Journal of Physics, 31 (1953) 15

[21] W. W. Mullins and R. F. Sekerka, Journal of Applied Physics, 35 (1964) 444.

[22] J. Jorstad, High Integrity Die Casting, chapter 3, North American Die Casting Association (NADCAL Rosemont, IL, USA, July 2008.

[23] H.V. Atkinson, Progress in IVhterials Science, 2005, v. 50, pp. 341-412.
[24] D. Apelian, G. Langford: Final Report, Drexel University, Philadelphia, USA, 1981.
[25] T. Okubo, J. Okamoto, A. Tsuchida: *Colloid Polym. Sci.*, 2009, vol. 289, pp. 645-57.
[26] G. Birsan, P. Ashtari, S. Shankar: *Int. J. Cast Met. Res.*, 2011, vol. 24-6, pp. 378-84.
[27] G. Birsan: Masters Thesis, McMaster University, Hamilton, Canada, 2009.
[28] A.A. Khalaf, P. Ashtari, S. Shankar: *Metall. Mater. Trans. B*, 2009, vol. 40B, pp. 843-49.

Shape Casting: 5th International Symposium 2014
Edited by: Murat Tiryakioğlu, John Campbell, and Glenn Byczynski
TMS (The Minerals, Metals & Materials Society), 2014

CORRELATION BETWEEN MELT QUALITY AND FLUIDITY OF A356

Baris Akkaya*, Emine Ertürk*, Derya Dispinar*

*Istanbul University, Metallurgical and Materials Engineering Department
Avcilar, 34320 Istanbul Turkey
baris.akkaya@gmail.com, emineee_erturk@hotmail.com, deryad@istanbul.edu.tr

Keywords: Al-Si, fluidity, bifilm, quality

Abstract

Aluminium and its alloys are prefer choice of automotive parts due to their low density and high strength ratios. These critical applications requires high quality castings to be produced. Melt quality can be measured by bifilm index using reduced pressure test. This index is a numerical indication of bifilm quantity of the melt. It has been shown that these bifilms affect many of the properties significantly. One of the limiting factor of casting methods is the fluidity of the liquid metal. The aim of the study is to investigate the effect of bifilms (i.e. melt quality) over the fluidity of A356 by using spiral test mould. Different mould temperatures were studied and it was found that as the bifilm index was increased, the scatter of the test results was increased.

Introduction

Aluminium alloys are used in the automotive and aerospace industry in critical places. Due to such critical applications, it is necessary to prepare high quality aluminium alloys. The first most important factor when casting aluminium alloys is to check the quality of the liquid. This is then followed by the optimised runner design and pouring the melt in a controlled way and fill the mould cavity with extreme caution not to generate any turbulence.

Turbulence generates bubbles which is should be avoided during casting processes. Turbulence has another effect on casting, particularly when the alloy is easily oxidized. In liquid form many metals have an oxide skin on its surface. The oxide formed on top of the aluminium works as a protection and prevents the liquid from further oxidization. As long as the oxide film remains on the surface, it does not generate a problem however when it enters into the liquid due to turbulence, the double oxide film or bifilm [1] is formed and after solidification is complete it will act as a stress generator leading to deteriorated properties.

These bifilms may seem harmless when they are folded and crumpled and their sizes are small. However, when solidifying these bifilms may open up (i.ie unravel) they became more dangerous for the mechanical properties. The decrease in the mechanical properties of the aluminium alloy have been attributed to hydrogen in the form of porosity. This is because the presence of bifilms were neglected.

It has been shown [2-8] that there is a good correlation between bifilms and mechanical properties (tensile and fatigue). Dispinar [9] proposed an index called bifilm index and shown that an increase in bifilm index results in decrease of tensile properties [10-13].

As metals quality effects the mechanical properties, it also has influence on the fluidity of the melt as well. The definition of fluidity used in the foundry is: "the distance covered by the flow of a metal within a certain form before it is solid". The factors that affect fluidity of aluminium alloys are temperature of melt and mould, inclusion content, alloying elements and porosity. In literature, it can be found that as the casting temperature is increased, fluidity increases [14-21]. Sabatino [17] has shown that the increase in inclusion content has resulted in the decrease of fluidity (Fig 1).

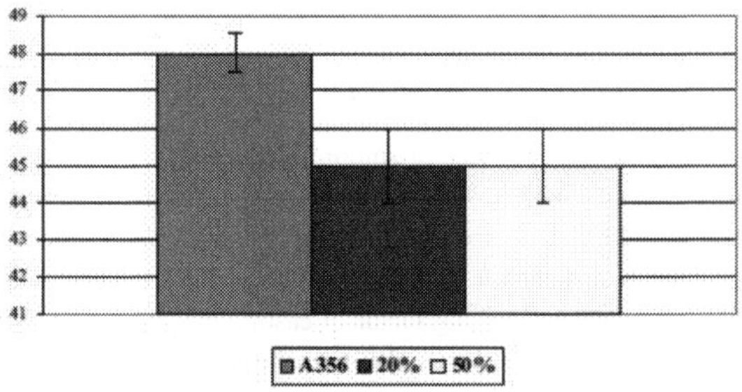

Figure 1: Fluidity change with scrap ratio [17]

In this work, A356 alloy was used and melt quality was measured by bifilm index and the correlation between melt quality and fluidity was investigated.

Experimental Study

The composition of A356 which is commercially provided is shown in Table I.

Table 1: A356 composition used in the experiments

Fe	Si	Ti	Zn	Cu	Mn	Sn	Mg	Pb	Al
0,15	6,7	0,10	0,004	0,02	0,35	0,05	0,35	0,02	rem

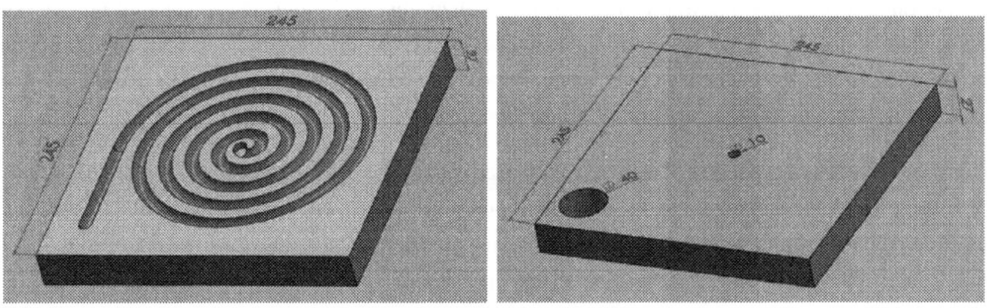

Figure 2: Dimension of the spiral fluidity mould

Alloys which are cut from ingots are melted in graphite crucibles at 750°C. They are poured into the moulds seen in Figure 2. The fluidity measurements are made both in room temperature and 200°C mould temperatures. After the fluidity experiments, metallographic examination is made on the samples' cross sectional areas and the image analysis was carried out to calculate dendrite arm spacing and the eutectic phase's ratio.

To calculate the bifilm index, reduced pressure test was used under a vacuum of 100 mbar.

Results

In this work the effect on the fluidity of bifilms A356 have been tested by using a spiral test method. The results of the fluidity of metals poured into a form after heating to 20°C and 200°C is shown in Figure 3 and 4.

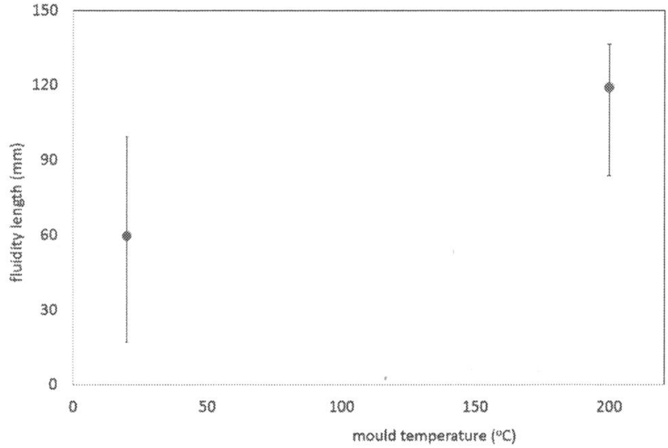

Figure 3: Fluidity lengths at different mould temperatures

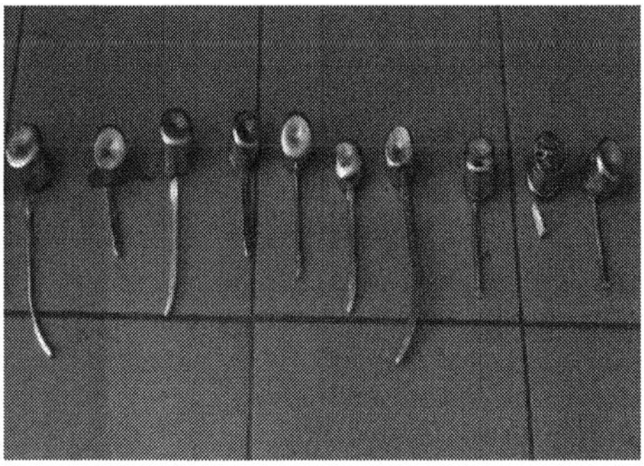

Figure 4: Fluidity test samples after solidification

The microstructure of the testing samples has been given in Figure 5 and 6 for 20°C and 200°C mould temperatures respectively. The same samples were subjected to image analysis for measuring eutectic percentage and dendrite arm spacing. It can be seen that there is no difference in the eutectic phase ratios. The metallographic work show that the dendrite arm spacing was 20 μm and 8 mm for mould temperature was 20°C and 200°C respectively.

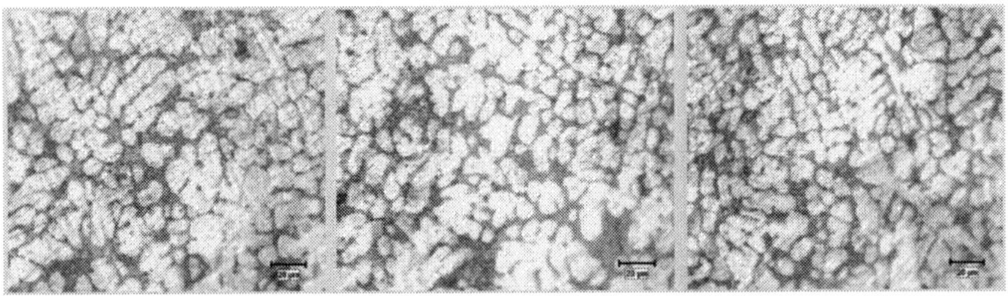

Figure 5: Microstructural images of the cross sections of fluidity test carried out in mould at 20°C

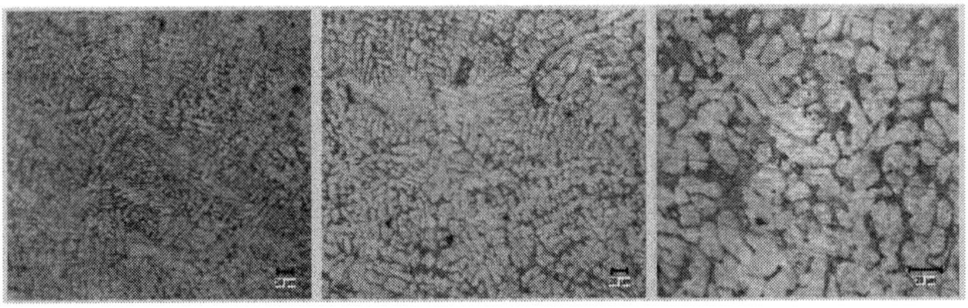

Figure 6: Microstructural images of the cross sections of fluidity test carried out in mould at 200°C

The bifilms index measurements have shown that the average index was to be at 96 mm. This is considered to be a bad quality melt with regard to Dispinar scaling of the index [13].

Discussion

One of the factors that affect the fluidity is the composition of the alloy. The fluidity is high when the metal is pure, free from alloying elements. As the content of the alloying elements increase, the fluidity is reduced. However, when the composition reaches a eutectic point, fluidity increases significantly. For example in Al-Si alloys, silicon levels at 17-18% result in increased fluidity. In the experiments done on Al-Si alloys with Mg, Fe, Ti and Sr, has resulted in that Mg's solution fluidity having an important effect and as the Mg increases the fluidity decreases [20].

Aside from this, with the addition of Ti, i.e. grain refinement, fluidity increases again. Especially with Al-Si alloys the modification with Sr you can increase fluidity [20].

As seen in the fluidity tests, as the mould's temperature rises, the fluidity also rises (Fig 3). The metal solidifies late and the grain size is smaller, that leads to increased flow of metal.

Other parameters that effect the flow of the metal are the eutectic phase and the length of the dendrite arms [15, 22]. Short dendrite arms create a particle thinning effect therefore increases fluidity.

The reason why the fluidity value scatters can be based on the bifilms in the metal which is one of the factors that affect the fluidity is the metal's quality. Bifilms that decrease the metal's quality also decrease the metal's fluidity.

Conclusion

As the mould's temperature increases, the fluidity of A356 increases.
Shorter the dendrite arms, higher the fluidity of A356.
Scatter of fluidity length is related with the presence of bifilms, confirmed by the bifilm index.

References

1. Campbell, J., *Castings*. 2003: Elsevier Science.
2. Green, N.R. and J. Campbell, *Influence of oxide film filling defects on the strength of Al-7Si-Mg alloy castings.* AFS Transactions, 1994. **102**: p. 341-347.
3. Mi, J., R.A. Harding, and J. Campbell, *Effects of the entrained surface film on the reliability of castings.* Metallurgical and Materials Transactions A, 2004. **35**(9): p. 2893-2902.
4. Nyahumwa, C., N.R. Green, and J. Campbell, *Effect of mold filling turbulence on fatigue properties of cast aluminum alloys.* AFS Transactions, 1998. **106**: p. 215-224.
5. Nyahumwa, C., N.R. Green, and J. Campbell, *The concept of the fatigue potential of cast alloys.* Journal of Mechanical Behaviour of Metals, 1998. **9**(2): p. 227-235.
6. Rezvani, M., X. Yang, and J. Campbell, *Effect of ingate design on strength and reliability of Al castings.* Transactions of the American Foundrymen's Society, Vol 107, 1999. **107**: p. 181-188
7. Runyoro, J.J., S.M.A. Boutarabi, and J. Campbell, *Critical gate velocities for film-forming casting alloys: a basis for process specification.* AFS Transactions, 1992. **100**: p. 225-234.
8. Yang, X. and J. Campbell, *Pouring basin desgin.* International Journal of Cast Metals Research, 1998. **10**(5): p. 239-299.
9. Dispinar, D. and J. Campbell, *Critical assessment of reduced pressure test. Part 1: Porosity phenomena.* International Journal of Cast Metals Research, 2004. **17**(5): p. 280-286.
10. Dispinar, D., et al., *Degassing, hydrogen and porosity phenomena in A356.* Materials Science and Engineering: A, 2010. **527**(16–17): p. 3719-3725.
11. Dispinar, D. and J. Campbell, *Metal quality improvement in recycling of aluminium.* Journal of Institute of Cast Metals Engineers, 2006. **180**(3640): p. 328-331.
12. Dispinar, D. and J. Campbell, *Effect of casting conditions on aluminium metal quality.* Journal of Materials Processing Technology, 2007. **182**(1-3): p. 405-410.
13. Dispinar, D. and J. Campbell, *Porosity, hydrogen and bifilm content in Al alloy castings.* Materials Science and Engineering: A, 2011. **528**(10–11): p. 3860-3865.
14. Dahle, A.K., et al., *Effect of grain refinement on the fluidity of two commercial Al-Si foundry alloys.* Metallurgical and Materials Transactions A, 1996. **27**(8): p. 2305-2313.
15. Di Sabatino, M. and L. Arnberg, *Effect of grain refinement and dissolved hydrogen on the fluidity of A356 alloy.* International Journal of Cast Metals Research, 2005. **18**(3): p. 181-186.
16. Di Sabatino, M., et al., *Fluidity evaluation methods for Al–Mg–Si alloys.* International Journal of Cast Metals Research, 2006. **19**(2): p. 94-97.
17. Di Sabatino, M., et al., *The influence of oxide inclusions on the fluidity of Al–7 wt.%Si alloy.* Materials Science and Engineering: A, 2005. **413–414**(0): p. 272-276.
18. Di Sabatino, M., et al., *An improved method for fluidity measurement by gravity casting of spirals in sand moulds.* International Journal of Cast Metals Research, 2005. **18**(1): p. 59-62.

19. Kwon, Y.-D. and Z.-H. Lee, *The effect of grain refining and oxide inclusion on the fluidity of Al–4.5Cu–0.6Mn and A356 alloys.* Materials Science and Engineering: A, 2003. **360**(1–2): p. 372-376.
20. Ravi, K.R., et al., *Fluidity of aluminum alloys and composites: A review.* Journal of Alloys and Compounds, 2008. **456**(1–2): p. 201-210.
21. Samuel, A.M., A. Gotmare, and F.H. Samuel, *Effect of solidification rate and metal feedability on porosity and SiCAl2O3 particle distribution in an Al-Si-Mg (359) alloy.* Composites Science and Technology, 1995. **53**(3): p. 301-315.
22. Sabatino, M.D. and L. Arnberg, *A REVIEW ON THE FLUIDITY OF AL BASED ALLOYS.* Metallurgical Science and Technology, 2013: p. 9-15.

Shape Casting: 5th International Symposium 2014
Edited by: Murat Tiryakioğlu, John Campbell, and Glenn Byczynski
TMS (The Minerals, Metals & Materials Society), 2014

FLUIDITY CHARACTERISTICS OF A356 ALLOY WITH VARIOUS THICKNESS SECTIONED NEW TEST MOULD

Murat Çolak*, Ramazan Kayikci**, Derya Dispinar***

*Sakarya University, Faculty of Technical Education, Metal Education
mcolak@sakarya.edu.tr
**Sakarya University, Faculty of Technology, Metallurgical and Materials Engineering
rkayikci@sakarya.edu.tr
***İstanbul University, Faculty of Engineering, Metallurgical and Materials Engineering
deryad@istanbul.edu.tr

Keywords: Fluidity, A356, modification, simulation

Abstract

Fluidity of alloys are typically carried out in a spiral test mould. Due to the variation of the results, several attempts have been made to optimise the test procedure. Pouring basin has been updated by addition of a stopper and using a defined quantity of melt to be poured. However, one of the discussed issues of the spiral test mould is its fixed size cavity where the liquid is transferred. Therefore, a mould was introduced by Campbell in which a single runner bar was used to feed different thickness sections. In this study, this mould design was used to investigate the effect of temperature (700oC, 725oC and 750oC) and modification (Ti grain refinement and Sr) on the fluidity of A356 alloy. The results were compared with the simulation (SolidCast) to optimise the characteristics of the alloy.

Introduction

Aluminium-Silicon alloys are mainly preferred in several applications due to their high strength/low density ratios. The simplest and economical way of producing parts from this alloy group is casting. One of the limiting factor of casting methods is the fluidity of the liquid metal. The metallurgical factors include composition, microstructure, modification, super heat, melt cleanliness and surface tension. Ravi [1] has reviewed many of these factors. Many of the researches [2-9] have focused on the microstructure where castability was considered as a function of feedability parameters such as eutectic ratio, dendrite arm spacing etc. Campbell [10] has defined five feeding mechanisms which depended upon solidification conditions. Typically, pure metals and eutectic alloys have the highest fluidity in a binary alloy systems as shown in Figure 1.

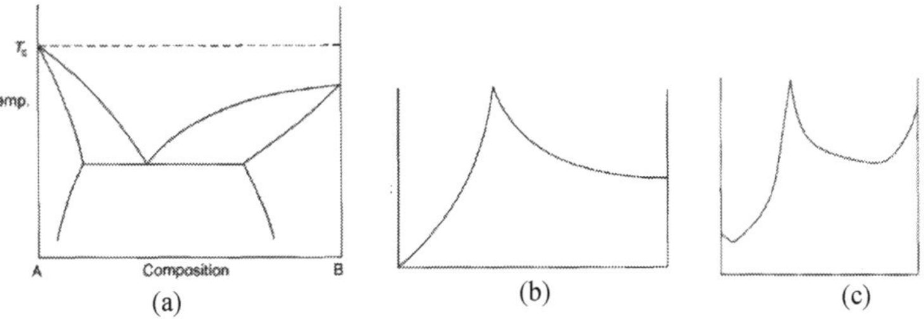

(a) (b) (c)

Figure 1: (a) a simple binary alloy with eutectic reaction, (b) fluidity with composition at zero super heat, (c) fluidity with composition certain super heat [10]

In addition, dendrite coherency (particularly secondary dendrite arm spacing), phases that are formed (eutectic ratio and/or intermetallics) and porosity play significant role on the fluidity of the A356. Therefore, the influence of Sr modification and grain refinement has been widely studied [2, 4, 7, 11].

In this study, the aim was targeted to investigate the effect of surface tension on the fluidity by means of using different section thickness mould.

Experimental work

Commercially available A356 alloy was used in this work. The composition of the alloy is given in Table 1.

Table 1: A356 composition used in the experiments

Fe	Si	Ti	Zn	Cu	Mn	Sn	Mg	Pb	Al
0,15	6,7	0,10	0,004	0,02	0,35	0,05	0,35	0,02	rem

Three different pouring temperatures were selected: 700, 725 and 750°C. The alloy was cast in three various composition: as-received, Ti-grain refined and Sr-modified. The fluidity tests were carried out in the revised mould design where the dimension is given in Figure 2.

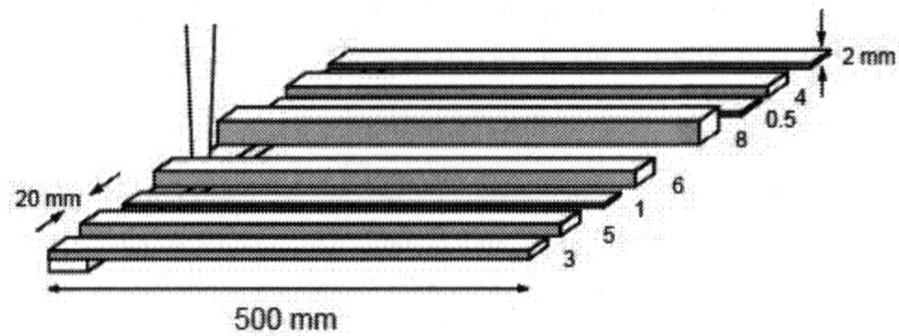

Figure 2: Dimension of the fluidity mould used in the experiments

The moulds were produced by sand moulding with CO_2 hardened AFS60 sands. The same experimental conditions were input to SolidCast simulation program and the results were compared with the casting trials. The basic assumptions used in the software were made according to the casting trial; such as, filling time was selected as 5 seconds. The flow was stopped after 609°C and the critical fraction of liquid (CFL) was set to 0.6.

Results

The fluidity test results were measured in millimetres and summarised in Figures 3-5.

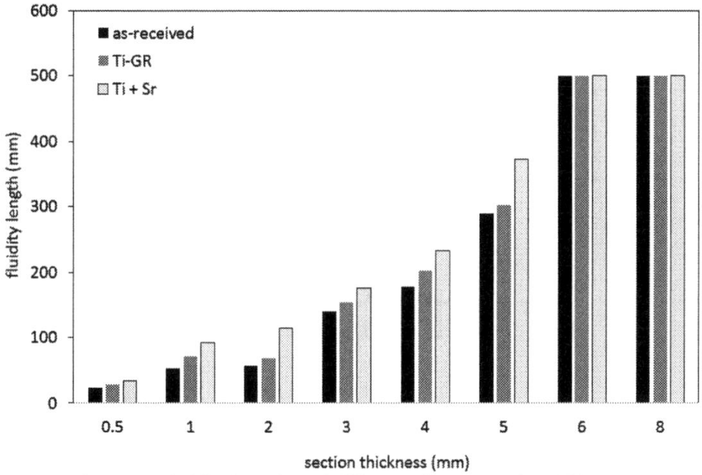

Figure 3: Fluidity length at 700°C according to section thickness

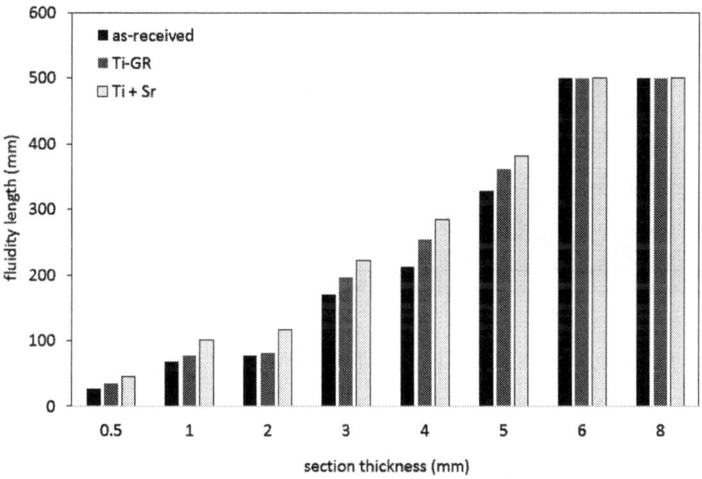

Figure 4: Fluidity length at 725°C according to section thickness

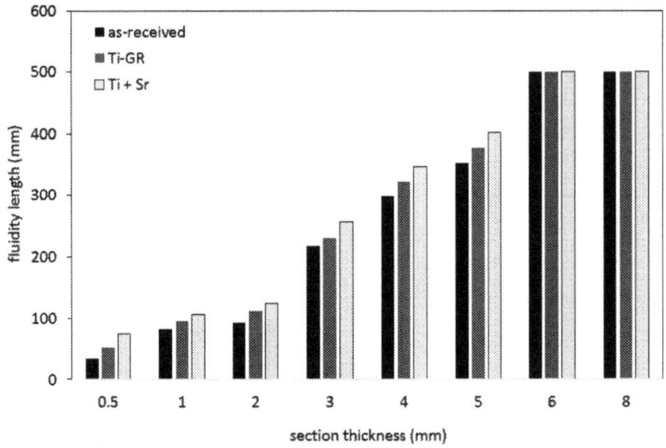

Figure 5: Fluidity length at 750°C according to section thickness

Some of the selected images that show the comparison of actual castings with the simulation is given in Figures 6-8.

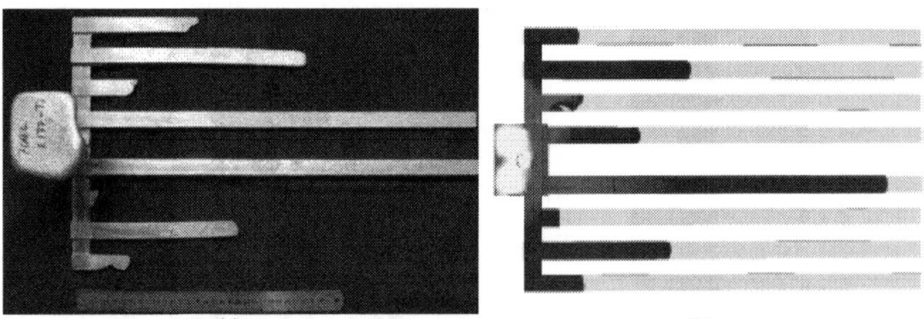

(a) (b)

Figure 6: Fluidity test results of Ti-grain refined A356 at 700°C (a) casting, (b) simulation

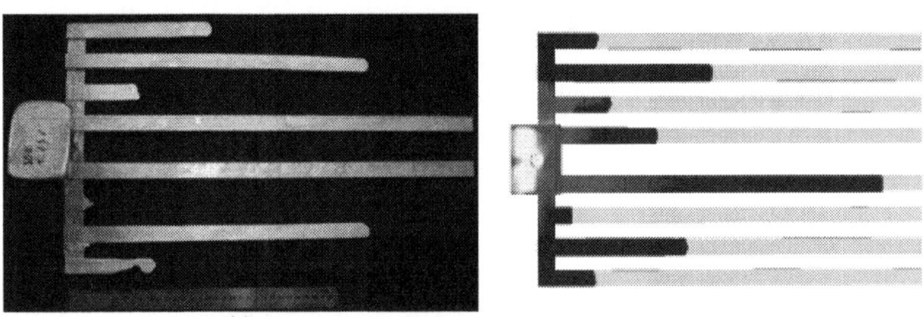

(a) (b)

Figure 7: Fluidity test results of Ti-grain refined and Sr-modified A356 at 700°C (a) casting, (b) simulation

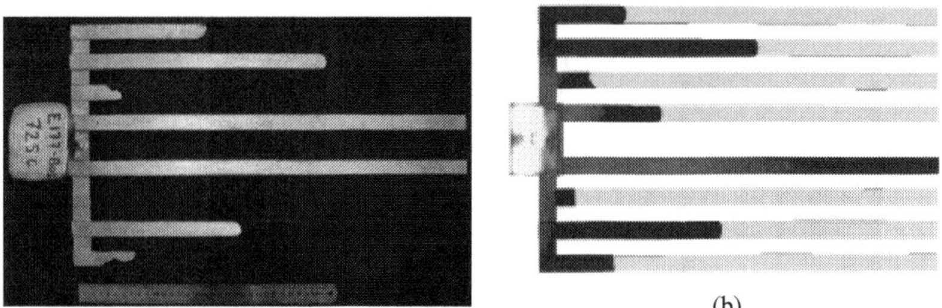

(a) (b)

Figure 8: Fluidity test results of as-received A356 at 725°C (a) casting, (b) simulation

All the castings that were produced under different conditions (i.e. 700, 725, 750 and as-received, Ti grain refined and Sr modified) were all subjected to metallographical examination in order to investigate the effect of microstructure on the fluidity. For simplicity, only few of the selected micrographs are given in Figure 9-11.

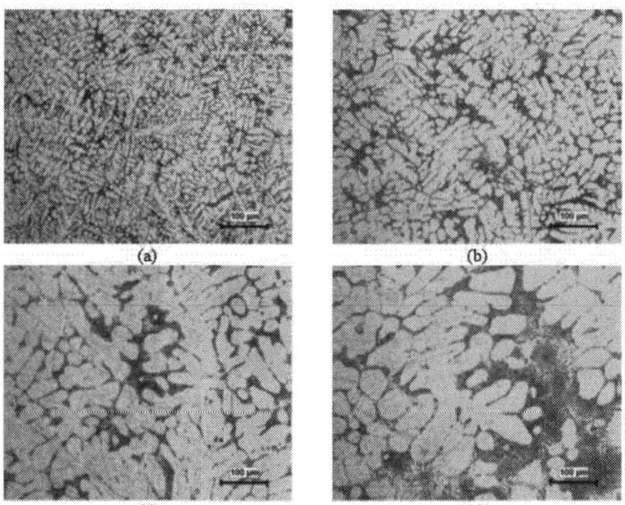

Figure 9: Sample cast at 700°C in as-received condition (a) 0.5 mm section, (b) 1 mm section, (c) 5 mm section, (d) 8 mm section

Figure 10: Sample cast at 700°C with Ti grain refined (a) 0.5 mm section, (b) 1 mm section, (c) 5 mm section, (d) 8 mm section

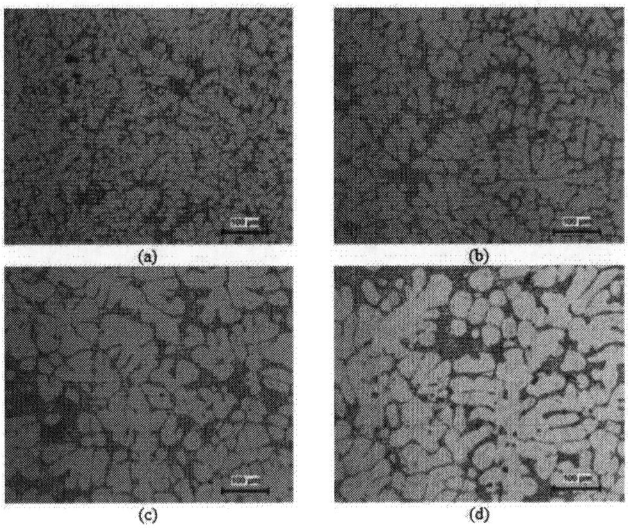

Figure 10: Sample cast at 700°C with Ti grain refined and Sr modified (a) 0.5 mm section, (b) 1 mm section, (c) 5 mm section, (d) 8 mm section

Discussion

As seen from Figures 3 to 5, in all of the tests, regardless of the microstructure (as-received, Ti grain refined or Sr modified) and casting temperature (700, 725, 750°C), 6 mm and 8 mm sections were all filled to the maximum which is 500 mm. On the other hand, 0.5 mm section was the lowest fluidity length with an average around 35 mm.

When the castings were made at 700°C, the fluidity length of as-received alloy and Ti-grain refined was quite close to each other at all section thicknesses. However, the fluidity was significantly increased when the alloy was modified with Sr; particularly at 2, 4 and 5mm sections. When the casting temperature was increased to 725 and 750°C, this apparent effect of Sr modification was less pronounced.

Kwon [4] had used a similar mould design but used 8 channels with same cross sectional areas. The dimension were 5x3x2. It was suggested that for A356 alloy, the grain refinement had increased fluidity even at lower melt temperatures. However, no such results were found in this work when the same cross sections were compared. As Ravi [1] concluded in their review, the influence of grain refinement on fluidity is remains trivial.

As expected, when the melt temperature was increased, the fluidity at all different sections of the mould was increased. However, there were no effect of the critical solid fraction over the experimental works conducted in this study. By Ti grain refinement, the dendritic coherency would have been delayed [3, 9], and as a results, feedability would be increased considerably. However, the fluidity lengths were not that different for the castings in the as-received and Ti-grain refined conditions. The establishment of the planer growth and fine eutectic structure by the addition of Sr was not as dramatic as expected for the thin sections. This may possibly be due to the effect of the melt cleanliness. It has been shown that the presence of oxide inclusions have noticeable effect over the fluidity [4, 8, 11, 12].

Conclusion

The fluidity of A356 increases with grain refinement and further increases with additional Sr modification.
Fluidity of A356 in sand moulds varies with section thickness and it is highest after 5 mm and it decreased exponentially down to 0.5 mm thickness.
Simulation results are in good agreement with the experimental work for the modified fluidity tests for A356.

References

1. Ravi, K.R., et al., *Fluidity of aluminum alloys and composites: A review.* Journal of Alloys and Compounds, 2008. **456**(1–2): p. 201-210.
2. Dahle, A.K., et al., *Effect of grain refinement on the fluidity of two commercial Al-Si foundry alloys.* Metallurgical and Materials Transactions A, 1996. **27**(8): p. 2305-2313.
3. Di Sabatino, M. and L. Arnberg, *Effect of grain refinement and dissolved hydrogen on the fluidity of A356 alloy.* International Journal of Cast Metals Research, 2005. **18**(3): p. 181-186.
4. Kwon, Y.-D. and Z.-H. Lee, *The effect of grain refining and oxide inclusion on the fluidity of Al–4.5Cu–0.6Mn and A356 alloys.* Materials Science and Engineering: A, 2003. **360**(1–2): p. 372-376.

5. Liu, L., et al., *Precipitation of β-Al5FeSi Phase Platelets in Al-Si Based Casting Alloys.* Metallurgical and Materials Transactions A, 2009. **40**(10): p. 2457-2469.
6. Mollard, F., M. Flemings, and E. Niyama, *Aluminum Fluidity in Casting.* JOM, 1987. **39**(11): p. 34-34.
7. Nabawy, A.M., et al., *Influence of additions of Zr, Ti–B, Sr, and Si as well as of mold temperature on the hot-tearing susceptibility of an experimental Al–2% Cu–1% Si alloy.* Journal of Materials Science, 2012. **47**(9): p. 4146-4158.
8. Samuel, A.M., A. Gotmare, and F.H. Samuel, *Effect of solidification rate and metal feedability on porosity and SiCAl2O3 particle distribution in an Al-Si-Mg (359) alloy.* Composites Science and Technology, 1995. **53**(3): p. 301-315.
9. Veldman, N.M., et al., *Dendrite coherency of Al-Si-Cu alloys.* Metallurgical and Materials Transactions A, 2001. **32**(1): p. 147-155.
10. Campbell, J., *Castings.* 2003: Elsevier Science.
11. Liu, L., et al., *Influence of oxides on porosity formation in Sr-treated Al-Si casting alloys.* Journal of Materials Science, 2003. **38**(6): p. 1255-1267.
12. Di Sabatino, M., et al., *The influence of oxide inclusions on the fluidity of Al–7wt.%Si alloy.* Materials Science and Engineering: A, 2005. **413–414**(0): p. 272-276.

Shape Casting: 5th International Symposium 2014
Edited by: Murat Tiryakioğlu, John Campbell, and Glenn Byczynski
TMS (The Minerals, Metals & Materials Society), 2014

Effect of Feeder Configuration on the Microstructure of Ductile Cast Iron

Nikolaj Kjelgaard Vedel-Smith[1], Niels Skat Tiedje[1]

[1]Technical University of Denmark, Department of Mechanical Engineering, Produktionstorvet Building 427A, 2800 Kgs. Lyngby, Denmark

Keywords: SG Iron, Color Etching, Feeder, Porosities, Dimensional Accuracy, Hot spot

Abstract

Feeding and microstructure of a test casting rigged with different feeder combinations was studied. Castings were examined and classified by soundness and microstructure. Subsequently the casting macro- and microstructure was analyzed to study how differences in solidification and segregation influence the soundness of different sections of the castings. Moreover, the microstructural changes due to variations in thermal gradients are classified, and the variations in the mushy zone described.

The paper discusses how solidification and segregation influence porosity and microstructure of ductile iron castings. The goal is to enable metallurgists and foundry engineers to more directly target mushy zone development to prolong the possibility to feed through this section. Keeping smaller section open for an extended period will make it possible to use fewer or smaller feeders, with reduced energy consumption and cheaper products as a result.

Introduction

Energy for melting is a significant expense and represents approx. one quarter of the production cost for cast iron foundries. New applications for ductile iron (DI) increase the requirements for improved mechanical properties, and some of these alloys show significant problems with shrinkage and porosities. Larger modulus feeders and better designs are required to successfully cast sound castings. The changes in alloy composition also entail a different behavior of the primary and secondary graphite expansion making it difficult to feed secluded sections of the casting.

This paper is the result of an ongoing project, involving several companies, working towards characterizing, quantifying and understanding the effect and functioning of various feeder applications.

On vertically parted molds it can be difficult to place feeders freely. Traditional feeders are geometrically restrained to the upper half of the parting line. Spot feeders, on vertically parted molds, enable feeding of secluded sections located away from the parting line. The spot feeders can be insulation, exothermic or a combination, and provide heat and melt to a given section.

Experimental Setup

Casting Geometry

The test geometry used for the experiments consist of a disc with an inner boss and an outer ring, separated by a thin plate like section. The geometry is designed to display the same problems as found in disc brakes, flywheels and other castings with combinations of small and large modulus sections. The casting layout and feeder location is seen in Fig. 1. For more details please refer to [1].

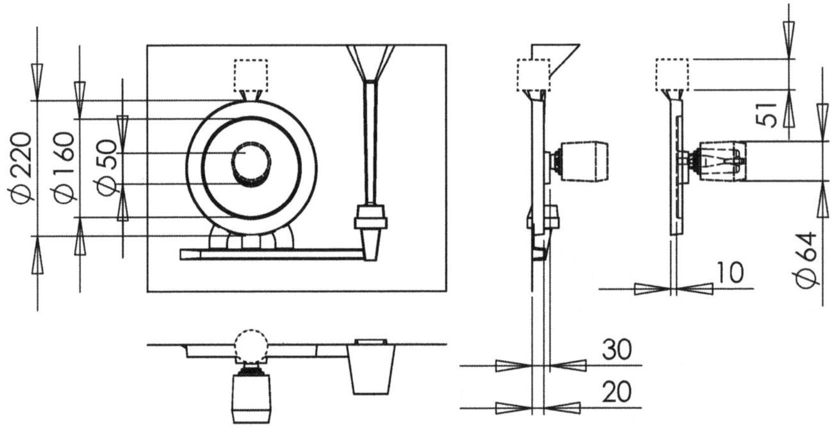

Figure 1: Overview of casting geometry and feeder placement. The dotted lines indicate the top feeder, and the dashed lines indicate the centre feeder. Measurements are in mm.

All feeders used where sleeved and had the same geometric modulus (M_g), calculated as $M_g = \frac{V}{S}$, where V is the volume of the feeder and S is the cooling surface. By changing the sleeve material, the thermal (or true) modulus (M_t) was increased without changing the geometry—thus keeping the ferrostatic pressure constant. The modulus of the feeders change because of the thermal properties of the chosen sleeve material. This multiply the size of the geometric modulus with a Modulus Extension Factor (MEF) specific for each sleeve material. Thus, the true modulus becomes $M_t = \text{MEF} \times M_g$.

Alloys and Combinations

Two different alloys for three different feeder combinations were cast in triplicates. The alloys were a pearlitic-ferritic EN-GJS-500-7 and a fully ferritic EN-GJS-450-10 [2]. Both alloys were spheroidal graphite irons (SGI), and their chemical composition is shown in Tab. 1(a). The combination of alloy, sleeve material and feeder modulus can be seen in Tab. 1(b). The castings are identified by alloy (α or β), feeder combination (1, 2 or 3) and triplicate copy (A, B or C). E.g. β3A is EN-GJS-450-10 fully ferritic alloy cast with a combined material feeder sleeve for the top feeder, and no center feeder.

Table A: *Dimension and M_g in mm*

	I	III	IV	V	VI	VII	IX
w	ø32	20	10	30	10	20	ø40
h	40	30	55	50	55	30	60
M_g	7	6	5	9	5	6	8

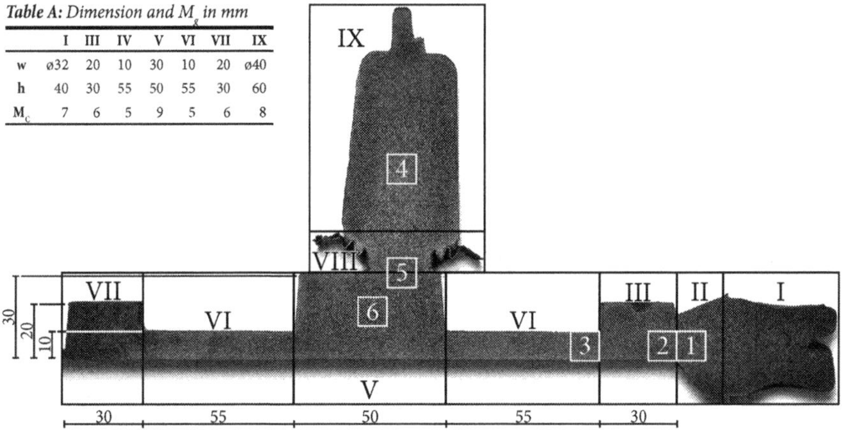

Figure 2: Overview of sectioned casting (α2A). Etched with 1% Nital for 600 s. The casting is divided into 9 non-overlapping areas (Roman numerals) for quantification and analysis of porosities. 6 sections (Arabic numerals) of 10 x 10 mm were cut and color etched. The inserted table (Table A) shows the dimensions and geometric modulus (M_g) of the different areas. All measurements are in mm.

Table 1: Alloy compositions and feeder combinations used.

(a) Alloy compositions in wt%. α is the EN-GJS-500-7 pearlitic-ferritic alloy, and β is the EN-GJS-450-10 fully ferritic alloy.

	CE	C	Si	Mn	P	S	Mg	Cu
α	4.6	3.69	2.75	0.50	0.015	0.005	0.044	0.25
β	4.5	3.35	3.48	0.34	0.017	0.003	0.046	0.10

(b) Sleeve materials: Exothermic (Exo), insulating (Ins) or combined (E/I). Parenthesis shows the true modulus (M_t) in mm.

	α1	α2	α3	β1	β2	β3
Top	E/I (10)	Ins (9)	E/I (10)	E/I (10)	Ins (9)	E/I (10)
Center	Exo (12)	Ins (11)	-	Exo (12)	Ins (11)	-

Production

α1 and α2 was cast at $(1401 \pm 5)\,°C$, α3 was cast at $(1408 \pm 5)\,°C$ and the β-series was cast at $(1392 \pm 5)\,°C$. The poured weight was 8 kg with a pouring time of 3.5 s and a casting weigh of 4 kg. All castings were made on the same vertical molding line. For more details please refer to [1].

Methods

Liquid Penetrant Test

The porosities in the sectioned castings were examined using liquid penetrant testing and classified accordig to the European Standard EN 1370-1:2011 [3]. The analysis of each casting was divided into 9 areas as seen on Fig. 1, and was evaluated for size and type of porosities. The process in fully described here [1].

Deformation Measurements

The deformation of the reverse side of the castings is in this paper described by its flatness value (f_v). In simple terms the f_v is the difference between the highest and lowest point on the surface. The f_v was measured using a Ziess OMC 850 mechanical Coordinate Measuring Machine (CMM) with a resolution of 0.2 µm. The measurements were made using a 3 mm prob which acted as a mechanical filter with respect to the surface roughness. For more detail see [1].

Etchings

Macro Etching All analyzed discs were sectioned using a cold saw, after which the sectioned piece was ground plane. The newly ground surface was etched for 600 s in a 20 °C 1% Nital solution—99 mL Ethyl alcohol and 1 mL Nitric acid (HNO_3). After etching the sectioned casting was cleaned in ethanol and left to dry in a hot air oven at 110 °C for 1800 s. The macro etched castings were analyzed using a magnifying glass—see Fig. 1.

Color Etching The cut out sections (1-6) from castings α1, α2 and β1 was color etched with a picric acid—50 ml Distilled water, 10 g Sodium hydroxide (NaOH), 40 g Potassium hydroxide (KOH) and 10 g Picric Acid ($C_6H_3N_3O_7$). After mixing the mounted and polished pieces were etched at 105 °C. Most pieces required an etching time around 330 s, but some required more. Each piece was analyzed in an optical microscope after etching, and then etching again if the etching was not fully developed.

Results

Porosities

For all 18 castings areas II, IV, VI and VII display no sign of porosities in any of the casting. Area II is the feeder neck of the top feeder, and areas IV and VI are the thin walled sections between the outer ring and the center boss. Area VII is the ring section at the bottom of the casting. Areas I and IX are the feeders and are excluded from the porosity analysis. Thus, the analysis focus on the remaining three areas—III, V and VIII—and the results are shown in Tab. 2.

All 6 castings that used the exothermic sleeves (α1 and β1) displayed no porosities in the areas analyzed. Of the three castings with insulating sleeves and the EN-GJS-500-7 alloy, the α2B showed the smallest category of non-linear isolated porosities (SP1). The same feeder configuration with the fully ferritic EN-GJS-450-10 alloy showed SP1 porosities in 2 of 3 feeder necks (VIII), a single more severe SP2 porosity at the boss (V) and a single SP1 porosity at the upper ring (III). The reference castings without center feeder all displayed porosity defects at the boss (V).

Casting Flatness

All 18 castings have a flatness value (f_v) between 0.33 and 0.50 mm. The results are shown in Tab. 2, displaying each casting configuration as separate bar graphs. The

Table 2: Feeder types and modules are shown in red, orange and yellow boxes. Soundness of individual castings is shown in the bottom section. Green means no porosities. SP indicate non-linear isolated porosities and CP indicate non-linear clustered porosities. The suffix indicate severity—higher is more severe [3]. The graph show the f_v of the different casting groups.

		EN-GJS-500-7									EN-GJS-450-10								
		α1			α2			α3			β1			β2			β3		
Top Feeder:	Material	E/I			Ins			E/I			E/I			Ins			E/I		
	M_t	10 mm			9 mm			10 mm			10 mm			9 mm			10 mm		
Center Feeder:	Material	Exo			Ins			N/A			Exo			Ins			N/A		
	M_t	12 mm			11 mm						12 mm			11 mm					
		A	B	C	A	B	C	A	B	C	A	B	C	A	B	C	A	B	C
Areas of the	III															SP1			
castings examined	V							CP1	CP1	SP1					SP2		CP2	SP1	CP3
for porosities	VIII					SP1								SP1	SP1				

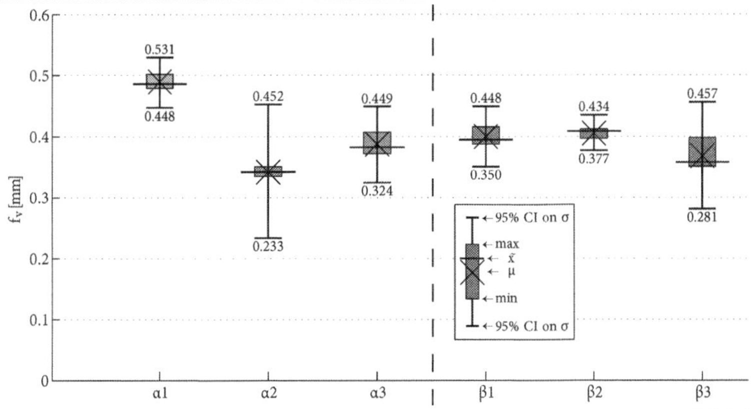

bar ends show the highest and lowest value measured. × mark the arithmetic mean of the population, and — mark the median value. The error bars mark a 95% confidence interval assuming a Gaussian distribution. Using an outliers factor of 3, no outliers were found. There was a greater variation in the pearlitic-ferritic α-alloy, compared to the fully ferritic β-alloy, but only α1 and β2 can be identified as statistically different. All other groups have overlapping confidence intervals and cannot be concluded to be different.

Etchings

Macro Etchings The macro etching had little or no effect on the fully ferritic β-alloy. The lack of reaction showed that these castings contained very little or no pearlite. The analysis of the pearlitic-ferritic α-alloy showed evenly distributed pearlite in all areas of the castings. Dendrite structures were visible to the naked eye in areas I, II, III, V, VIII and IX. Areas IV, VI and VII do not have visible dendritic structures. All α-castings showed clear signs of directional dendrite growth across areas I, II and III. The dendrites in areas V, VIII and IX did not show a clear direction of solidification. α2A after etching with Nital can be seen in Fig. 1.

Color Etchings Comparing the microstructures of the α- and β-alloys it was seen that the β-alloy display a greater nodule count. All three castings displayed good nodularity in all of the etched sections. The thin walled section (3) showed directional solidification from the edge and towards the center of the section. Dendrites were found in sections 3, 4, 5 and 6 for all three castings. Dendrites were not identified in etchings of section 1 and 2.

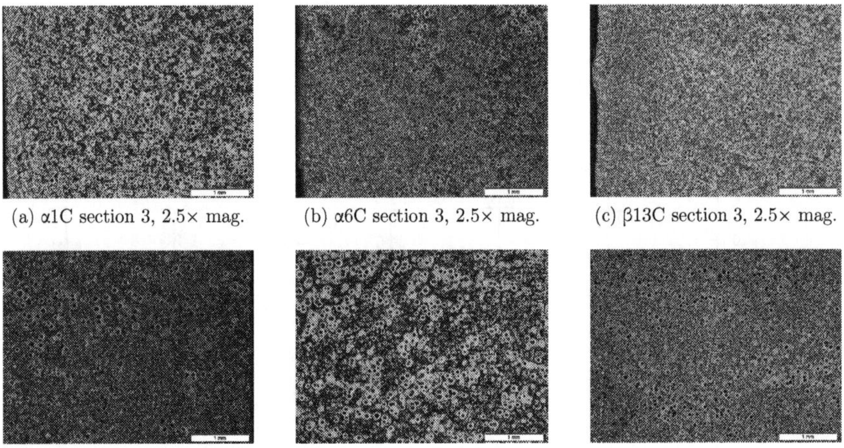

(a) α1C section 3, 2.5× mag. (b) α6C section 3, 2.5× mag. (c) β13C section 3, 2.5× mag.

(d) α1C section 6, 2.5× mag. (e) α6C section 6, 2.5× mag. (f) β13C section 6, 2.5× mag.

Figure 3: Color etchings: Si segregation gives a blue tint. Brown areas are low Si regions.

α1C and β1C with the exothermic sleeves contained a large fraction of low Si eutectic segregation in section 6. α2C with the insulating sleeves, showed a large fraction of low Si eutectic segregation in sections 3, 4 and 5. Furthermore, sections 4 and 5 for castings α1C and β1C showed alignment of the graphite nodules according to the dendritic structure, while the same section for α2C was less orderly and displayed a larger fraction of non-linearized nodules. For section 6 none of the three castings showed a high degree of linearized nodules.

Discussion

Porosities and Deformation

First of all the results show that the test casting could not be cast porosity free without the center feeder. All six castings from α3 and β3 displayed porosities at the boss (V). The porosity analysis also showed that the β-alloy was more prone to porosities, as also the β2 configuration displayed porosities in the boss (V), the top ring (III) and on two occasions at the feeder neck (VIII) close to the boss. In comparison the α2 castings that had the same feeder configuration showed only a single porosity defect, namely at the feeder neck (VIII). Porosities in the feeder neck do not deem the casting itself unsound, but it indicates that the given feeder is close to the limit of how much it can feed. For both the α- and the β-alloy the feeder configuration using exothermic sleeves showed no sign of porosities at the examined areas.

There were no direct correlation between the flatness value f_v of the castings and the porosities. Comparing the two alloys it was shown that the pearlitic-ferritic α-alloy displayed a greater variance than the fully ferritic β-alloy. The higher Si content of the β-alloy increase the austenite to ferrite transformation temperature [4], which result in an earlier graphite expansion, which again occur at a time where the casting has regained less strength compared with the low Si α-alloy. The pattern plate was measured to have a f_v of 16.4 μm—approx. 20-30 times smaller than the flatness measured on the castings and is thus concluded to have no significant influence on the results.

Solidification and Microstructure

The greater nodule count in the β-alloy cannot be contributed to the difference in alloying elements. Increased Mn content can increase the nodule count, but the β-alloy have a lower Mn content than the α-alloy, and should thus have a lower nodule count. However, the two alloys were cast at different dates and the inoculation procedure may have changed in between. The castings with the exothermic sleeves (α1C and β1C) both showed a large fraction of low Si eutectic solidifying in between the nodules at the boss (6). The insulating sleeves showed the opposite effect, having low Si eutectic at the feeder neck and in the feeder it self. Before making final conclusions in this matter, further color etchings should be made. The etchings are very sensitive with respect to etching time, and the lack of brown areas in section 6 of β6C may be due to under etching. Assuming the etchings are reliable, the phenomena may be explained by the delayed graphite expansion of the exothermic feeders. The boss section (V) will have expanded and pushed out a portion of the low Si eutectic, after which the graphite expansion in the feeder (IX) will push back the last to freeze melt—the low Si eutectic.

The greater tendency to shrink found in the β-alloy may be, in part, related to an improved inoculation procedure, as has previously been shown in gray cast iron [5]. While under-inoculation will result in a decreased graphite expansion, over-inoculation can advance the occurrence of the graphite expansion resulting in a decreased expansion later in the process. Another phenomena that influence the movement of the last to freeze melt through the mushy zone is the thermal center. The movement of the thermal center during solidification—a so called migrating hot spot—has been shown to play a significant role in the development and location of porosities [5, 6]. A migrating hot spot can also be part of the explanation for the different locations of the last to freeze melt, as the not only the different M_t of the different feeders will influence the development and movement of the global and local hot spots, the differences in sleeve material will likewise also influence the migration of the hot spots.

Conclusions

The results showed that the high Si β-alloy was more prone to porosities than the low Si α-alloy, and that the exothermic feeder sleeves could feed both alloys while the insulating sleeves was insuficient for the β-alloy. The measurement of the castings showed that the two alloys had close to the same average f_v, but that the β-alloy displayed a smaller variance. This is suspected to be related to the elevated austinite to ferrite transitions temperature controlled by the Si content. Finally, the color etchings indicated that the low Si segregation of the last to freeze melt were located differently in the castings with exothermic feeders, as compared to the castings with insulating feeders. It is suspected that this is related to the shift in graphite expansion due to the changed M_t and the effect of the migrating hot spot.

Acknowledgments

This work is funded by the PSO funds from the Danish Government, and is performed in collaboration with FOSECO Ltd., MAGMA GmbH, DISA Industries A/S, Vald. Birn A/S and the Technical University of Denmark, Department of Mechanical Engineering. A special thanks to Professor Attila Diószegi and Sadaf Vazehrad at Jönköping University, School of Engineering, Sweden, for their kind assistance with color etching the cut sections.

References

[1] N. K. Vedel-Smith, et al. Quantification of Feeding Effects of Spot Feeding Ductile Iron Castings made in Vertically Parted Molds. *AFS Proceedings*, (1310), 2013.

[2] EN 1563:2012-3—Founding: Spheroidal Graphite Cast Irons, 2012.

[3] EN 1371-1:2011—Founding: Liquid Penetrant Testing—Sand, Gravity Die, and Low Pressure Die Castings, 2011.

[4] Ductile Iron Data. *Ductile Iron Data - for Design Engineers.* Rio Tinto Iron & Titanium Inc., 1990.

[5] Lennart Elmquist, et al. On the Formation of Shrinkage in Grey Iron Castings. *Key Engineering Materials*, 457:416–421, 2011.

[6] Lennart Elmquist and Attila Diószegi. Shrinkage Porosity and its Relation to Solidification Structure of Grey Cast Iron Parts. *International Journal of Cast Metals Research*, 23(1):44–50, 2010.

Shape Casting: 5th International Symposium 2014
Edited by: Murat Tiryakioğlu, John Campbell, and Glenn Byczynski
TMS (The Minerals, Metals & Materials Society), 2014

ANALYSIS OF HETEROGENEOUS NUCLEATION IN DUCTILE IRON

Simon N. Lekakh

Missouri University of Science and Technology, 1400 N. Bishop, Rolla, MO 65409

Keywords: Ductile iron, Graphite, Heterogeneous nucleation, 3-D quantitative technique

Abstract

A combination of an automated SEM/EDX analysis, a special 2D-3D converter of nodule size distribution, and adaptive thermal analysis were used for the study of heterogeneous nucleation in ductile iron. The special quenching technique was applied to increase probability to reveal non-metallic heterogeneous nuclei in small graphite nodules. Ternary diagrams were developed to compare statistics of oxide and sulfide compositions in graphite nodules and metal matrix. Thermodynamics of heterogeneous nucleation of graphite phase in Mg-treated cast iron is discussed based on the novel experimental data.

Experimental and Simulation Methods

Shape, size, quantity, and distribution of the graphite phase, developed during heterogeneous solidification in high carbon iron alloys (cast irons), are some of the most important microstructural parameters because the graphite phase influences the physical and mechanical properties of the final castings. In this article, developed approaches, including (i) an automated SEM/EDX analysis of graphite nodule heterogeneous nuclei chemistry[1-2] in quenched specimens and (ii) a special algorithm to convert two-dimensional to three-dimensional graphite nodule size distribution[3-5], were used in combination with (iii) an adaptive thermal analysis[6].

Heterogeneous nucleation statistic. Heterogeneous nucleation plays an important role in stable graphite eutectic solidification to avoid metastable cementite formation and associated shrinkage defects[7]. A vast variety of *FeSi*-based inoculants are used in ductile iron industrial practices. The thermo-chemistry of heterogeneous nucleation formation during melt inoculation was described by authors[8,9]. An experimental assessment of inoculation efficiency can be done on the basis of knowledge of heterogeneous nuclei chemistry. However, large graphite nodules (20-50 μm) in casting, relative to the size of potential nuclei (< 2 μm), makes it difficult to analyze potential nuclei chemistry in random section. Fig. 1 illustrates the calculated probability of visible nuclei in random section depending on graphite nodule size.

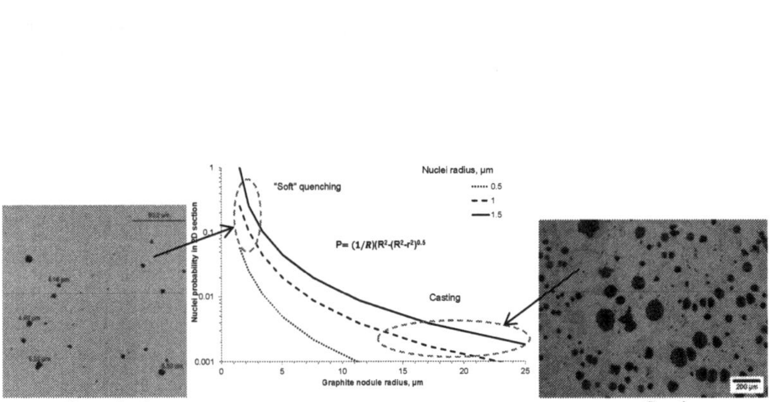

Quenched Casting

Fig. 1. Probability of nuclei view in random section of graphite nodules in cast structure and in quenched *Mg*-treated ductile iron.

A suggested technique for an evaluation of heterogeneous nuclei inside graphite nodules included two steps: (i) melt quenching to develop small graphite nodules (3-6 μm) with increased probability to reveal non-metallic heterogeneous nuclei and (ii) using automated SEM/EDX analysis to search potential nuclei in the center of small graphite nodules. In this study, samples were collected directly from the melt in the ladle after *Mg*-treatment using a submerged sampler with two internal steel chillers at 4 mm apart. A typical microstructure contained 3-6 mm diameter graphite nodules (Fig. 1).

Statistics of non-metallic inclusions were studied using an automated inclusion analyzer system, the Aspex PICA-1020. The applications of the automated analysis for control of non-metallic inclusions in iron matrix are described elsewhere[1,2]. In this study, the special search routine and rule files were developed to separately analyze non-metallic inclusions located in the metal matrix and inside graphite nodules. Fig. 2 shows the examples of different observed chemistries of heterogeneous nuclei inside small graphite nodules: (a) complex silicate with *MgS*, (b) *Si-Mg* oxide with small amount of *S*, and (c) complex *Mg-Ca* sulfide with *Mg-Si* oxide.

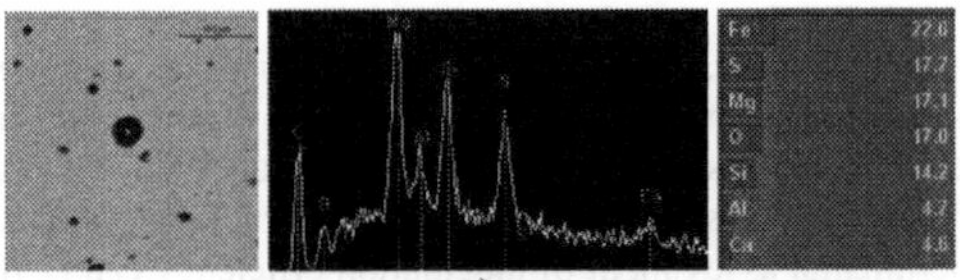

a)

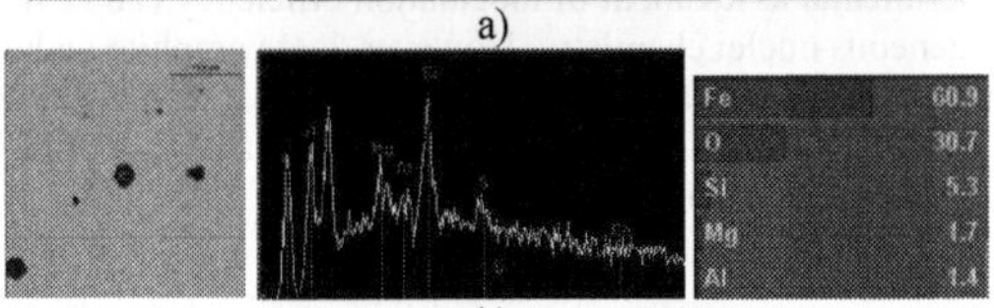

b)

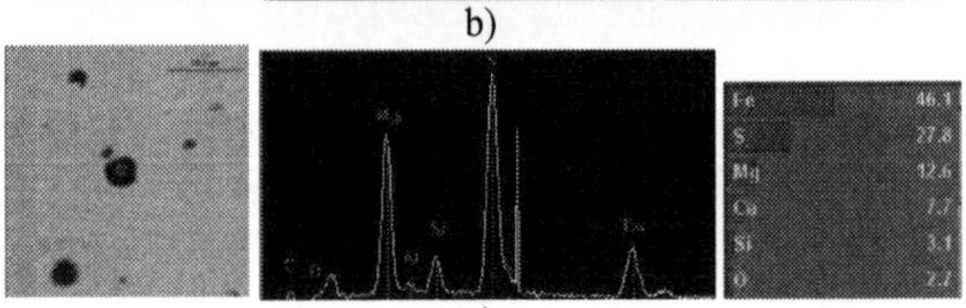

c)

Fig. 2. Different types of non-metallic inclusions located inside graphite nodules.

Ternary diagrams were used (see below) to present experimental statistics of the graphite nuclei chemistry and for comparison to chemistry of inclusion located in the metal matrix.

Three dimensional quantitative analysis of ductile iron structure. Because micro-structure parameters obtained from the optical image of random planes have significant limitations[3,4], an automated SEM/EDX analysis was used to collect nodule size statistics in a random section. EDX analysis of each particle was used to separate the carbon containing phase from non-metallic inclusions. After that, two-dimensional graphite nodule diameter statistic (2D) was converted into three-dimensional spheres diameter (3D) statistic and volume population density methods using the suggested method[3,4].

Adaptive thermal analysis. ATAS (Novacast) thermal analysis was used for the determination of the set of solidification parameters which characterize inoculation efficiency[6].

Structure and Properties of Two Ductile Irons

It is well known that charge materials could effect on the chill tendency of produced ductile irons. Two ductile irons (Table 1) were melted in the laboratory 200 lbs induction furnace targeting the same basic chemistry (3.7 %*C*, 1.7 %*Si*); however, using different industrial charges to verify its effect on structure and properties of ductile irons. In both cases, melt was *Mg*-treated in the ladle by 1.6% alloy (*Fe*-46%*Si*-6%*Mg*-1%*Ca*) and inoculated by 0.3% of *Fe*-75%*Si*-1%*Al*-1%*Ca*.

Table 1. Casting chemistry in two experimental heats (wt. %).

Heat	*C*	*Si*	*Mn*	*S*	*Cr*	*Ni*	*Ti*	*La*	*Mg*
1	3.56	2.86	0.20	0.005	0.026	0.023	0.006	0.010	0.036
2	3.63	2.84	0.22	0.005	0.022	0.025	0.006	0.005	0.047

Adaptive thermal analysis of *Mg*-treated and inoculated melts from the ladle just before pouring castings (Fig. 2) showed that these two ductile irons had the large differences in solidification parameters, which typically are linked to nucleation efficiency[6] (highter TE_{low} and *GRF1*, lower recolecence *R* in Heat 1).

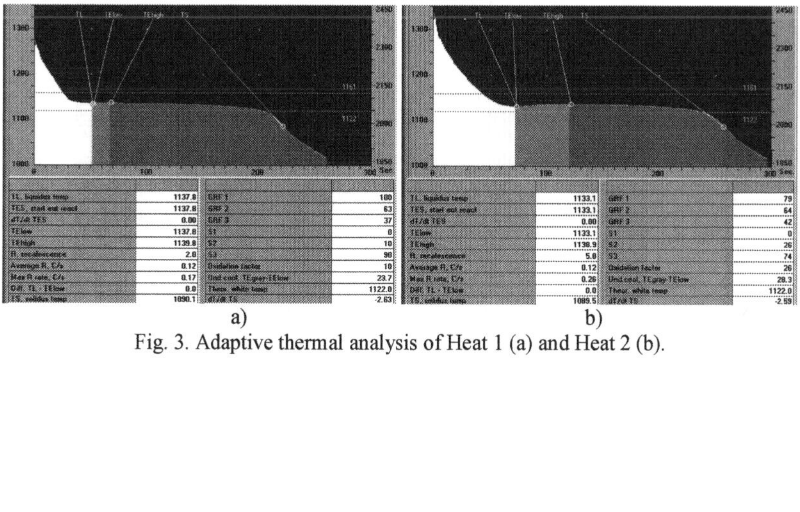

a) b)

Fig. 3. Adaptive thermal analysis of Heat 1 (a) and Heat 2 (b).

Microstructure in step plates (Fig. 4) also showed significant differences in the number of graphite nodules and related structure of metal matrix. Heat 1 had larger nodules number and less ferrite in 30 mm casting section.

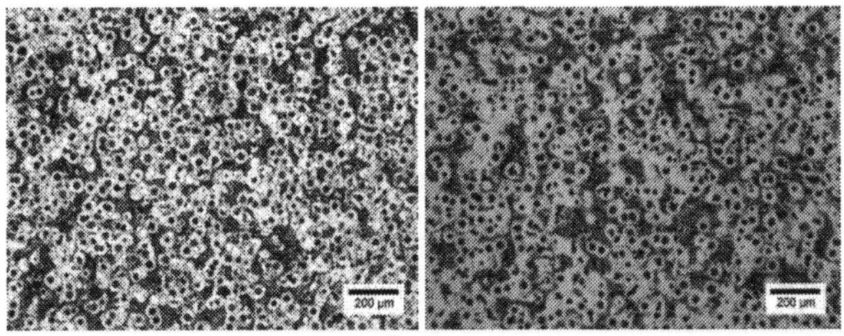

5 mm

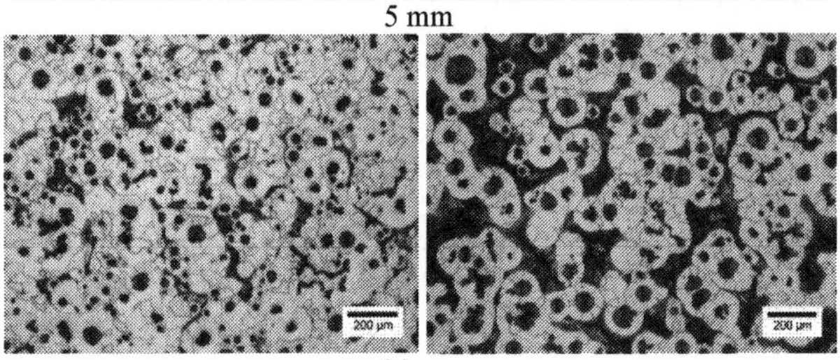

30 mm

Heat 1 *Heat 2*

Fig. 4. Etched microstructure of castings in 5 mm and 30 mm thicknesses of step plate.

Three-dimentional statistic of graphite nuclei showed (Fig. 5) that the ductile iron in Heat 1 was better nucleated (highter nodule count in 1 mm^3 and less 3D-diameter) when compared to casting from Heat 2.

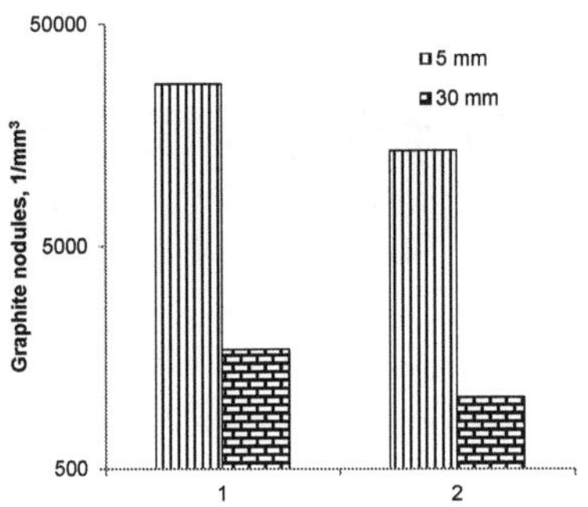

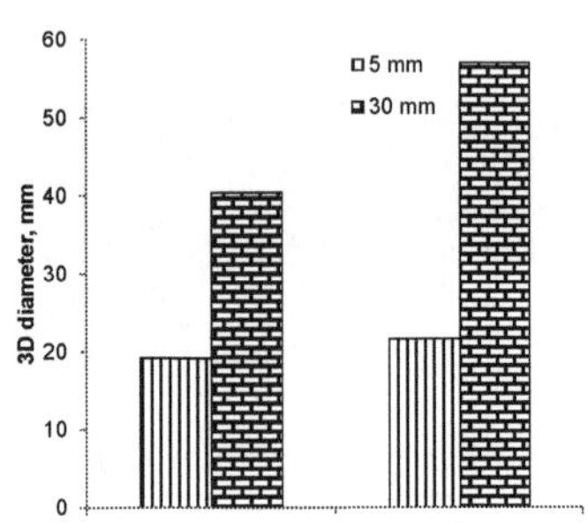

a) b)

Fig. 5. Graphite nodules volume population density (a) and average 3-D diameter (b) in in step plates (5 and 30 mm thickness) from Heat 1 and Heat 2.

These ductile irons also have a different 3-D diameter statistic. Ductile iron from Heat 1 had bi-nodal 3-D diameter (Fig. 6a) and volume population density (Fig. 6b) distributions with a significant part of small graphite nodules.

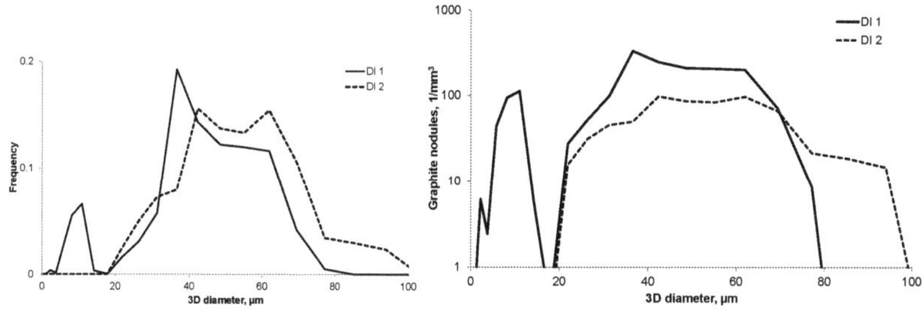

a) b)

Fig. 6. Graphite nodule 3-D statistic: a) diameter and b) volume population density in 30 mm thickness step.

Described differences in graphite nodule dimensional statistic and metal matrix structure influenced on mechanical properties of these two ductile irons. Ductile iron from Heat 1 had higher elongation, toughness and, so-called", quality index Q=UTS^2*El.

Table 2. Mechanical properties of two ductile irons.

Ductile iron	UTS, KPa	YS, Kpsi	El, %	BHN	Q=UTS^2*El	CVN, ft-lbs
1	71.4	49.7	14.9	162	76.1	10.1
2	75.7	51.1	12.2	167	69.8	6.6

Analysis of Heterogeneous Nuclei

Two types of charge materials were used for production of the described ductile irons. In this article, the observed nucleation efficiency were linked to the actual chemistry of graphite nuclei. It could be observed that the majority of heterogeneous nuclei contained a complex of *Si-Mg-Ca* oxides and *Mg-Ca* sulfides. A limited amount of mono- *Mg* and *Ca* oxides and sulfides were found inside graphite nodules. Pure alumina was not found inside graphite nodules. The majority of inclusions in the metal matrix were complex *Mg-Si-Ca* oxide with significantly less concentration of *S* and *Ca* when compared to inclusions found inside graphite nuclei (Fig. 7).

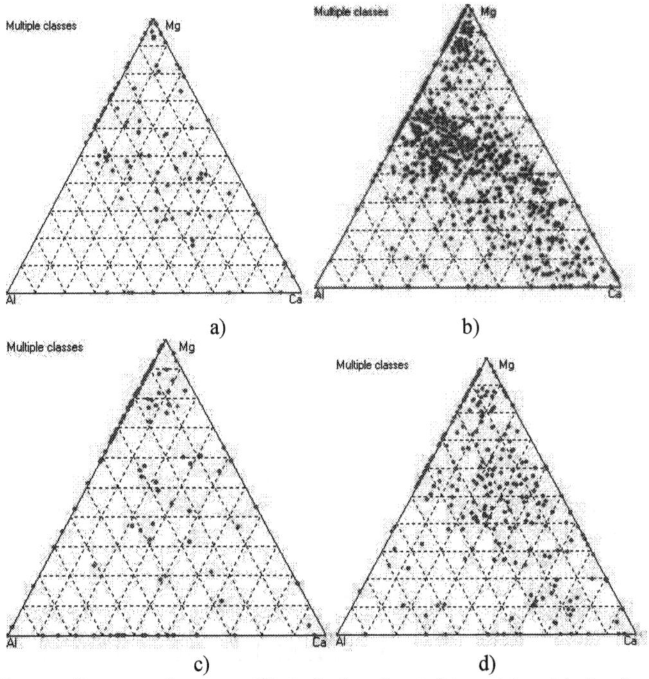

Fig. 7. Ternary diagrams of non-metallic inclusions located in metal matrix (a, c) and inside graphite nodules (b and d) in quenched specimens from Heat 1 (a, b) and Heat 2 (c, d).

Analysis showed significant partitioning of non-metallic inclusion between metal matrix and graphite nucleus (Table 3). Inclusions, which can be considered as potential heterogeneous nuclei, had a higher amount of *Ca* and *S* in both ductile irons. However, graphite nodules in ductile iron from Heat 1 showed more Ca and S when compared to Heat 2.

Table 3. Average chemistry of non-metallic inclusions in metal matrix and inside graphite nodules.

Heat	Location	Chemistry, wt. %				
		Mg	*Al*	*Si*	*S*	*Ca*
1	Nodules	22	10	41	19	12
	Matrix	25	5	49	17	4
2	Nodules	22	8	42	17	9
	Matrix	23	7	57	8	4

The thermodynamic simulation (FACTSAGE software) was used to predict the possible sequences of reactions which could take place in inoculant dissolution regions[8,9]. In these super-saturated regions, active elements, such as *Mg, Ca, Si* and *Al*, react with dissolved in the melt impurities *S* and *O*. In this article, different simulated scenarios included variations in a sulfur and calcium in *Mg*-treated melt. At very low residual *S* and *Ca* in the melt, *MgO* and more complex oxides such as forsterite (Mg_2SiO_4) and merwinite ($Ca_3MgSi_2O_8$) could be formed at a high temperature 200-300^{0}C above ductile iron solidification interval (Fig. 8a). These inclusions can be easily agglomerated and floated out from the melt. From the stand point of stability of very fine inclusions in the melt, the scenario with some residual *S* and *Ca* in the *Mg*-treated melt looks more promising to enhance heterogeneous nucleation (Fig. 8b). The possible sequence of reactions will include early forming *MgO* and complex *Mg-Si-O* oxides (forsterite, for example) with sequential "coating" inclusions by sulfides (*CaS*) just before graphite eutectic solidification.

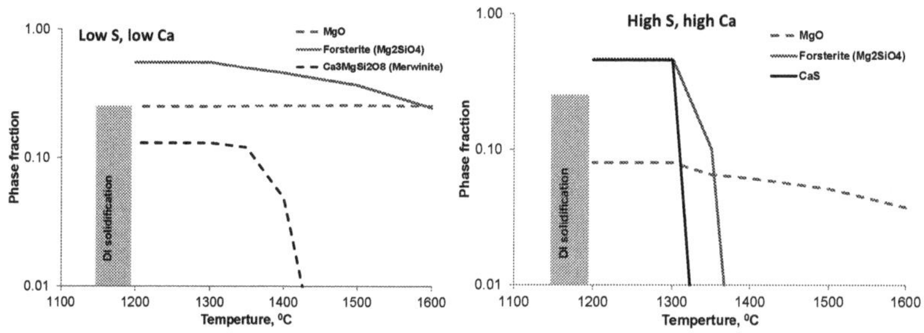

a) b)

Fig. 8. Thermodynamic predictions of oxi-sylfide formation in *Mg*-treated melt with different levels of residual *S* and *Ca*.

Nucleation potential depends on several factors, including requirements of minimal interfacial energy between nuclei and growing phase which could be less for sulfides when compared to oxides. Important also is the temperature-time condition for inclusions precipitation in the melt: small and fresh formed inclusions with a minimal interfacial to graphite energy will have a higher graphite phase nucleation potential. Statistics of non-metallic inclusions inside graphite in the metal matrix showed a higher probability of complex (*Ca*+*S*)-containing inclusions inside graphite nodules. The experimental data support these thermodynamic considerations. Direct high resolution observation also showed the complex nature of heterogeneous nuclei in ductile iron[7]. Active heterogeneous nuclei provide continuous nucleation of spherical graphite during solidification, resulting in bi-modal three-dimensional graphite nodule distributions in the casting. Fig. 9 illustrates the effect of nuclei chemistry on the size of graphite nodules in "soft quenched" specimens. In all cases, *S-Ca*-contained graphite nodules had significantly larger diameters, which indicated evidence of early nucleation.

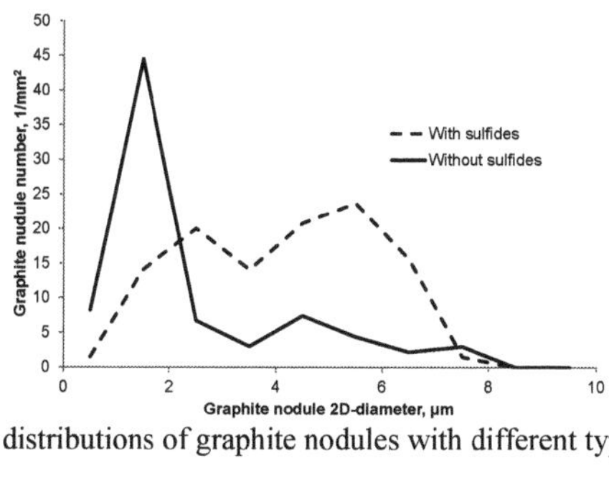

Fig. 9. Size distributions of graphite nodules with different types of nuclei.

Summary

The applied approaches including an automated SEM/EDX analysis of graphite nodule nucleation sites and a special algorithm to convert two-dimensional to three-dimensional graphite nodule size distribution were used to evaluate the effect of charge materials on heterogeneous nucleation in two ductile irons with similar main chemistries. It was shown that partitioning of *Ca* and *S* containing non-metallic inclusions between metal matrix and inclusions enclosed into graphite nodules takes place. Thermodynamic analysis was used to predict non-metallic inclusion formation sequences and correlate it with observed inclusion partitioning. Melting "history" plays a significant role in nucleation efficiency of inoculants and the methodology, described in this article, could be used for optimization of ductile iron treatment.

Acknowledgments

Dr. Von Richards, Dr. Kent Peaslee, and PhD student Jingjing Qing for help and support.

References

1. Singh V., Lekakh S., and Peaslee K., "Using Automated Inclusion Analysis for Casting Process Improvements", *Proceedings of the Steel Founders' Society of America 62th Annual Technical and Operating Conference*, Chicago (2008).
2. Lekakh S., Richards V., and Peaslee K., "Thermo-chemistry of Non-metallic Inclusions in DI", *The Carl Loper Cast Iron Symposium Proceedings*, AFS (2009).
3. Lekakh S., Qing, J., Richards V., Peaslee K., "Graphite Nodule Size Distribution in Ductile Iron", *AFS Proceedings* (2013).
4. Lekakh S., Thapliyal V., and Peaslee K., "Use of an inverse algorithm for 2d to 3d particle size conversion and its application to non-metallic inclusion analysis in steel", *AIST Proceedings* (2013).
5. Lekakh S., Harris M., "Novel Approaches to Analyze Structure of Ductile Iron", *Keith Millis Symposium on Ductile Cast Iron Proceedings*, AFS (2013).
6. Novacast: ATAS, http://www.novacast.se
7. Skaland T, "Nucleation Mechanism in Ductile iron", *Proc. AFS Cast Iron Inoculation Conference* (2005).
8. Lekakh S. and Loper C. R., "Improving Inoculation of Ductile Iron", *AFS Proceedings* (2003).
9. Lekakh S., Robertson D., and Loper C. R., "Thermochemistry and Kinetics of Iron Melt Treatment", *World Foundry Congress Proceedings*, UK (2006)

Shape Casting: 5th International Symposium 2014
Edited by: Murat Tiryakioğlu, John Campbell, and Glenn Byczynski
TMS (The Minerals, Metals & Materials Society), 2014

The External and Internal Shrinkages in Aluminum Gravity Castings

Fu-Yuan Hsu, Shin-Wei Wang and Huey-Jiuan Lin

Department of Materials Science and Engineering, National United University,
No.1 Lein-Da, Kung-Ching Li, MiaoLi, 36003 Taiwan, R.O.C.

Keywords: shrinkage, micro-porosity, modeling, aluminum gravity casting.

Abstract

In aluminum gravity castings, the external and internal shrinkages (micro-porosity) are more prominent than those in ferrous castings. In order to understand the relationship between these two shrinkages during solidification, three castings with various tilting angles (i.e., 0°, 45° and 90° designated as H- T-, and S- castings respectively) were designed.

Two pouring temperatures (700 and 760°C) were conducted in the real aluminum casting experiments. The volume change of external surface sinks was quantified by the reverse engineering scanning method, while that of internal porosities was measured by Archimedes method. Numerical models for simulating the shrinkages of aluminum alloy casting were suggested and validated by these experiments. The influence of tilting angle of castings was reported.

A linear function between maximum micro-porosity and pore formation pressure in the modeling result was proposed to predict the negative pressure within castings during solidification. Two feeding (liquid and mass feeding) mechanisms of the solidification in the real aluminum casting were clarified by the modeling results.

1 Introduction

In the solidification of liquid metal, Campbell [1] concluded five feeding mechanisms. Among five feeding mechanisms, many authors suggested the most of aluminum solidification defects, such as micro-porosity and surface sink, are resulted from the inter-dendritic feeding mechanism at the third stage of solidification. Using D'arcy law, Kao *et al* [2 3] derived the equation of local pressure drop (ΔP) of A206 and A356 aluminum alloys. Chiesa [4] found that the distribution of micro-porosity in aluminum alloy casting is related to the hydrogen content of the alloy. Kubo、Pehlk [5] believed that the formation of micro-porosity is resulted from the inter-dendrite, surface tension and gas precipitation. They also used a numerical model to predict the content of micro-porosities within the casting. Campbell [6] discovered that intrinsic double-oxide (bifilm) defects within liquid aluminum are main sources of heterogeneous nucleation sites for formation of hydrogen molecule (gas porosity). The inflated bifilms are the micro-porosities found in aluminum castings.

The purpose of this study was to understand which feeding mechanism results in the external and internal shrinkages during the solidification of aluminum melt. Real casting and numerical modeling were employed in this investigation.

2 Methods

Three types of aluminum alloy castings with equal volume, which are V-, T- and H-types, were designed for this study (Figure 1a). For these castings, their geometries having angles against to gravity direction (-z direction) are 0, 45 and 90 degrees, respectively. Because the breadth of the three castings is equal length (300mm) and the insulated material (i.e., glass fiber cotton) used at the two ends of the casting in real casting experiments (shaded regions in Figure 1a), the solidification of these castings became two-dimensional condition. The thermal transferring in the y-direction in Figure 1a is therefore ignored in this study.

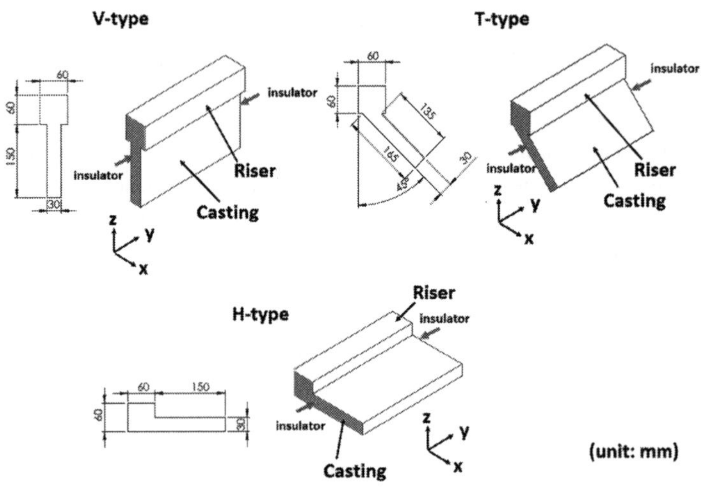

Figure 1 Three types of casting geometries. (noted that shaded regions are insulators at two ends of the castings)

In the casting experiment, the bottom-gating runner systems were designed for the three casting types. They are the same total head height of 300mm to provide the equal static pressure to the casting cavity. 10ppi ceramic foam filter was included within the runners to reduce gating velocity under the critical velocity of 0.5 m/s [7]. The sand mold material was silica sand with the mean particle size of ASTM sieve no. 70~85 mixed with Phenolic-Urethane Cold Box (PUCB) binder. Aluminum alloy, A356 (Al-7%Si-Mg), was melted at 680±10°C and degassed by pure argon for two hours. After degassing, liquid aluminum then was heated up to two designated pouring temperatures (700 and 760 ±5°C). The pouring lasted approximately 1~3 seconds.

After solidification, the external profile of the castings was measured and scanned by the three-dimensional scanning machine (Breuckmann opto Top-530). The difference between the scanned data and its original 3D CAD file was evaluated and the external shrinkage of the castings was estimated.

Samples with thickness of 10 mm was cross-sectioned from the center of the three castings. The sample was wax sealed and its bulk density was measured by Archimedes method. The internal shrinkage (or micro-porosity) within the sample was thus estimated by the difference between theoretical and the measured densities.

The commercial computational fluid dynamics (CFD) code, Flow-3D™, was utilized for the solidification modeling. Two-dimensional condition of heat transfer was considered in these three casting geometries. The physical data of A356 aluminum alloy applied in the computational modeling. For the latent heat of fusion, 4.29×10^5 (Joule/Kg) was used. During the solidification of this liquid aluminum, the critical solid fraction f_c and the coherent solid fraction f_{co} are 0.68 and 0.23 respectively.

A solidification model for solid-to-liquid phase change can be constructed using the porous media drag concept. By neglecting volume changes associated with a phase change and assuming that solid aluminum is at rest with respect to the computational mesh, approximate solidification processes (i.e., the state of zero flow velocity) is modeled by using a drag coefficient $\boldsymbol{K}$ that is a function of the local solid fraction $\boldsymbol{F_s}$. The D'Arcy-type drag model for the liquid phase flow in the mushy zone given by Equation 1:

$$K = D\frac{F_s^2}{(1-F_s)^3}$$

Equation 1

when F_s is equal to critical solidification, the drag coefficient K becomes effectively infinite. At intermediate states, the drag is an intermediate value. To estimate the modeling of micro-porosity, two values applied for the critical pressure at which gas pores can form were -1.0×10^2, -1.5×10^5 and -2.0×10^5 Pa. The negative value implied the gas pore can only form when the pressure is lower than one atmosphere (i.e. pressure difference).

3 Results

The 2D casting profile in the center of the casting geometry can represent the whole 3D casting since the two-dimensional solidification condition was considered in this study. In Figure 2 the real castings and their silhouette in their center plane were demonstrated. Apparent external shrinkages in the riser and some places of the castings were observed (as the arrows indicated in Figure 2a).

Figure 3a shows the total external shrinkage as function of the tilting angle. The external shrinkage increases as the tilting angle is increased. The external shrinkage of V type is the greatest among all casting types at a pouring temperature. At low pouring temperature (700°C) the change of increasing volume of the shrinkages against to the tilting angle (the slope of the trend line in Figure 3a, 7×10^{-5})

is more prominent than that of high temperature (1×10^{-5}).

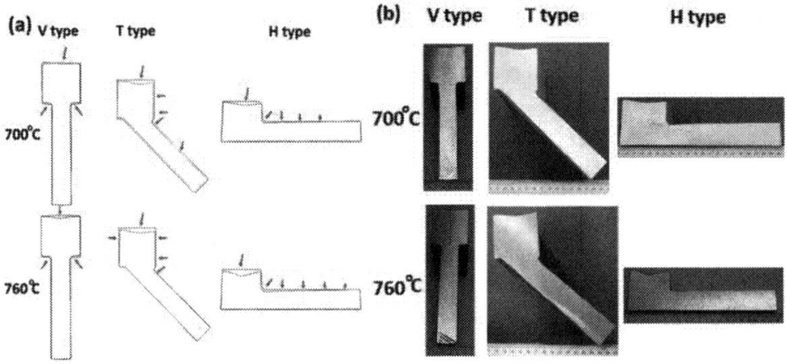

Figure 2 The external shrinkages profile of three casting types at two pouring temperatures (a) (note: the outline of the shrinkage profile is the original cross-sectional area of the castings) and the real castings sectioned in the center (b).

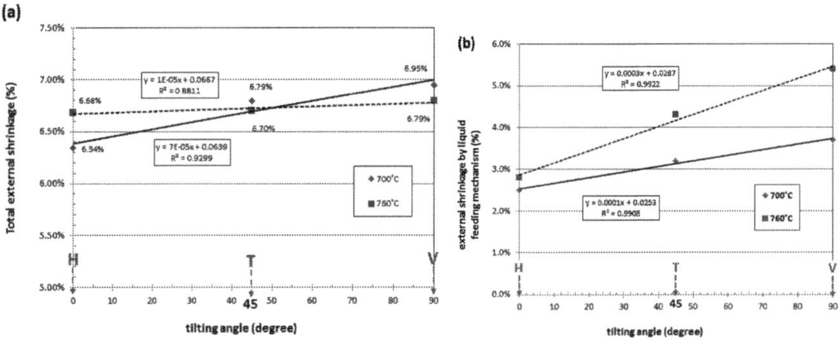

Figure 3 The total external shrinkages of three casting types against its tilting angle (a) and external shrinkage by liquid feeding mechanism in the modeling results (b). (the error range ±0.15%) .

The volumes of the internal shrinkages or micro porosity for these samples and their percentage were shown in Table 2. The micro porosities for three casting types are almost the same at a constant pouring temperature. At low pouring temperature (700°C), the micro porosity is approximately 1.8%, while 1.9% is for the high temperature (760 °C). It implies that the internal shrinkage is not influenced by the tilting angle of the casting geometry. But it may affected by the pouring temperate.

Table 2 internal shrinkage (micro-porosity) for the three castings

Pouring Temperature (°C)	Type	Volume of Micro Porosity, mm^3	Micro Porosity (%)
700	V	41956	1.80
	T	41046	1.78
	H	44234	1.88
760	V	45166	1.94
	T	44749	1.90
	H	44925	1.94

The modeling results of the micro porosity for three casting types were shown in Figure 4a. To have a better illustration, two color scales with different maximum porosity percentages (1.86% and 1.50% i.e., the red in color scales) were placed on the top of the column header. In the modeling, the maximum micro-porosity is dependent on the pore formation pressure. Lower the pressure (-1.5×10^5 Pa) smaller the maximum percentage of micro-porosity (1.5%) within castings. Figure 4b shows this linear tendency between the pore formation pressure and the micro porosity, predicted by the modeling.

(a)

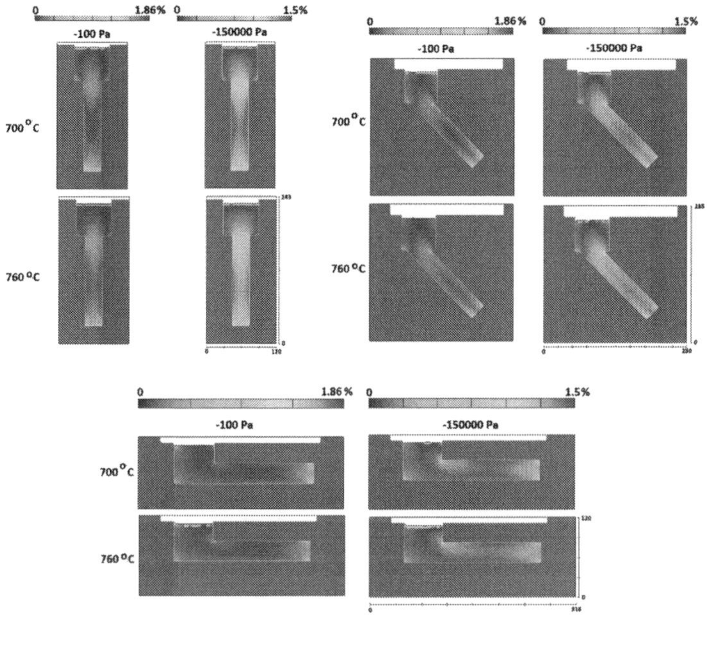

(b)

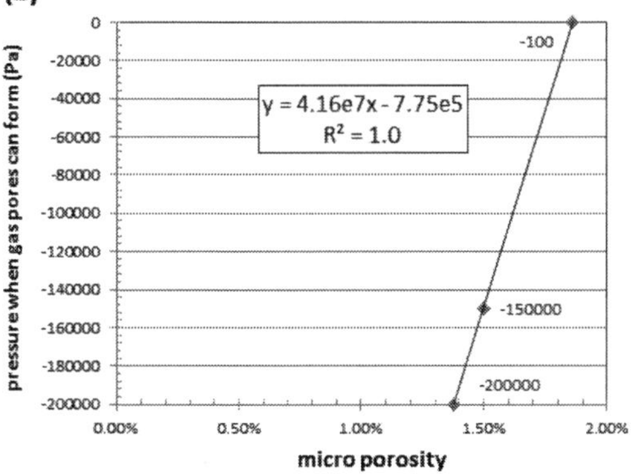

Figure 4 The micro porosities of three casting types at two initial liquid temperatures against to two pressures when gas pore can form (a) (note that the row header is initial melt temperature while the column is pore formation pressure. The maximum of micro porosity percentage are different in two color scales.) and the linear relationship between the pore formation pressure and the maximum porosity in the modeling results (b).

In the modeling, the external shrinkage only appeared at the first solidification mechanism, liquid feeding mechanism (lower the liquid level on top surface of the risers in Figure 4a). After the shrinkage by liquid feeding mechanism, no apparent external shrinkage was shown in the modeling result. However, the regions of micro-porosity in the castings were predicted. Figure 3b shows only the external shrinkage by liquid feeding mechanism in the modeling results. The linear trend lines demonstrate that the external shrinkage of liquid feeding mechanism increases as the tilting angle is increased. At the high temperature (760˚C), the slope (0.0003/degree) is higher than that (0.0001/degree) at the low temperature (700 ˚C). This is resulted from the thermal expansion of liquid aluminum considered in the modeling (Table 1).

4 Discussion

Averaged total external shrinkages are 6.69% and 6.72% for pouring temperatures of 700 ˚C and 760 ˚C respectively, while averaged internal shrinkages are 1.82% and 1.93%. The sum of external and internal shrinkages is 8.51% at 700 ˚C and that is 8.65% at 760 ˚C.

Comparing to the slopes of the trend lines in the real casting (Figure 3a), that of the external shrinkage by liquid feeding mechanism in the modeling result (Figure 3b) is more pronounced. Thus, the external shrinkage at liquid feeding stage is more influenced by the geometry of castings related to the direction of gravity. It is also affected by the initial melt temperature. At 760˚C, the slope is 3 times than that at 700 ˚C (Figure 3b).

The actual casting profiles (Figure 2a) are more concaved than those (Figure 4a) predicted by the modeling. It suggests that the actual casting is not only affected by the liquid feeding mechanism but also some other mechanisms. Figure 5 shows the difference between the actual casting and the modeling one. In Figure 5a, the arrows indicated are the regions of external shrinkages in this casting. These regions are related to the places of the maximum micro-porosities illustrated in the modeling results (i.e., the red color in Figure 5b). These regions are the places in very large negative pressure (-1.5×10^5 Pa, more than one atmosphere). At the temperature range between liquid and solid states, the strength of actual aluminum casting is weak enough to cause the mass and Inter-dendritic feeding mechanism.

Therefore, the pre-solidified aluminum skin attached on the surface the mold is peeled away from it. That results in the smaller size of the shape similar to the geometry of the mold cavity as mass feeding mechanism takes place. Although the modeling result did not show exactly profile of the actual casting, the negative pressure predicted by the modeling can explain the cause of the final casting shape.

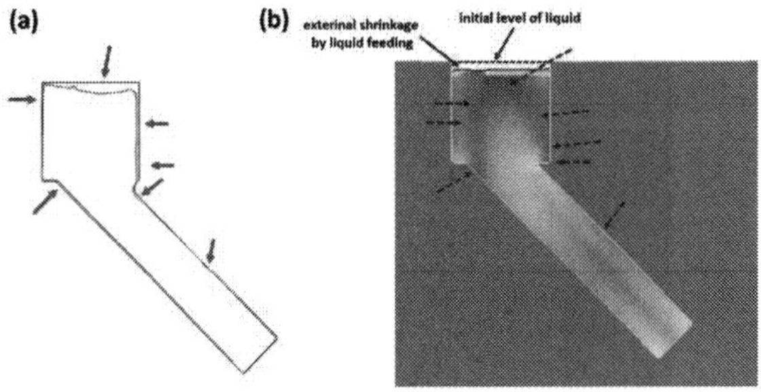

Figure 5 the external shrinkage profile comparison between the actual casting (a) and the modeling one (b). (note: the dashed arrows indicate the potential regions of external shrinkages could result from the large negative pressure within the casting)

Considering the tilting angle, total external shrinkages of the actual castings are more likely affected by the pouring temperature (Figure 3a). At 700˚C the increment of shrinkage percentages per tilting angle is 7 times larger than that at 760 ˚C. The change of shrinkages per degree is less at the high temperature. Against to the tilting angle, the trend line of the pouring temperature 760 ˚C shows nearly a horizontal straight line (independent the tilting angle) in Figure 3a. It implies that the total external shrinkages are less influenced by the tilting angle at the high temperature. The total external shrinkage will be nearly the same regardless of various tilting angles. Thus, the shortage of total external shrinkages resulted from liquid feeding mechanism (Figure 3b) is compensated by the mass feeding mechanism. At the high temperature, the mass feeding becomes more evident.

During the transformation of liquid-solid phase, the maximum of volume shrinkages $\mathbf{S_{max}}$ would be obtained from the density difference between solid and liquid and critical solid fraction $\mathbf{f_c}$ (0.68) of the aluminum alloy as shown in the following equation.

$$S_{max} = \frac{(\rho_{sol}-\rho_{liq})}{\rho_{sol}}(1\text{-}f_c)$$

Equation 2

where ρ_{sol} is solid density of aluminum alloy A356 (2570 Kg/m^3), while ρ_{liq} is the liquid density (2420 Kg/m^3). In the modeling, D'Ary type drag is considered in liquid flow through the partial solidification (such as the mushy zone, Equation 1).

The region of maximum micro-porosity is thus calculated as 1.87%. Once a pore formation pressure was set in the modeling (Figure 4a), S_{max} is a constant value. Figure 4b shows the linear relationship between S_{max} and pore formation pressure. The linear equation of this trend line can be used for predicting the total negative pressure in the real castings as the total micro-porosity of castings is acquired. Based on the linear equation in Figure 4b, the negative atmosphere pressure of -25,082 Pa is calculated, as the averaged micro porosity is 1.8% at 700˚C (Table 2). In comparison, as averaged micro porosity is 1.9% at 760 ˚C, the castings are experiencing the positive pressure of +16,561 Pa during their solidifications. If maximum micro porosity is 1.87% the pressure is close to zero Pa (one atmosphere). At the high pouring temperature, additional gas content may be consumed in the aluminum casting. Therefore the positive pressure at which pore can form in the casting is predicted by the linear equation.

Weak strength of partial solidified skin on the surface of the mold at the high temperature is easily sucked away from it. Strong skin strength at low temperature is less affected by the negative pressure at mass-feeding step. During solidification the negative pressure at the high temperature can be compensated by the skin deformation (external shrinkage). As a result no more negative pressure exists in the casting. At low temperature, negative pressure is developed due to lack of skin deformation.

5 Conclusion

1. For three casting types, the external shrinkage at liquid-feeding step is affected by casting tilted angle. The total external shrinkage at the high pouring temperature is almost the same and it is independent on the tilting angle.
2. At the high pouring temperature, mass-feeding mechanism becomes significant in the solidification. Weak strength of aluminum skin on the surface of the mold is more easily deformed. As a result, there is no negative pressure at this temperature. At low temperature the strong skin is less affected by the negative pressure at mass-feeding step. Negative pressure is therefore created during solidification.
3. Micro porosity is not influenced by the tilting angle of castings but it is affected by pouring temperature. The linear equation between maximum micro porosity and pore formation pressure was proposed to forecast the possible additional gas content and the region of negative pressure in aluminum casting.

Acknowledgements

F-Y Hsu acknowledges the help of Dr. Yen-Hung Chen, Department of Mechanical and Automatic Engineering, Chung Chou University of Science and Technology in Taiwan.

References

1. Campbell, J., *Castings*. 2003: Butterworth-Heinemann.
2. Kao, S.T., E. Chang, and Y.W. Lee, *Role of Interdendritic Fluid Flow on the Porosity Formation in A206 Alloy Plate Castings*. Materials transactions-JIM, 1994. **35**: p. 632-632.
3. Kao, S.T. and E. Chang, The role of the pressure index in porosity formation in A356 alloy castings. Cast Metals, 1995. **7**: p. 219-219.
4. Chiesa, F., R. Fuoco, and J.E. Gruzleski, Porosity distribution in directionally solidified test bars sand cast from a controlled A356 melt. Cast Metals, 1994. **7**: p. 113-113.
5. Kubo, K. and R.D. Pehlke, *Mathematical modeling of porosity formation in solidification*. Metallurgical and Materials Transactions B, 1985. **16**(2): p. 359-366.
6. Campbell, J., *An overview of the effects of bifilms on the structure and properties of cast alloys*. Metallurgical and Materials Transactions B, 2006. **37**(6): p. 857-863.
7. Fu-Yuan Hsu, Huey-Jiuan Lin, Foam filters used in gravity casting, Metallurgical and Materials Transactions B: Volume 42, Issue 6 (2011), Page 1110-1117.

Shape Casting: 5th International Symposium 2014
Edited by: Murat Tiryakioğlu, John Campbell, and Glenn Byczynski
TMS (The Minerals, Metals & Materials Society), 2014

EFFECT OF Si AND Cu CONTENT ON THE SIZE OF INTERMETALLIC PHASE PARTICLES IN Al-Si-Cu-Mg-Fe ALLOYS

T. Sivarupan[1], J. A. Taylor[1] and C. H. Cáceres[1, 2*]
[1]CAST Co-operative Research Centre
[2]ARC-Centre of Excellence for Design in Light Metals
Materials Engineering, School of Mechanical and Mining Engineering,
The University of Queensland, Brisbane, QLD 4072, Australia

Keywords: Secondary dendrite arm, sandcasting, Fe-rich intermetallics, aluminium silicon alloy, alloy 319

Abstract

Quasi-directionally solidified plates were sand cast using unmodified Al-xSi-yCu-0.1Mg-0.5Fe alloy with two Si (x = 4.5 or 9 mass%) and three Cu (y = 0, 1 or 4mass%) contents, and the size of the intermetallic phase particles (β-Al_5FeSi and Al_2Cu) assessed at constant secondary dendrite arm spacing (SDAS) using optical microscopy and back scattered electron images. Increasing the concentration of Si alone or in combination with Cu refined the β-Al_5FeSi platelets, whereas increasing Cu at constant Si shows an SDAS and Si level dependent effect.

Introduction

Recent work on Sr modified Al-Si-Cu-Mg-Fe-Mn alloys indicates that increasing Si content increases the ductility of high Fe containing alloys [1]. The enhanced ductility was explained with the help of the ternary Al-Si-Fe phase diagrams [2]: a higher level of Si shortens the solidification path and refines the β-Al_5FeSi platelets. The effect was observed in the presence of Cu, Mg, Mn and Sr. Further experiments aimed at verifying this hypothesis in ternary Al-Si-0.8Fe alloys contradicted the earlier conclusions, i.e., that increased Si led increased the size of the Fe-rich intermetallics [3]. It was thus suggested that the refining effect of Si upon the Fe-rich intermetallics was related to the formation of low meting point Cu-based eutectics rather than just due to the sortening of the solidification path of the Al-Si base alloy. In the present work the effect of Cu (with levels of 0, 1 and 4 %) on the size of β-Al_5FeSi platelets was studied. The Fe content (0.5%) was the same as those of reference [1]. Unmdified alloys were studied. The commercial software package, ThermoCalc^{TM}, was used in conjunction with the experiments to help understand the solidification sequence and the formation of intermetallics.

Materials and experimental methods

Six alloy compositions, listed in Table I, were studied. The casting followed the description of Refs [4, 5]. Briefly, Al ingot (99.8% purity: main impurities 0.03% Si, 0.12%Fe) was molten in a 20kW induction furnace along with commercial purity Si. Cu, Fe as AlTab Fe 75% compacts, and Mg were subsequently added in the order; the liquid was degassed for about 20 minutes with Ar at a rate of 2 l/min (3.33×10^{-5} m^3/s), and poured into the mould after 2 minutes of holding.

* Corresponding author email: c.caceres@uq.edu.au, (C H Cáceres)

Samples for chemical analysis were taken prior to each pouring. Pouring was done at 1003 K (730°C) into a sand mould assembly, with heavy steel chills at the far ends, designed to produce three plates at a time, each 160x120x15 mm^3 in size. The mould contained the chills, had dimensions of 32x40x140 mm^3 and their surface was smooth. The mould assembly was inverted right after pouring to allow the plates to solidify in a quasi-directional manner upwards away from the chill, towards the risers. Furher details of the mould assembly can be found in Ref [4].

Longitudinal sections were cut from each cast plate and metallographically polished. Back scattered electron images were taken at predetermined locations selected using data from reference [4] to ensure constant SDAS in order to compare the size of Fe- or Cu-rich intermetallics. The SDAS selected for metallographic obseravations (~50 and ~30 μm), correspond to cooling rates of <1 °C /s and >5 °C /s, respectively. The cooling rates calculated between liquidus and solidus; CR1 cooling rate as mentioned in the Ref. [4]

Results

THERMOCALC™ CALCULATIONS

The onsets of precipitation of each phase in the Cu-free alloys (whose composition matches the alloys of the prior studies of Refs. [3, 5]) are shown in Fig 1 as per the values listed in Tbale 2. The β-Al_5FeSi phase precipitates in the pre-eutectic stage for all of the alloys except for the high Si ones where it forms in the post-eutectic stage. When present, Cu reduces the onset temperature of all reactions, namely, β-Al_5FeSi phase and Al-Si eutectic temperatures, as well as a slight reduction in the values of $T_{\beta-eut}$, and $T_{liq-eut}$. The later two, however, were more than halved by the increase in Si content from 4.5 to 9.0%. i.e., increased Si level reduced the solidification range, $(T_{liq} - T_{sol})$.

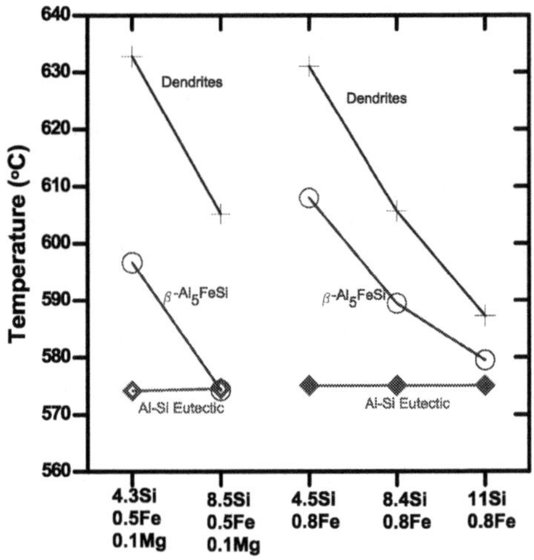

Figure 1: The temperatures at the onset of relevant reactions for the Cu-free alloys studied as per Table II. (Note that the Cu-free alloys of the work match those of the prior studies of Refs [3, 5]).

Table I: Target (left paired column) and actual (right, **bold** column) chemical compositions (mass %) of the alloys studied. The alloys nomenclature follows that of Ref. [4], i.e.: the superscripted/subscripted elements indicate higher/lower level)

Alloy	Chemical Composition†							
	Si		Cu		Mg		Fe	
Al_{SiMg}^{Fe}	4.5	**4.3**	0.0	**0.01**	0.1	**0.08**	0.5	**0.49**
Al_{SiCuMg}^{Fe}	4.5	**4.4**	1.0	**0.96**	0.1	**0.08**	0.5	**0.48**
$Al_{Si}{}^{Cu}{}_{Mg}{}^{Fe}$	4.5	**4.3**	4.0	**3.7**	0.1	**0.08**	0.5	**0.42**
$Al^{Si}{}_{Mg}{}^{Fe}$	9.0	**8.5**	0.0	**0.01**	0.1	**0.13**	0.5	**0.46**
$Al^{Si}{}_{CuMg}{}^{Fe}$	9.0	**8.7**	1.0	**0.96**	0.1	**0.13**	0.5	**0.45**
$Al^{SiCu}{}_{Mg}{}^{Fe}$	9.0	**8.5**	4.0	**3.79**	0.1	**0.12**	0.5	**0.43**

Table II: ThermoCalc™ values for the onset of precipitation and the temperature range of key reactions. All values are in [°C].

Alloy	Liquidus (T_{liq})	Al solid solution (dendrites) (T_{den})	β-Al_5FeSi (T_{β})	Al-Si eutectic (T_{eut})	($T_{liq-eut}$)	($T_{\beta-eut}$)
Al_{SiMg}^{Fe}	632.96	632.76	596.66	574.16	58.6	22.5
Al_{SiCuMg}^{Fe}	629.96	629.26	591.76	570.06	59.9	21.7
$Al_{Si}{}^{Cu}{}_{Mg}{}^{Fe}$	621.96	621.56	579.86	560.16	61.8	19.7
$Al^{Si}{}_{Mg}{}^{Fe}$	605.96	605.16	574.26	574.56	31.4	-0.3
$Al^{Si}{}_{CuMg}{}^{Fe}$	600.96	600.36	571.76	572.46	28.5	-0.7
$Al^{SiCu}{}_{Mg}{}^{Fe}$	591.96	591.66	565.36	566.76	25.2	-1.4

BACKSCATTERED ELECTRON IMAGES

Back scattered electron images were obtained at an SDAS of ~ 30 µm and ~ 50 µm are shown in (Fig. 2 (a-f)) and (Fig. 3 (a-f)) respectively. The Cu-free, Fig 2 a and d, and 0.1 Mg alloys showed a similar effect as in the alloys of Ref. [5] (the Fe content was ~0.8 mass % in Ref. [5] against 0.5mass% in the present alloys): increasing the level of Si from 4.5 to 9 reduced the size of Fe intermetallics.

In low Si alloys at small SDAS (Fig. 2 a-c), the Cu-free alloy (Fig 2a) showed a large amount of long Fe-rich platelets, the size of which decreased when the Cu content increased to 4 mass% (Fig.2c). In the 1 mass% Cu alloy, the Fe-rich platelets exhbit a lgreater degree of interconnection than in the Cu-free alloy.

A high Si content (Fig. 2(d-f) and Fig.3(d-f)), refined the intermetallics at small SDAS regardless of the presense of Cu. At large SDAS, Cu does increased the size of the intermetallics (i.e., the effects were opposite to those of the low Si alloys).

† All other common impurities, such as Mn, Pb, Ti, Sn, Cr, Zn, Ni, etc., levels were well below 0.01 w.t.%

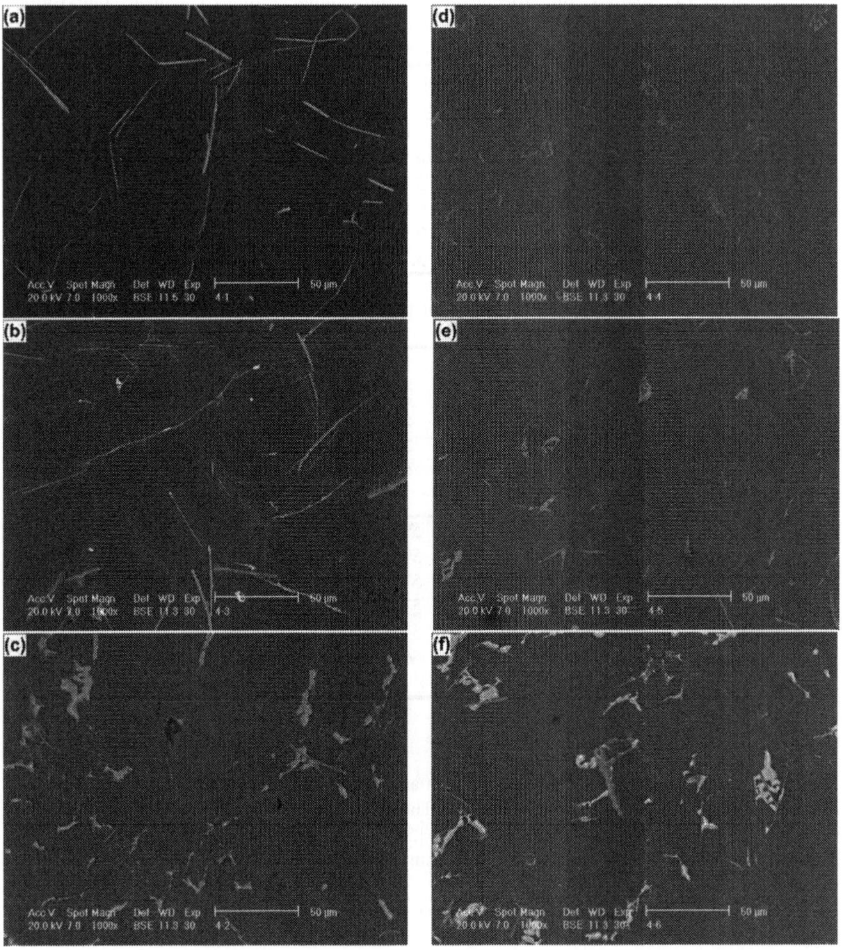

Figure 2: Back Scattered Electron images taken near chill end on the (left) low Si and (right) high Si alloys with (a & d) 0 Cu, (b & e) 1Cu, and (c & f) 4Cu. Average SDAS ~ 30 μm.

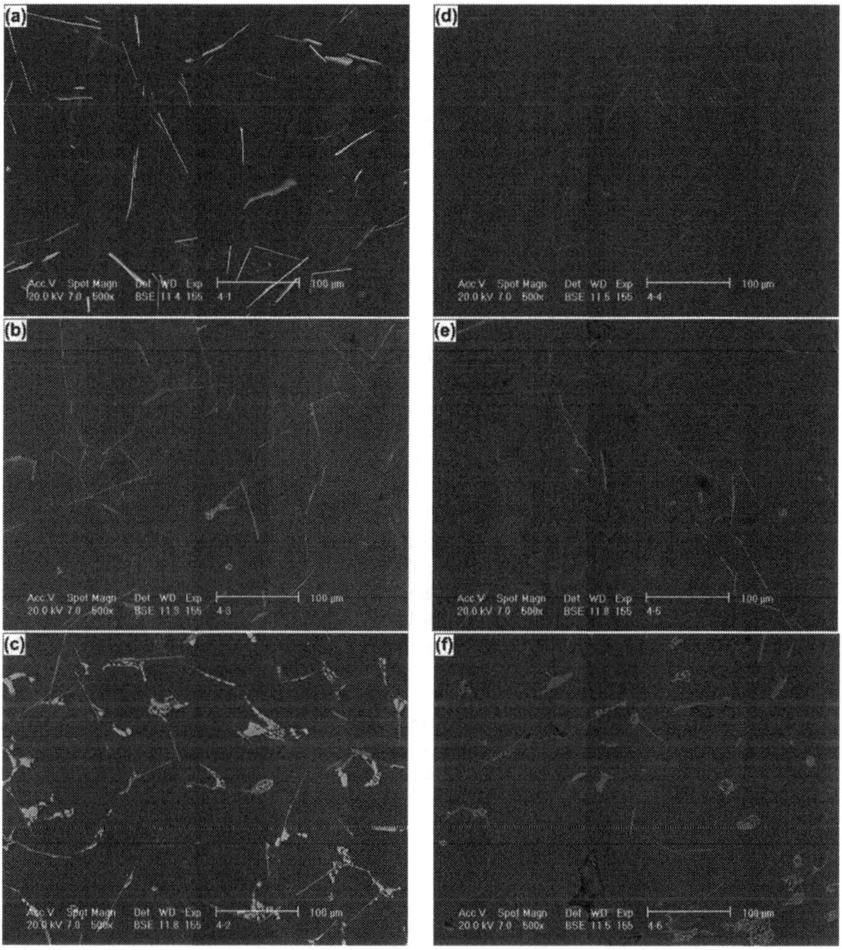

Figure 3: Back Scattered Electron images taken near riser end on the (left) low Si and (right) high Si alloys, with (a & d) 0 Cu, (b & e) 1Cu, and (c & f) 4Cu. SDAS ~ 50 µm.

Discussion

All of the alloys with low Si (4.5 wt. %) precipitate the β-Al_5FeSi intermetallics in the pre-eutectic stage (Table II and Fig 1). Introduction of Cu or increasing the Cu content reduces the $T_{\beta-eut}$ slightly (Table II) and possibly following the suggestion made in Ref. [3], the Cu rich pools of liquids isolate the β-Al_5FeSi plates preventing them from forming long interconnected clusters. Alternatively, it may be speculated that if the Al_2Cu particles nucleate on the β-Al_5FeSi plates, then the lengthening process of the latter may somehow be disturbed by the newly nucleated particles; hence the overall size reduction of Fe-rich intermetallics. The multiple Cu pools lead to extensive comminution and this is the dominant effect, especially at small SDAS.

Prior work showed that increased Si increases the size of the β-Al_5FeSi plates at 0.8 Fe level [3] whereas the more detailed study of ref. [5] made eveident the a cooling rate dependent evolution of α-Al_8Fe_2Si and β-Al_5FeSi phases at this high Fe level. In the present study, with only 0.5 Fe, the β- plates did not increase in size with increasing Si level (against Ref. [3]'s main conclusion); instead, it shows preferential formation of α-Al_8Fe_2Si particles at small SDAS and of β-Al_5FeSi plates at large SDAS. These observations can be easily rationalised with reference to different onset of precipitation temperature (Figs. 1 and 2, and Table II). The β-Al_5FeSi plates evolve in the pre-eutectic stage in the low Si alloys with 0.5 Fe, and even in high Si alloys with 0.8 Fe. In the high Si alloy with 0.5 Fe, the β-Al_5FeSi plates form in the post-eutectic stage, hence size refinement for increased Si occurs (Fig. 3). In the high-Cu low-Si alloys the β-Al_5FeSi plates form in the pre-eutectic stage and no refining should be expected,still the Figs 2 and 3 showed a size refining effect compared to the low-Si (Cu-free) alloys. In fewer words, further growth of the pre-eutectically nucleated β-Al_5FeSi plates in the post eutectic stage is prevented by the liquid Cu-rich pools.

Increasing the Cu content to the low Si alloys reduces the size of the β-Al_5FeSi plates in small SDAS, but the effect requires a critical level of Cu which is a function of the cooling rate and the Fe and Si content. The critical level of Cu seems determined to the solubility in Al, which is around 1mass%. That is, an excess of Cu above this concentration is required for the refining of Fe-platelets to occur (ie. Formation of Al_2Cu is required).

At large SDAS in the low Si alloys, increased Cu, actually increased the size of β-Al_5FeSi plates. Since the Fe-intermetallic particles nucleate in the pre-eutectic stage and due to the slow cooling rate there would be much time and space for the independent growth of these plates to grow larger before the nucleation of the detrimental Al_2Cu particle which could prevent the lengthening of these plates. In the slow cooling condition, i.e. SDAS >50 μm, Cu forms both blocky and fine eutectic Al_2Cu in the low Si alloys. In high Si alloys it only forms very fine eutectic Al_2Cu. Increasing Si increases the size of the β-Al_5FeSi plates in Cu free alloys with 0.8 mass% Fe [3, 5]. Increasing Si with Cu reduces this effect. Hence, addition of Cu to high Si alloys not only improve its strength but also be beneficial by improving ductility especially for alloys made using slow cooling processes such as sand castings. Again Cu addition to low Si alloys (which form with SDAS of < 25 μm) is also beneficial for refining the β-Al_5FeSi plates. The reason for the refining effect of high level of Si on the Al_2Cu is, increasing Si in low cooling condition provides more nucleation sites, post eutectically formed β-Al_5FeSi plates[6], for the eutectic Al_2Cu and the formation of this fine eutectic precipitation prevent the lengthening of β-Al_5FeSi plates in the later part of solidification.

Conclusions

- High level of Si and/(or) Cu decreased the amount and size of the β-Al_5FeSi platelets in Al-Si-Cu-Mg-Fe casting alloys, especially at small SDAS if the Fe level is below the Si dependent critical level for the formation of the pre-eutectic platelets (in the post eutectic stage).
- The Fe containing intermetallics become script-like, alpha-Al_8Fe_2Si, at small SDAS while at large SDAS they take a plate-like, β-Al_5FeSi, shape.
- At small SDAS, in the low Si (4.5%) alloys, Cu actually leads to decreased size of the interemetallics whereas in the Cu-free, high-Si alloy, the plates are replaced by mixture of the irregular alpha-$Al_8Fe_2Si$and β-Al_5FeSi platelets. ie. in the Cu containing alloys, plates are formed even at small SDAS.

Acknowledgement

The CAST Cooperative Research Centre was established under and was supported in part by the Australian Government's Cooperative Research Centre's Programme.

References

[1]C.H. Caceres, I.L. Svensson, and J.A. Taylor: *International Journal of Cast Metals Research* 2003, vol. 15, pp. 531-543,

[2]C.H. Caceres and J.A. Taylor: *Shaping Casting: The John Campbell Symposium* 2005, vol. 1, pp. 245-254,

[3]C.H. Caceres, B. Johannesson, J.A. Taylor, A. Canales-Nunez, M. Cardoso, and J. Talamantes: *Shape Casting: 2nd International Symposium* 2007, vol., pp. 135-141,

[4]T. Sivarupan, C.H. Caceres, and J.A. Taylor: *Metallurgical and Materials Transactions A* September 2013, vol. 44 A pp. 4071-4081, DOI: 10.1007/s11661-013-1768-x.

[5]T. Sivarupan, C.H. Caceres, and J.A. Taylor: *Materials Science Forum* 2013, vol. Light Metals Technology 2013, pp. 107,

[6]M.C. Eva Tillová, Lenka Hurtalová, Peter Palček: *Acta Metallurgica Slovaca - Conference, Vol. 3, 2013, p. 196-201* 2013, vol., pp. 10.12776/amsc.v3i0.127.

Shape Casting: 5th International Symposium 2014
Edited by: Murat Tiryakioğlu, John Campbell, and Glenn Byczynski
TMS (The Minerals, Metals & Materials Society), 2014

Efficient Use of Titanium and Boron in Grain Refining of a Die-cast Strontium-modified A356

Sebastian F. Fischer[1], Roman Boras[1], Andreas Bührig-Polaczek[1], Matthias Bünck[2]

[1]Foundry-Institute, RWTH Aachen University; Intzestraße 5; 52072 Aachen, Germany
[2]Access e.V.; Intzestraße 5; 52072 Aachen, Germany

Keywords: Grain Refining, Modification, Interaction Effect, AlSi7Mg

Abstract

In the Al-Si alloy-system, titanium, boron and strontium are very effective refining primary aluminum and eutectic silicon, thus enhancing casting and mechanical properties. In literature there are different statements which titanium and boron contents are the most effective for grain refining and optimal mechanical properties. In addition some studies showed an interaction between boron and strontium which would lead to a decreased modification of the eutectic silicon. To evaluate these contrary effects in detail, the influence of titanium and boron on setting, microstructure and mechanical properties of die-cast strontium-modified A356 are shown by systematically varying the titanium and boron contents in the present study. Titanium in combination with a low boron content (50 ppm) leads to the highest eutectic undercooling and maximum tensile strengths. By the use of 150 ppm boron in a titanium-free A356 melt the smallest grain size and highest elongation at fracture could be achieved.

Introduction

Due to their excellent casting and mechanical properties Al-Si alloys are most commonly used casting lightweight aluminum parts [1-5]. Grain refining of these alloys is very important to realize fine, equiaxed grain structures and thus enhanced casting and mechanical properties as well as a good machinability [1,6,7]. For grain refining usually Al-Ti-B master alloys are introduced into an aluminum melt leading to the presence or formation of Al_3Ti, TiB_2, AlB_2 and AlB_{12}, particularly [2]. The TiB_2 particle supports grain refinement when solute titanium is additionally present, e.g. by using an Al-5Ti-1B master alloy (Ti/B ratio > 2.2:1). In combination with silicon the solute titanium segregates onto the TiB_2/melt interface leading to the formation of an Al-Ti-Si phase which acts as a nucleant for the α-aluminum [2,5,8]. When boron is in excess, using an Al-3Ti-3B master alloy for example, a boron-rich layer forms at the TiB_2/melt interface leading to the nucleation of α-aluminum as well [2]. According to different statements in literature, the Ti/B ratio in a master alloy has the biggest effect on grain refinement [e.g. 6,9,10]. However, according to [5] the grain refining effect of some titanium-boron master alloy variants in Al-Si alloys is overestimated, because the effect is mostly only investigated in relation to the titanium content neglecting the boron content. In his review article, J. Spittle shows that most investigated master alloys with different titanium and boron content are added to the aluminum melt in a Ti/B ratio of > 2.2:1 [5]. Moreover, in a lot of studies the absolute titanium and boron content of the analyzed aluminum melts is not clearly indicated. Chemical compositions of the base melt show that the Al-Si melts often already contain some amounts of titanium and/or boron before adding the Ti-B master alloys.

Even master alloys of the type Al-Ti-B are commonly used for grain refinement of Al-Si alloys, they have significant disadvantages. The higher density of Al_3Ti and TiB_2 in relation to pure aluminum (4.5 and 3.35 g/cm³ to 2.7 g/cm³) leads to a segregation of potential nucleant particles with an increasing melt holding time [3,11,12]. Another disadvantage of the application of titanium as part of a grain refining agent is the possible formation of insoluble flake- or block-like AlTiSi phases, which decrease the casting and mechanical properties of Al-Si melts and castings, significantly [13]. Additionally the formed $TiSi_2$ and Ti_5Si_3 particles, which exhibit a relatively poor crystallographic matching with α-aluminum, can coat Al_3Ti particles thus decreasing their nucleation ability [14,15].

As shown by [16] and [17] master alloys containing only boron (Al-B and Si-B; the former is usually used in pure aluminum for electrical applications to eliminate transition elements) are promising grain refining agents in Al-Si alloys. With boron additions in excess of 200 ppm using e.g. an Al-4B master alloy, the melt's undercooling is completely eliminated and the nucleation temperature is remarkably increased in relation to the growth temperature proving the effectivity of the boron grain refining treatment [3]. The analyses of Al-3B treated nearly titanium-free samples with 200 ppm boron indicate a fine globular grain with an average grain diameter of less than 200 μm [18]. By the use of Al-B master alloys, AlB_2 and AlB_{12} particles are present in the Al-Si melts. However, only AlB_2 particles achieve a clear refinement of the microstructure [16]. According to [3] and [13] the AlB_2 crystal exhibits a small disregistry with the aluminum matrix (4.96 %) in contrast to AlB_{12} with aluminum (151 %) thus indicating AlB_2 as a very effective α-aluminum nucleus [3]. Because this nucleus is built at the eutectic point of the Al-B system at approx. 660 °C, AlB_2 can rather act as nucleant in aluminum alloys than in pure aluminum due to their lower liquidus temperature [19]. It cannot be excluded that even at a very low titanium level in the melt titanium transforms AlB_{12} particles to an AlTiB compound which is subsequently eventually transformed to TiB_2. These TiB_2 particles can be encapsulated by AlB_2 and thus act as nuclei for the α-aluminum [3]. In addition, there is diffusion controlled slow transformation in the melt of AlB_{12} to AlB_2 which extends the period of the grain refining effect of Al-B master alloys [20]. When the titanium content of the melt is 0.12 wt.%, Al-B master alloys lose some of their outstanding refining effect, which is comparable or lower than that of Al-5Ti-1B or Al-3Ti-3B, respectively [18]. The boron content in the melt should not be increased to any level; not only to limit the production costs but also to avoid agglomerated intermetallic boron particles in the casting's microstructure [3]. In contrast to the TiB_2 and Al_3Ti phases the AlB_2 phase shows a lower density difference to aluminum and is not poisoned by silicon thus making this phase very attractive for refining nearly titanium-free Al-Si alloys [3].

A disadvantage of the use of only boron as grain refining agent is its reaction with strontium, which is usually added to die-cast aluminum-silicon alloys to modify the eutectic silicon [12]. Besides the refinement of the grain structure a modification of the eutectic silicon is usually conducted to transfer the flake-like silicon to a fibrous morphology thus enhancing the tensile strength and ductility. For modification of die-cast aluminum-silicon alloys the addition of up to 200 ppm strontium via an Al-Sr master alloy is a common melt treatment [7, 21-23]. In the presence of boron, strontium can form Sr-B rich particles which may settle in the crucible lowering the dissolved strontium content in the melt for the eutectic silicon's modification [24]. The addition of boron does not directly influence the eutectic growth temperature and the morphology of the eutectic silicon [25,26].

Due to the continuous downsizing of components, casting alloys have to be optimized with respect not only to mechanical properties but also to the production costs. A key to meet this challenge is the improvement of the grain refinement and the modification treatment. Up to now no optimum refinement agent was introduced for the shape casting of Al-Si alloys since Al-Ti-B grain refiners show different grain refining effects within one casting, a fading effect, and an agglomeration and settling of potential nucleant particles [3]. In addition, the interactions between the grain refinement and modification agents, namely boron and strontium, limit the effect of both melt treatments. To contribute to the finding of the optimum titanium and boron content different titanium and boron levels where set in a 200 ppm strontium containing A356 melt by Al-Ti and Al-B master alloys. The particular effects of titanium and boron as well as the interactions between titanium, boron and strontium on setting behavior, microstructure and mechanical properties of a die-cast A356 are shown by thermal and metallographic analyses and tensile tests. These results allow the adjustment of the grain refining according to a maximum tensile strength by a minimal size of the eutectic silicon, a maximum ductility by a minimal grain size or an agreement of both.

Experimental Procedure

Production, Testing and Analyses of the Specimens

The A356 base material (Ti: 50 ppm; B: 4 ppm; Sr: 4 ppm; Fe: 0,1 wt.%), prepared and supplied by RHEINFELDEN ALLOYS GmbH & Co. KG (Rheinfelden, Germany), was melted in a SiC crucible (Aug. Gundlach KG, Grossalmerode, Germany) using a 5.5 kW 7 kg Nabertherm resistance furnace (Lilienthal, Germany) at a furnace bulk temperature of 750 °C. The modification agent was added as Al-10Sr rods (KBM Affilips B.V., Oss, Netherlands) after the melt reached a temperature of 740 °C. After 20 minutes, depending on the test run, the titanium master alloy (Al-10Ti) and/ or boron master alloy (Al-5B) were/ was introduced to the melt in form of waffle plates (TECHNOLOGICA GmbH, Bad Homburg, Germany). After grain refinement, the hydrogen content of the melt was measured with a partial pressure-density testing device (mk Industrievertretungen GmbH, Stahlhofen a. W., Germany) and, on reaching a density index over 1.5 %, the melt was degassed by argon 4.6 using a graphite lance (HASCO Hasenclever GmbH + Co KG, Lüdenscheid, Germany). According to [27] degassing after grain refinement increases the dispersity of the potential nuclei thus enhancing their nucleation effect. When the hydrogen content and the particular casting temperature was stable, the melt was stirred and then cast with the aid of a pouring spoon into a Diez die, which was preheated to 190 °C with the aid of an oil heating system. The die was coated with graphite in the area of the specimen and with an insulating coat in the area of the feeders according to norm P 372 of the Association of German Foundrymen (VDG). 45 seconds after casting, the specimen was removed. The quality of the melt was controlled by thermal analyses and spectrometer measurements.

From 4 cast rods, $d_0 = 8$ mm tensile specimens were machined according to DIN EN 50125 and tested with the aid of a 8033 Instron tensile testing machine using a cross head speed of 0.35 mm/s according to DIN EN 10002.

Metallographic sections were prepared from two different positions of one sample per test run representing two different cooling rates by embedding the specimens in Araldit combined with the hardener Ren HY 956 (both from Huntsman, Germany). The specimens were grinded using abrasive paper (320 to 1000 grades) and polished using a VibroMet machine (Buehler, Düsseldorf, Germany). Optical micrographs were taken at different magnifications at four defined positions per sample using the Axio Imager A1 m light microscope (Carl Zeiss, Oberkochen, Germany). These images were analysed with respect to the number and the area of eutectic silicon particles using the image analysis software Axiovision (Zeiss, Oberkochen, Germany). For a sound analyses the detected particles were divided into four classes according to their area (< 1 µm², 1-5 µm², 5-10 µm², >10 µm²). The areas of other phases were included in the measurement of the silicon phase since these phases, on the one hand, occurred in same amount and size in every sample (e.g. Mg_2Si), and the different phases could not be separated due to the low grey level difference between them and silicon. However, with a maximum of 0.25 wt.% titanium and 200 ppm boron no intermetallic phases - are detectible by the image analyses software - had to be expected. The DAS from selected samples was manually determined from the 50x magnified micrographs according to the Association of German Foundrymen's (VDG) Norm P 220 and indicated a cooling rate of 20 and 10 K/s at position 1 and 2, respectively. The average grain size was manually measured from the 50x magnified micrographs taken from metallographic sections which were etched according to Barker.

Setting of the Parameters and Design of Experiments

In the present work the titanium and boron contents were systematically varied to show the interaction effect of these parameters on the solidification, microstructure and mechanical properties of A356. The boron contents were set at 50, 150 and 250 ppm. According to [3] an amount of 250 ppm is adequate for an entire refining of the microstructure. Due to the lower settings the range of an optimum boron level related to the particular titanium content of the A356 melt can be shown. The low titanium content's setting was set to zero to realize a better visualization of the boron effect. Since the maximum solubility of titanium in aluminum is 0.15 wt.% [28], the mean titanium content was set at this level. In [7] is was shown, that under consideration of the single effect a titanium content of 0.20 wt.% is effectual to gain the best mechanical properties of A356. Due to this result the maximum titanium level was set at 0.25 wt.%. The strontium content was fixed in all melts at a common level of 200 ppm [23].
The titanium and boron contents of the A356 melts were varied according to a full factorial design and each melt composition was repeated three times hence preparing and casting 27 melts. From each melt a minimum of nine specimens was cast. In addition two melts without titanium, boron and strontium were cast to evaluate the properties of the basis material resulting in an overall sample quantity of approx. 260. The results of the thermal analyses, tensile tests and the microstructure measurements were analyzed and evaluated with the aid of the analyses of variance (ANOVA) indicating the statistical relevance of each parameter and parameter setting [29]. For this, the p-value test and a tukey test were conducted.

Results and Discussion

In figure 1a) the difference between the higher and lower liquidus temperature of the not grain refined melts is divided by the same difference of the grain refined melts. High values represent an effect of the focused parameter. The thermal analyses indicate that the addition of boron without titanium and a mean boron content in combination with a mean titanium addition show the biggest effect. At a titanium level of 0.25 wt.% an increase of the boron content leads a decreased liquidus temperature ratio.

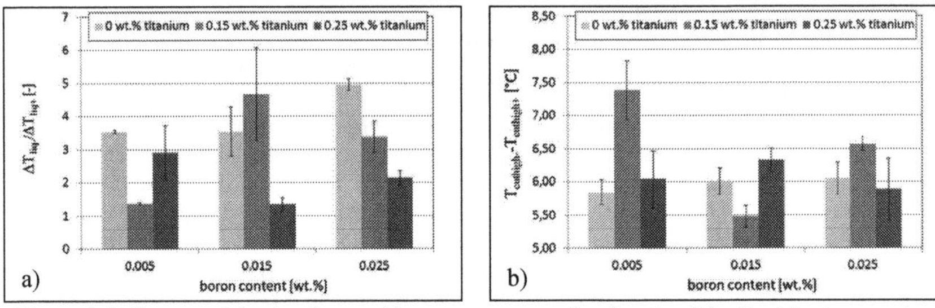

Figure 1. Influence of the titanium and boron content on the a) ratio of the liquidus temperature differences and b) the difference of the high eutectic temperatures.

Even the eutectic silicon is not directly influenced by boron and titanium, an interaction between titanium, boron and strontium should result in a change of the eutectic solidification and silicon's size. The difference between the high eutectic temperature of the not grain refined and the grain refined melts is shown in figure 1b). The higher this difference the higher should be the influence

of the focused parameter on the silicon's modification. The temperature difference is statistically significant increased by the addition of a mean titanium amount in combination with a boron content of 50 ppm and 250 ppm, respectively.
In figure 2 the influence of titanium and boron on the grain size at position 1 and 2 are shown. Due to the higher cooling rate the untreated specimens exhibit at position 2 a smaller grain size than at position 1. At a cooling rate of 20 K/s a boron content of 50 ppm shows no statistical significant effect on the A356's grain size (figure 2a). In combination with any boron content the mean titanium content of 0.15 wt.% has no statistically significant influence on the A356's grain size, too. If the cooling rate is decreased to 10 K/s the grain size is increased by a low boron content and by the combination of 50 or 150 ppm boron and a mean titanium content (figure 2b). Independent of the cooling rate the smallest grain size is produced with a mean boron content of 150 ppm (figure 2a) confirming the findings of [3]. The addition of titanium to a melt containing 150 boron leads to a significant increase of the grain size, what was also reported by [18]. If a low boron content of 50 ppm is used, a titanium content of 0.25 wt.% produces a small average grain. At 0.25 wt.% titanium, the grain size is increased when the boron level is increased to 150 ppm or, at a cooling rate of 20 K/s, 250 ppm.

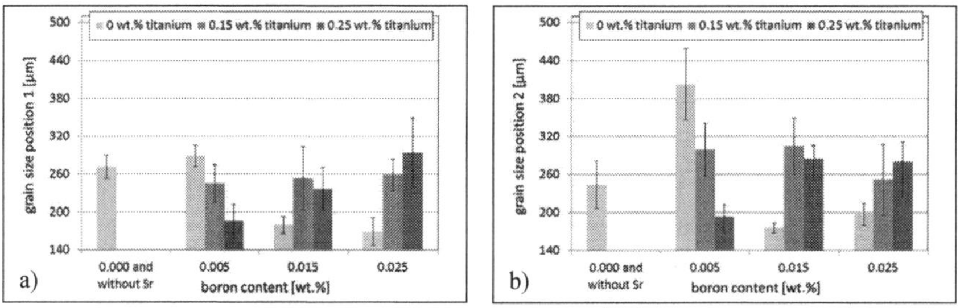

Figure 2. Influence of the single and interaction effect of the boron and titanium content on the A356's grain size at a cooling rate of a) 20 K/s and b) 10 K/s.

The distribution of the intermetallic particle classes' ratio in relation to the titanium and boron content at position 1 is shown in figure 3. As known from literature the addition of strontium refines the eutectic silicon (figure 3a) thus leading to an increase of the number of particles < 1 µm² and to a decrease of the number of particles with an area of 5-10 µm² and > 10 µm².

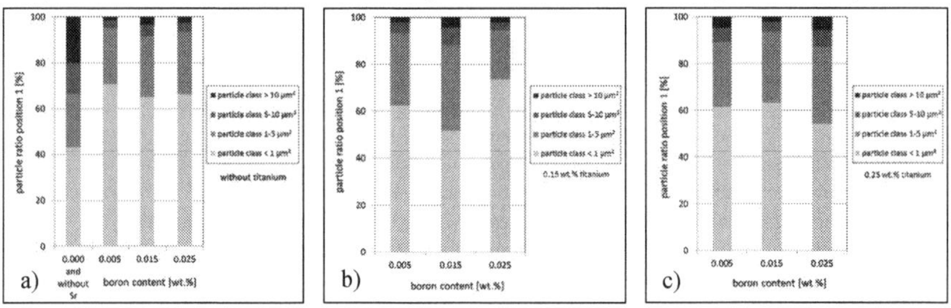

Figure 3. Influence of the boron addition on the size of the intermetallic phases at position 1 at a titan level of a) 0 wt.% b) 0.15 wt.% and c) 0.25 wt.%

The number of particles with an area of 1-5 µm² is only slightly influenced by a strontium addition. With the addition of 0.15 wt.% titanium, the number of particles with an area of 5-10 µm² is increased at a low and mean boron level (figure 3b). The finest particles are produced by the addition of 250 ppm boron and 0.15 wt.% titanium or 50 ppm boron without a titanium addition.

The effect of the varied titanium and boron contents on the mechanical properties of an A356 are shown in figure 4. In general, the untreated basis material shows relatively good mechanical properties maybe resulting from the low die temperature of 190 °C used for the production of the samples. According to [7] the die temperature is the most important factor influencing the mechanical properties of a die-cast A356. Due to the modification with strontium the tensile strength and the elongation at fracture are increased. In relation to the tensile strength the strontium's effect on the elongation is more distinctive.

The highest tensile strength is always achieved by a mean to high titanium content independent of the boron level in the specimens (figure 4a). Boron shows no statistically significant single effect and interaction with titanium according to the tensile strength.

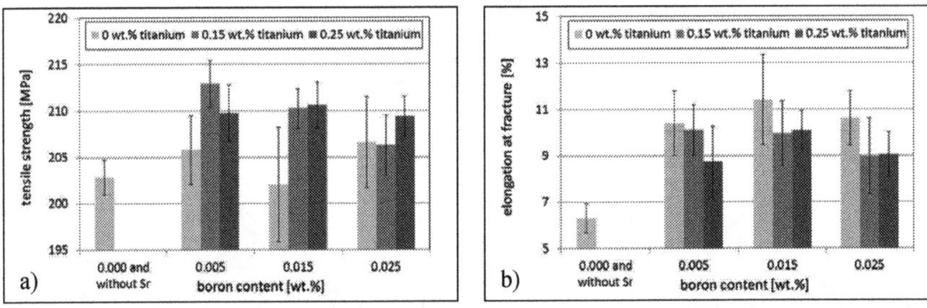

Figure 4. Influence of the single and interaction effect of the boron and titanium content on the A356's a) tensile strength and b) elongation at fracture.

As indicated by figure 4b), the best elongation at fracture can be realized by a mean boron content which produced the smallest grain size (figure 2). In general the elongation at fracture is higher when no titanium is present in the boron-treated samples. There is no statistically significant difference between the effect of 0.15 and 0.25 wt.% titanium. This indicates that there is no modification and grain refinement over-treatment by the maximum strontium, titanium and boron additions, which would lead to a clearly decrease of the elongation at fracture due to the formation of intermetallic phases. Within the current study it was not possible to proof this assumption by scanning electron microscopy and energy dispersive X-ray spectroscopy.

Summarizing Conclusions

In the present study the single and interaction effect of titanium and boron on setting, microstructure and mechanical properties of a die-cast strontium-modified A356 were shown. In general the used base material exhibits a comparatively high tensile strength probably due to the low die temperature of 190 °C during the production of the samples. At low boron content and when adding titanium, the tensile strength can even be increased in contrast to the elongation at fracture. As indicated by thermal analyses, the addition of titanium supports the decrease of the eutectic temperature and thus the modification of the eutectic silicon by strontium. However, the present analyses of the silicon's size can only rudimentarily confirm this assumption. An improved description of the silicon's modification grade would enhance the examination of the titanium's influence on the silicon size. The highest elongation at fracture and smallest grain size can be achieved by the use of only boron in a nearly titanium-free A356. The influence of boron on the liquidus temperature can only slightly be connected with its effect on the grain size. Besides the use of only boron, what is in the agreement with its effect on the elongation at fracture and grain size, the lowest undercooling occurs with the application of a mean boron and a mean titanium content. The presented effects of titanium and boron on the microstructural and mechanical properties of an A356 leading to the assumption that, for the present experimental conditions, the tensile strength is mostly governed by the eutectic silicon's size and the elongation at fracture by the grain size. Since for the present conditions no minimal silicon phase size can be realized at the same time with a minimal grain size by the use of titanium, boron and strontium, the utilization of other grain refining and modification agents is promising to avoid interactions between the element additions and thus to achieve maximum mechanical properties [21].

Acknowledgements

The authors would like to thank Thorsten Hackel from the TECHNOLOGICA GmbH, Bad Homburg, Germany, for the in-kind donation of the titanium and boron master alloys, Ingo Braun, Andreas Gruszka, Dirk Freudenberg and his staff, Claus Groten, Elke Schaberger-Zimmermann, and Elke Breuer for their support during the production, preparation and testing of the metallic specimens.

References

1. D.G. Mallapur, K. Rajendra Udupa and S.A. Kori, "Influence of grain refiner and modifier on the microstructure and mechanical properties of A356 alloy", *International Journal of Engineering Science and Technology*, 2 (9) (2010), 4487-4493.
2. P.S. Mohanty and J.E. Gruzleski, "Grain refinement mechanisms of hypoeutectic Al-Si alloys", *Acta Materialia*, 44 (1996), 3749-3760.
3. S. Nafisi and R. Ghomashchi, "Boron-based refiners: Implications in conventional casting of Al-Si alloys", *Materials Science and Engineering A*, 452-453 (2007), 445-453.
4. S.F. Fischer. M. Thielen, R.R. Loprang, R. Seidel, C. Fleck, T. Speck and A. Bührig-Polaczek, "Pummelos as Concept Generators for Biomimetically Inspired Low Weight Structures with Excellent Damping Properties", *Advanced Engineering Materials*, 12 (12) (2010), B658-B663.
5. J. Spittle, "Grain refinement in shape casting of aluminium alloys", *International Journal of Cast Metals Research*, 19 (4) (2006), 210-222.
6. T. Wang, H. Fu, Z. Chen, J. Xu, J. Zhu, F. Cao and T. Li, "A novel fading-resistant Al-3Ti-3B grain refiner for Al-Si alloys", *Journal of Alloys and Compounds*, 511 (2012), 45-49.
7. S.F. Fischer, V.F. Groten, J. Brachmann, C. Fix, T. Vossel, and A. Bührig-Polaczek, "Influence of die and casting temperatures and titanium and strontium contents on the technological properties of die-cast A356 in the as-cast and T6 condition", *Light Metals 2013*, ed. B. Sadler (Hoboken, NJ: John Wiley & Sons, Inc., 2013), 1031-1036.
8. P. Schumacher, "Nucleation Mechanisms during Grain Refinement of Al-Si-Alloys" (in German), *Giesserei-Rundschau*, 50 (2003), 228-230.
9. N.S. Reddy, A.K. Prasad Rao, J. Krishnaiah, M. Chakraborty and B.S. Murty, "Desgin of an Ideal Grain-Refiner Alloy for Al-7Si Alloy Using Artificial Neural Networks", *Journal of Materials Engineering and Performance*, 22 (3) (2013), 696-699.
10. D. Boot, P. Cooper, D.H. StJohn and A.K. Dahle, "A comparison of grain refiner master alloys for the foundry", *Light Metals 2002*, ed. W. Schneider (Warrendale, PA: TMS, 2002), 909-916.

11. C. Limmaneevichitr and E. Eidhed, "Fading mechanism of grain refinement of aluminum-silicon alloy with Al-Ti-B grain refiners", *Materials Science and Engineering A*, 349 (2003), 197-206.
12. L. Lu and A.K. Dahle, "Effects of combined additions of Sr and AlTiN grain refiners in hypoeutectic Al-Si foundry alloys", *Materials Science and Engineering A*, 435-436 (2006), 288-296.
13. R. Ghomashchi, "The evolution of AlTiSi intermetallic phases in Ti-added A356 Al-Si alloy", *Journal of Alloys and Compounds*, 537 (2012), 255-260.
14. P. Schumacher and B.J. McKay, "TEM investigation of heterogeneous nucleation mechanisms in Al–Si alloys", *Journal of Non-Crystalline Solids*, 317 (2003), 123-128.
15. D. Qiu, J.A. Taylor, M-X. Zhang and P.M. Kelly, "A mechanism for the poisoning effect of silicon on the grain refinement of Al-Si alloys", *Acta Materialia*, 55 (2007), 1447-1456.
16. G. Sigworth und M. Guzowski, Grain Refining of Hypoeutectic Al-Si Alloys, *AFS Transactions*, (1985), 907-912.
17. P.A. Tøndel, G. Halvorsen and L. Arnberg, "Grain refinement of hypoeutectic Al-Si foundry alloys by addition of boron containing silicon metal", *Light Metals 1993*, ed. S.K. Das (Denver, CO: TMS, 1992), 783-789.
18. Y. Birol, "Performance of AlTi5B1, AlTi3B3 and AlB3 master alloys in refining grain structure of aluminium foundry alloys", *Materials Science and Technology*, 28 (4) (2012), 481-486.
19. Y. Birol, "Grain refinement of pure aluminium and Al-7Si with Al-3B master alloy", *Materials Science and Technology*, 28 (3) (2012), 363-367.
20. T. Wang, Z. Cheng, H. Fu, J. Xu, Y. Fu and T. Li, "Grain refining potency of Al-B master alloy on pure aluminum", *Scripta Materialia*, 64 (2011), 1121-1124.
21. J.H. Li, S. Suetsugu, Y. Tsunekawa and P. Schumacher, "Refinement of Eutectic Si Phase in Al-5Si Alloys with Yb Additions", *Metallurgical and Materials Transactions A*, 44 (2013), 669-681.
22. S.-Z. Lu and A. Hellawell, "The Mechanism of Silicon Modification in Aluminum-Silicon Alloys: Impurity Induced Twinning", *Metallurgical Transactions A*, 18 (1987), 1721-1733.
23. M. Timpel, N. Wanderka, R. Schlesiger, T. Yamamoto, N. Lazarev, D. Isheim, G. Schmitz, S. Matsumura and J. Banhart, "The role of strontium in modifying aluminium-silicon alloys", *Acta Materialia*, 60 (2012), 3920-3928.
24. K. Nogita, S.D. McDonald and A.K. Dahle, "Effects of Boron-strontium Interactions on Eutectic Modification in Al-10mass%Si Alloys", *Materials Transactions*, 44 (4) (2003), 692-695.
25. K. Nogita and A.K. Dahle, "Effects of boron on eutectic modification of hypoeutectic Al-Si alloys", *Scripta Materialia*, 48 (2003), 307-313.
26. G. Huiyuan, L. Yanxiang, C. Xiang and W. Xue, "Effects of boron on eutectic solidification in hypoeutectic Al-Si alloys", *Scripta Materialia*, 53 (2005), 69-73.
27. P. Feng, J. Tang, Y. Jin, S. Li and D. Zeng, "Influences pf Preparation Conditions and Melt Treatment Procedures on Melt Treatment Performance of Al-5Ti-B and Al-10Sr Master Alloys", *Journal of Materials Science Technology*, 22 (1) (2006), 50-54.
28. G. Sigworth, The Grain Refining of Aluminium and Phase Relationships in the Al-Ti-B System, *Metallurgical Transactions A*, 15 (1984), 277-282.
29. S. Kumar, P. Kumar, H.S. Shan, "Optimization of tensile properties of evaporative pattern casting process through Taguchi's method", *Journal of Materials Processing Technology*, 204 (2008), 59-69.

Shape Casting: 5th International Symposium 2014
Edited by: Murat Tiryakioğlu, John Campbell, and Glenn Byczynski
TMS (The Minerals, Metals & Materials Society), 2014

STUDYING ON THE EFFECTS OF QUENCHING RATE ON RESIDUAL STRESS IN Al-5Mg and Al-Mg-Cu ALLOYS

Ibrahim Halil Kalkan, Derya Dispinar
Istanbul University, Faculty of Engineering, Metallurgy and Materials Engineering
kalkan.i.halil@gmail.com, deryad@istanbul.edu.tr

Keywords: Residual stress, DSC Analyses, Precipitation Hardening, Quenching Rate, Al-5Mg Alloys

Abstract

In aluminium alloys, Cu and Mg are added in order to increase the strength. Obtained alloys are subjected to solution treated and precipitation hardening (T6) processing. Phases are formed with simultaneous diffusions of Cu and Mg in these alloys. The purpose of this study is the modelling of elements, which are in Al-Cu, Al-Mg and Al-Cu-Mg alloys, solution treatment times into alloys both one by one and in an interrelated way and their microstructural changings. Alloys are produced by pouring to both sand and metal mould. Then, these samples are subjected to solution treated and precipitation hardening processing in different temperatures and for different periods of time. Solution treated speeds of obtained samples are examined with microstructure and image analyses through metallographic examination and a model is designed. Microhardness analyses are also made. On the other hand, residual stress of alloys are examined with hardness, DSC and microstructure analyses.

Introduction

In order to increase mechanical properties, aluminium alloys must exhibit a solvus line for T6 heat treatment method that consists of "solution treated, quenching and aging". An aluminium alloy that is solution treated is heated until a temperature between upon the solvus curve and below the eutectic point and is held for a certain time. Then a homogeneous microstructural single phase is obtained. Alloy elements which are dissolved in aluminium matrix, provides hardness for the material by precipitating. This process is called as "precipitation hardening". In order to practise the precipitation hardening, alloying element that is supposed to be dissolved in matrix should be confined. Quenching process is held for this purpose. Thus, dissolved elements can stay in microstructure. While quenching process provides hardness and resistance for the material, it also causes material to have residual stress due to the rapid cooling. Material can be held in water, chemical composited water or in open-air. One of the most important points in this study is to examine "effects of cooling in oil quenching to residual stress". Therefore, effects of oil and water quenching on the residual stress, hardness and microstructure were compared. Besides, microstructural differences and similarities between casting into sand mould and metal mould are presented.

Quenching

After the heat treatment, in quenching process, three phases occur during the cooling processing: steam phase, boiling phase and convection phase. When heat treated specimen is immersed in the water, a steam film occurs on the surface of specimen during the steam phase. Transition temperature between steam phase and boiling phase is called "Leindenfrost Temperature". The shorter the steam phase occurs, the lower rate of risk of cracks and stress a material can have. However, the longest phase is the steam phase during quenching. When transition to the boiling phase, heat is moved away from specimen by breaking steam film. Quenching rate is maximum level at this stage. Finally, heat transfer has a form of a convection on the convection phase. The temperature of the heat transfer coefficient in the transfer of our cooling and also quenching rate of specimen depends on "Leindenfrost

Temperature". The higher the Leidenfrost Temperature is, the lower the cooling rate is. At the same time, the temperature of water, geometry and temperature of specimen are other factors which are affecting the rate of quenching [2]. A single-phase structure is formed at the end of quenching process. It means that matrix includes entirely a single-phase aluminium structure. Incompletely saturated solid solution, which is aluminium matrix alloy element, is imprisoned form. This structure is not stable at the room temperature. For the stabilizing of structure, alloy elements must be diffused from saturated aluminium matrix [3]. Hence, soluted alloy element which is imprisoned in aluminium matrix is precipitated via diffusion from space of structure. Correlation between the spaces of structure and the quenching rate is as it is shown in Figure 1 and Figure 2.

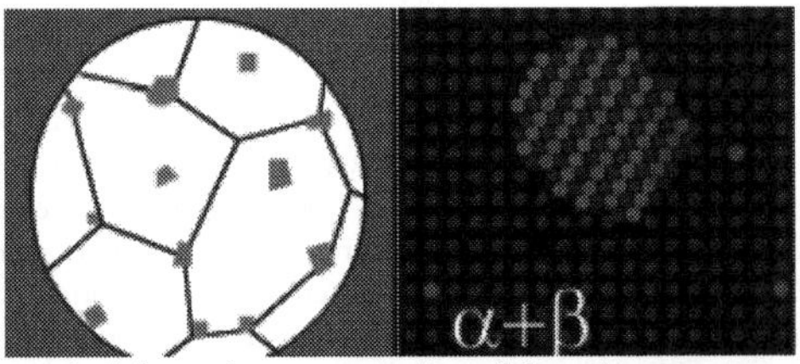

Figure 1. Schematic representation of the microstructure and atoms resulting from slow cooling [4]

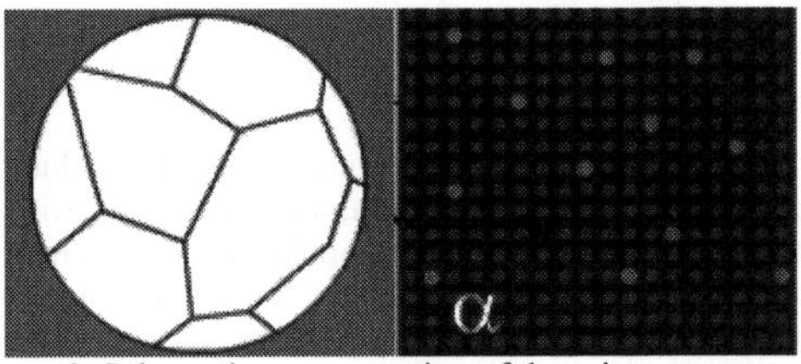

Figure 2. Schematic representation of the microstructure and atoms resulting from rapid cooling [4]

Quenching rate depends on the environment that is selected during the process of the quenching. If water is used in the process of quenching is at room temperature, it is called "rapid cooling". If boiling water or air cooling methods are chosen instead of water at room temperature, it is referred to as "slow-cooling" operation. Due to rapid cooling, the concentration of space is more and during the aging diffusion of second phase in matrix occurs easily. In addition to this, in order to imprison the second phase of the atom as dissolved in matrix, the process of rapid cooling is also needed. As it is shown in Figure 2, vacancy concentration is much more than slow-cooling which is as shown in Figure 2. As a result, a more rigid structure is obtained at the end of the aging. But the rapid cooling reduces the toughness and ductility.

In this study, the main focus is the avoiding of the residual stress that occurs during the rapid cooling. Thus, slow-cooling of oil environment is used besides the rapid cooling of water environment as being the most frequently used method generally. As mentioned above, the temperature of the quenching rate depends on the "Leindenfrost Temperature". Leidenfrost Temperature in water environment is less than in the oil environment. For this reason the oil cooling process's quenching rate will be slower than the water cooling process's.

Residual Stress

The biggest problem encountered during the process of rapid cooling is residual stress which occurs in internal structure of material after the cooling process. Residual stress is very important engineering failure for the material constitute. Microcracks, macrocracks and fatigue fracture which occur during using the material, are caused by residual stress as the material is associated directly with it.

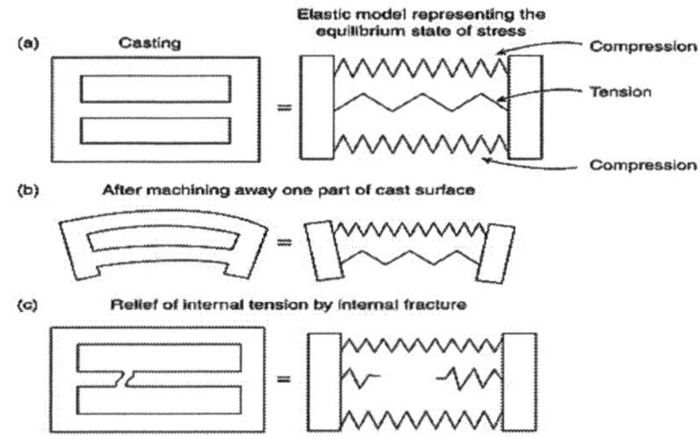

Figure 3. Heyn's model in rapid cooling after the internal strain [5].
(a) The rapid cooling of the material with high compressive stress and strain
(b) The internal structure after processing material distortion
(c) The result will consist of internal stress with cracking

Campbell's Analysis

We can calculate the stress of change when aluminium casting is quenched in water which is at the room temperature. Strain Exchange (ε) equals total temperature (ΔT) multiplied by the thermal expansion coefficient (α):

$$\varepsilon = \alpha \cdot \Delta T \quad (1)$$

According to this result, α value for aluminum is approximately 20×10^{-6} K^{-1} Temperature of the aluminium heat is 640 K in the experiment [6]. In figure 4, aluminum alloys' failure stress and accordingly the resulting residual stress is compared after the heat treatment process. And Figure 5 shows quenching, aging, and the relationship of the strain\stress of material. As can be seen in the figures, resulting residual stress which is generated after the rapid cooling is above the yield point before the heat treatment process. Then residual stress decreases yield point in approximately between 70% and 30% as a result of heat treatment process. Used materials are seen as safe to have 50% of this stress. But still the quality of the materials are influenced badly on the formation of residual stress [6].

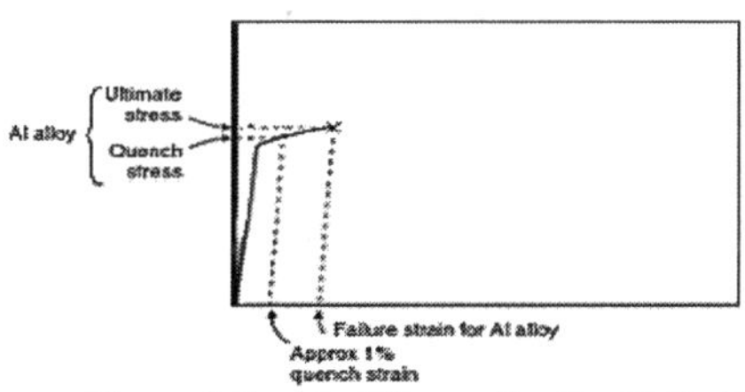

Figure 4. Residual Stress after the Al Casting [6]

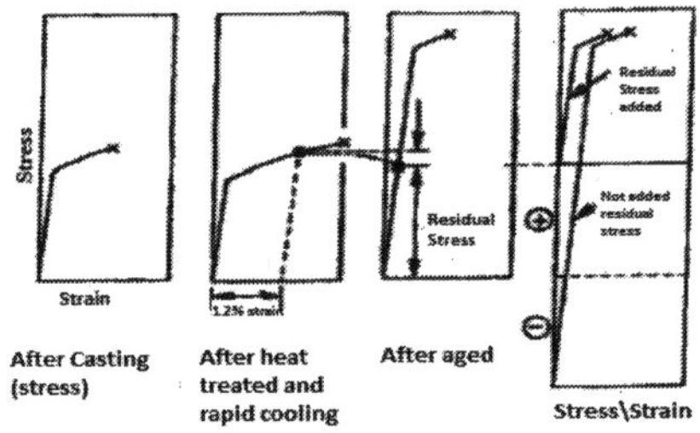

Figure 5. Stress\strain relate in during quenching in water [6,7]

The method which will be used in order to reduce the residual stress should be slower cooling method. For this, because of being rather slower, oil cooling media can be preferred.

Experimental Study

First, Al-5Mg was melted at 750°C and cylindrical bars were produced in both permanent mould and sand mould. Permanent mould temperature was about at 200 °C. 10 samples were quenched in oil media and water media. Each specimen were solution treated at 370 °C for 5-10-15-20-25-30 minutes. The specimens were quenched in water environment and oil environment after the homogenization process. They were aged for 30 days at room temperature and hardness values were measured on a regular basis every week.

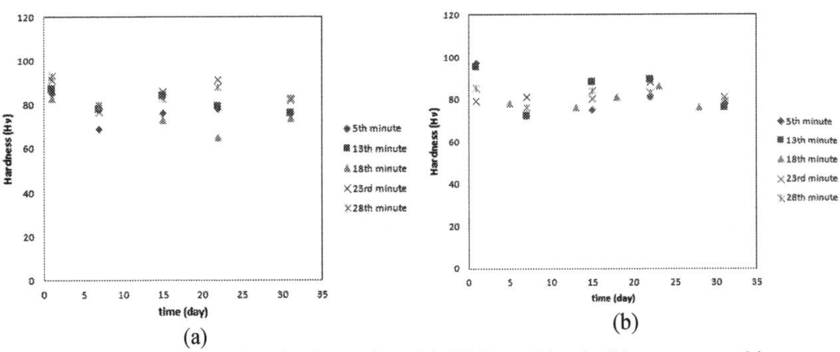

Figure 6. Time-dependent hardness chart (a) Oil Quenching in (b) water quenching (permanent mould casting)

Considering the temperature of homogenization in oil quenching, average hardness of the 1st specimen in Figure 6 is calculated by basing on aging time. Depending on this factor, 5th and 10th minute were chosen for oil quenching, and same choice was made for water quenching for metallography.

As in the Figure 6 and 7, hardness of 10 specimens were calculated and aging time periods of getting together were obtained with graphic result sand metallographic study was made with obtained specimens.

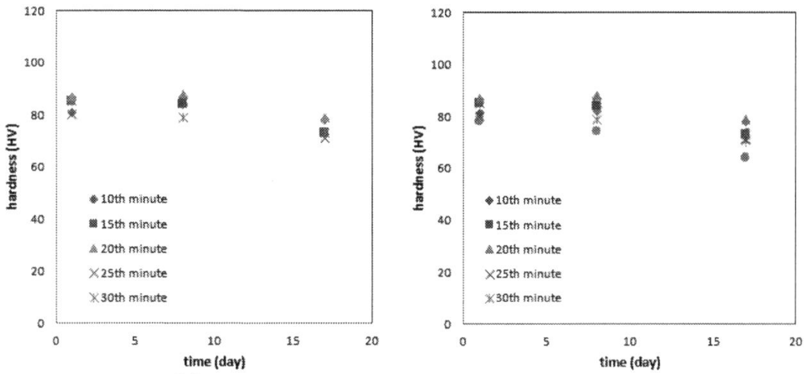

(a) (b)

Figure 7. Time-dependent hardness chart
(a) Oil Quenching in (b) water quenching (sand mould casting)

Discussion

The microstructure castings are shown in Figure 8. Dendrite arm spacing (DAS) was calculated with these images. Permanent mould casting microstructures and sand casting microstructures were compared. It was seen that there is a huge difference in DAS of sand and permanent mould castings. These specimens are heat treated, quenched and aged at the room temperature.

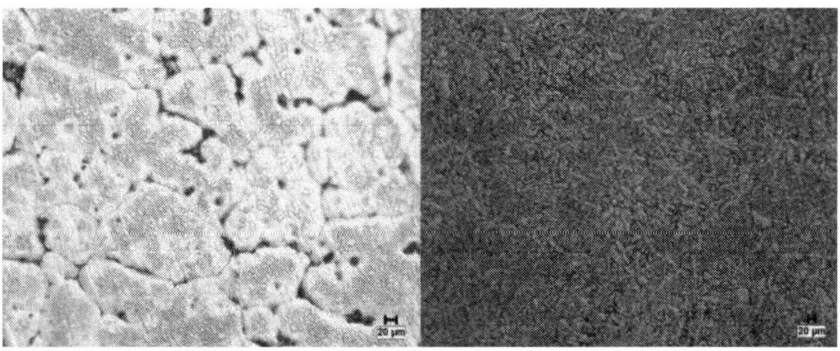

Figure 8. (a) Sand mould casting's microstructure.
(b) Metal mould casting's microstructure.

In Figure 8, average size of grain size of sand moulds was calculated as 12 μm. This was 3.25 μm for permanent mould casting. However, it is observed that hardness of these samples exhibit no difference. In addition to these, different quenching medium (oil or water) does not have an impact on hardness values. In both oil and water cooling, the same hardness values have been reached after a period of time.

Then, in order to determine the effect of residual stress, DSC test was performed and the results are shown in the Figure 9.

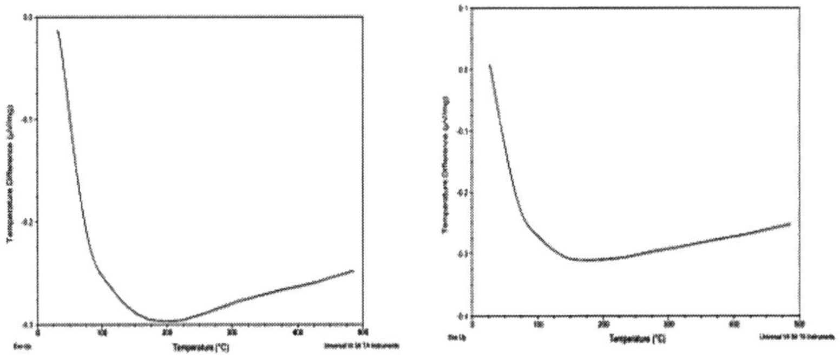

(a) (b)

Figure 9. DSC test results (temperature-microvolts \ mg chart):
(a) Water quenched Al-5 Mg alloy, (b) Oil quenched Al-5 Mg alloy.

As seen in Figure 9, residual stress was not significant in both oil and water quenched Al-5Mg alloys. Apparent residual stress was detected for water quenched Al-Mg-Cu alloy which is shown in Figure 10. The residual stress was detected approximately at 290 °C.

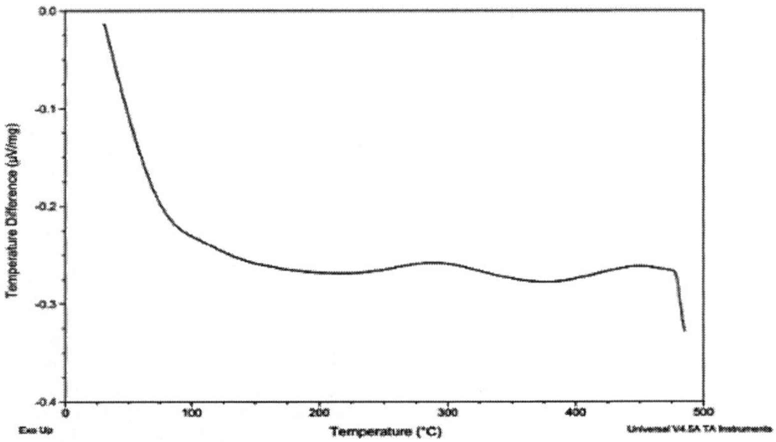

Figure 10. The residual stress is detected approximately at 290 °C

According to the work of Ma [8], Al-Mg alloys' hardness values do not depend on duration of heat treatment or microstructure; it depends on percentage of Mg. In the same study, it is found that residual stress increases depending upon the rate of the water quenching.

Conclusion

As a result of the experiments that are held in this study, it is detected that residual stress depends on the rate of quenching. But it is observed that microhardness measurements do not measure these residual stresses. With DSC method, it is detected that residual stress can be seen in the water quenching, while it cannot be seen in the oil and air quenching.

Acknowledgement

Authors would like to gratefully acknowledge the help and support of Begum Ozturk.

References

1) C. Demir, E.B: Mentese, M.A. Togay, Quenching and residual stresses in aluminum alloys, 3rd Aluminum Symposium, Istanbul, 2007, Turkey
2) HM Tensi, P Stitzelberger-Jakob, GE Totten, Quenching Fundamentals Quenching of Aluminium: Importance of Uniform Surface Rewitting, Quench Tek, Technical report
4) http://aluminium.matter.org.uk/content/html/eng/?catid=35&pageid=370721592 , 21.05.2013
5) John Campbell, Complete Casting Handbook, Elsevier, 2011
6) John Campbell, Castings Practice: The 10 Rules of Castings
7) Scott Mckenzie and George Totten, Handbook of Aluminum: Vol. 1: Physical Metallurgy and Processes. 2003: Taylor & Francis.
8) Shuhui MA, A Methodology to Predict the Effects of Quench Rates on Mechanical Properties of Cast Aluminum Alloys , Worcester Polytechnic Institute, PhD Thesis, 2006

SHAPE CASTING:
5th International Symposium 2014

Mechanical Properties

Shape Casting: 5th International Symposium 2014
Edited by: Murat Tiryakioğlu, John Campbell, and Glenn Byczynski
TMS (The Minerals, Metals & Materials Society), 2014

Metallurgy without Bifilms: No More Fractures

J. Campbell
Emeritus Professor of Casting Technology
Department of Metallurgy and Materials
University of Birmingham, UK

Abstract

There is growing evidence that the failure by cracking of nearly all solid metals, whether cast or wrought, is probably in most cases initiated by casting defects; entrained oxides known as bifilms which *are* the Griffith cracks necessary to initiate failure. Because bifilms are practically universally present in metals, the few examples of metals and alloys without bifilms, such as ductile iron, Hadfield Manganese Steel and Ni superalloy single crystals, appear to have unusually good properties, being unusually resistant to failure by cracking. Boron containing steels, and carbon steels deoxidised with CaSi might also be candidates. They indicate the potential for all metals and alloys if cast well. Current metallurgical thinking is required to recognize the current near-ubiquitous presence of bifilms in metals and the future casting technologies required for their avoidance. Uncracked and uncrackable metals having high strength, together with high ductility and toughness, should and could be the norm.

Keywords: Oxide bifilms; Defects; Failure; Cracks; Superalloy Single Crystals

Introduction

Although cracking failures in cast alloys has been attributed to such mechanisms as hot tearing, and so-called brittle inclusions and second phases, which are thought to be solidification defects, there is plenty of evidence that they are not solidification defects, but casting defects, and so potentially avoidable by use of an appropriate casting process.

The reason for this behavior is the near-universality of bifilms, the 'cracks' in liquid metals resulting from the surface oxide film on the liquid metal entrained into the bulk melt. Entrainment occurs by surface turbulence during such actions as the stirring or pouring of liquid metals during the melting and casting processes. The surface oxide is necessarily entrained in a double form, with the dry faces facing each other, and the original undersides, in perfect atomic contact with the liquid, become the outer surfaces of the double film. These features make the bifilm not only a crack but also a substrate for the formation of intermetallics and second phases. The presence of the central crack gives the appearance of these precipitates being brittle, giving rise to the ubiquitous phrase "brittle grain boundary precipitate" which is almost certain untrue in general, since most intermetallics and second phases are known to be strong and crack resistant [1].

A notable feature of some alloys which has come to light only recently is a natural propensity for freedom from bifilms. These alloys are the exception, but enjoy remarkable properties. We shall list some candidates of this special category of unusually reliable metal alloys.

In most metals, however, large bifilm populations appear to be easily created but less easily reduced. Efforts to reduce the bifilm content can be carried out by either mechanical or metallurgical means. These techniques are discussed below, illustrating the central importance of bifilms in the achievement of good properties, and castings in particular. The development of improved melting and casting techniques is beginning to reveal the potential for bifilm-free (i.e. crack-free) metallurgy, conferring the benefits of high toughness combined with high strength.

Mechanical Techniques

The most common mechanical method for the removal of inclusions of all types is filtration. However, because entrained films start their life in suspension as compact, convoluted bundles, ravelled into compact scrambled forms because of the action of bulk turbulence in the liquid, these compact forms can often pass through filters without problem. After the bifilms arrive in the casting, and turbulence in the liquid decays, the bifilms can unfurl, expanding once again to their full original size, impairing properties to a maximum degree. Thus filtration has limited effectiveness in arresting the transfer of bifilms [1].

Flotation; the bubbling of an inert gas through liquid Al alloys has been widely used as a technique assumed to be for the reduction of dissolved hydrogen, but acting more importantly for the reduction in the content of the bifilms suspended in the liquid. In this case the large bifilms, which are mainly formed by the primary oxide skins of the charge constituents of the melt, are easily floated out. This is a key benefit of the gas bubble flushing technique. However, oxides which are of similar size or smaller than the bubble size are not removed: they simply follow the flow lines around the rising bubble, never coming into contact with the bubble. Thus the very damaging large oxides, measured in fractions of meters across, are easily floated out and skimmed from the surface. The oxides smaller than the bubble dimensions, of the order of 5 mm or less, remain in solution. The smaller oxides may in fact be increased by the action of oxidation by air as bubbles of inert gas burst at the surface of the melt. In terms of chemistry, the large oxide skins are, in general, mainly spinels ($Al_2O_3.MgO$) whereas the final population of small oxides are generally pure alumina (Al_2O_3). It is possible that the overall population of bifilms may be increased by such techniques, although, regrettably, such facts are not currently known, and some careful research would be welcome.

Other mechanical techniques involve such methods as density separation (sedimentation or flotation depending on the relative densities of the defects and the liquid) of bifilms prior to casting, preferably followed by casting techniques specially designed to avoid re-entrainment of the surface.

Most *melting* and *transfer* operations in foundries are lamentably turbulent, so that the quiescence required for sedimentation or flotation is rare. For Al alloys, an exceptional system specifically designed for efficient sedimentation was the Cosworth Process for melting by (i) the avoidance of pouring, and (ii) the provision of a quiescent holding furnace to encourage sedimentation of oxides [3].

A notable *casting* technique which achieves the near-unique ability to achieve the filling of molds without the entrainment of the surface oxide is *counter-gravity casting*, in which the mold is filled uphill, against gravity, and in which the filling velocity is controlled below the critical velocity of 0.5 m/s so that the liquid metal never has sufficient energy to enfold its surface. This technique is used in a number of casting industries, but deserves universal application [4]. In contrast with the near-perfection of filling conditions conferred by counter-gravity casting, there are improved gravity filling techniques which are easily introduced to existing foundries, which although far from perfect, are significantly better than most currently used gravity pouring methods [5].

An unusual *solidification* technique which is successful to achieve the separation of oxide bifilms from the casting is employed in the manufacture of single crystal turbine blades. Here the slow upward progress of the freezing front not only allows plenty of time for oxide bifilms to float out, but also tends to push ahead those that are very slow or hindered from floating. The result is a product low in bifilm defects, explaining the excellent toughness and creep life of the single crystal in comparison with equiaxed products, in which the relatively rapid and multi-directional freezing traps bifilms between grains. These open during creep, giving the impression of weak grain boundaries (which incidentally, are now known from molecular dynamic studies to be strong; in some cases actually as strong as the matrix) [2].

Metallurgical Techniques

Metallurgical techniques to reduce or eliminate bifilms include a number of very different approaches which are listed below.

(i) Casting in a vacuum or inert gas

The melting and pouring of metals in a 'protective' atmosphere is widely employed in an attempt to reduce the oxidation of the surface of the melt. However, although the oxidation is, of course, generally reduced, the vacuum conditions are not sufficiently good to avoid all oxidation. The result in vacuum cast products is the presence of extremely thin bifilms which are therefore more difficult to detect, but continue to act efficiently as cracks, and as substrates for the precipitation of intermetallics and second phases. The precipitation of carbides in Ni and Co based superalloys seems always to occur on bifilms, explaining their tendency to exhibit cracks and therefore appear brittle [6]. It is necessary to conclude that vacuum melting and casting alone is usually not an adequate technique for achieving cast products of high reliability. Freedom from entrained bifilm cracks has to be sought by other techniques.

(ii) Naturally occurring *liquid* oxide surface films

Naturally occurring liquid surface films are not common in metallurgy, but in those instances where they do occur they are of key significance. In the case of the entrainment of liquid surface films, instead of solid films coming together to form an unbonded interface, effectively creating a crack, the liquid films come together and mutually assimilate, subsequently spherodizing to droplets and finally floating out, leaving no significant remnant defect. The formation of a liquid oxide film is therefore a major strategy for the manufacture of defect-free products. Some alloys in which this effect occurs naturally are listed below.

Hadfield manganese steels (13Mn steels) were, until recently, cast with copious quantities of air, entrained at the trumpet entry into the sprue when the steel was poured from bottom-teemed ladles. It led to eruptions of gas flames and smoke from the mold and to a mass of entrapped bubbles on the cope surfaces of castings, so that extensive dressing of the casting was required. To solve this problem contact pouring has been introduced, excluding air from entering the mouth of the trumpet. Contact pouring had an immediate and dramatic effect on the casting process: no gases or combustion products were seen to erupt from the mold. But most importantly of all, bubbles on the cope surfaces of castings immediately disappeared, greatly reducing the cost of dressing of the castings [7].

From the point of view of this study, it is most noteworthy that Hadfield manganese steel castings had been employed for years for rail crossings and points which are known to experience the most punishing conditions of impact and wear, but very rarely fail in service. This unusually high reliability in such a harshly testing environment is a key pointer to the fact that bifilms are likely to be absent from these products, despite the clear evidence that copious quantities of air bubbles had been systematically entrained. In most normal alloys and steels such entrainment of air would have been expected to result in masses of oxide bubble trails acting as long bifilms throughout the casting. The fact that such defects were necessarily absent, as testified by the reliability of the castings to withstand failure by cracking in this uniquely testing service condition, points to the character of the oxide film on high Mn steels. This surface oxide must be a liquid at steel casting temperatures. On reflection, this is predictable from the low melting point of the eutectic between MnO and FeO which is expected to be in the region of 1200 C or below. The liquid surface film on the liquid metal makes it impossible to generate bifilms.

B-containing steels have been renowned for their toughness and strength properties. This is perhaps surprising in the sense that carbon steels are chemically so similar. The alloying effects of carbon and boron being highly analogous, both forming similar compounds, carbides and borides, so that it would be reasonable to expect closely similar behavior. However, the carbon steels, in general, have a gaseous oxide, CO, but solid oxide film, usually Al_2O_3 from the deoxidation practice, whereas boron containing steels have the low melting point liquid borides as a surface film. Recent work on 9Cr1Mo steel weldments indicates that B improves cracking resistance [8] and displays many fewer ductile dimples (sites of bifilm embrittlement of inclusion) in tensile tests [9]. Steels with sufficient B should be free from bifilms. The level of B required for this transformation of behavior requires to be defined by further work. The significant effect of B on hardenability of steels may also have some contribution from the lack of bifilms, since the absence of a dense population of such cracks would normally be expected to increase the thermal conductivity of the solid metal.

High phosphorus cast irons, much beloved by Victorian casters of intricate ornamental street furniture, would be expected to fall to pieces because of the intrinsic brittleness of the phosphide eutectic in this alloy. However, the longevity of these castings testifies to their rather surprising resistance to fracture, which is likely to be the result of the liquid phosphide on the surface of the liquid iron, which would assist the casting to be free from entrained cracks, and so be more resistant to fracture than might otherwise be expected.

(iii) **Artificially created liquid surface films**.

In some circumstances alloys can be manipulated by minor alloy additions (within the chemical specification of the alloy) to achieve a liquid surface oxide film. The outstanding example of this situation is the deoxidation of steels and monels, in which a liquid surface film can be encouraged by careful design of the deoxidation process.

Deoxidation and degassing of steels and other high temperature alloys such as Ni-base alloys including Monel, has traditionally been preoccupied with the avoidance of gas evolution during solidification which might lead to porosity in the cast product.

In the author's experience, any form of degassing may usually be avoided by a modest delay, simply holding the liquid metal for a time to allow it to equilibrate with its environment. Provided the environment is sufficiently dry (for instance, the furnace lining is not too damp) the melt will evaporate away its hydrogen, and if a steel, might also lose some oxygen and nitrogen. A simple check by pouring a sample of a few grams into a small test mold is quicker and more informative than sophisticated solid state electrolyte probes. If the sample casting rises as a muffin, some degassing of some sort is required. This may consist of only waiting some extra time and testing once again. If the sample shrinks down to give a conical pipe, no degassing is required, and pouring can be carried out immediately. This practice has been most useful for the production of pore-free monel castings, which are otherwise impaired by the traditional but misplaced use of deoxidation by Mg and Li which leads to severe bifilm formation, possible cracking during subsequent working, and poor properties.

In monels, if deoxidation is required, the avoidance of Mg and Li, and the use of Fe and Mn at their highest alloy levels in the specification, together with modest levels of Si, will create a liquid Fe/Mn Silicate on the melt surface, so that pouring can be carried out safely without the formation of damaging bifilms. Too much silicon will over-ride the presence of Fe and Mn, creating a solid silicate (SiO_2) film which is usually a disaster for a turbulently poured alloy. Research is required to define optimum impurity and alloy levels for the different monels.

In steel melting, as the term deoxidation implies, the reduction of oxygen in solution in the melt is usually the principal target for attention. It can of course be tackled by such sophisticated techniques as an argon-oxygen-degassing (AOD) treatment in which carbon can also be reduced to low values by the CO reaction with no contamination from residual added elements. However, most steel foundries do not have the luxury of an AOD vessel, and reaction with 'deoxidisers' such as Mn, Si, Al, and Ca etc are most common, forming oxides in the melt, some of which float out, and some of which remain in suspension.

A highly counter-productive complication arises after this initial deoxidising action. There is usually an excess of the deoxidizing elements which have been added. Unfortunately, these remain in solution in the melt, but naturally react with the air during the pouring of the steel, forming an oxide film which becomes entrained into the plunging stream during pouring. These unwelcome reactions are known as reoxidation reactions.

If the main deoxidiser is Al, deoxidation of the steel is extremely effective, lowering oxygen in solution to harmless levels. However, the slight excess of Al remaining after this desired duty is highly undesirable. The excess residual Al reacts with the air to form a thin film of alumina on the surface of the pouring steel. This solid ceramic film becomes entrained, folded into the melt, dry-side-to-dry-side, so as to form alumina bifilms that act as cracks in the liquid. The bifilm population after a pour can be immense, and is slow to float out as a result of the high viscous drag of these thin double films. For shaped castings, filters can be blocked by the huge amounts of bifilms created during the plunge into the conical pouring basin and down the sprue. Furthermore, that portion of the population of bifilms that finds its way past the filter can enter the casting, unfurl to its original entrained size and shape, and be subsequently frozen into the casting, greatly impairing properties.

The bifilms resulting from reoxidation as a consequence of an over-enthusiastic deoxidation with Al (often together with Si and Mn) therefore act as cracks, weakening the product, with consequences such as hot tearing or cracking on solidification, or the expansion apart of their two films to create gas or shrinkage porosity (avoiding the any difficulty of the nucleation of these defects), or cracking during subsequent plastic working such as forging or extrusion, or even failure by cracking during its service life.

Alternatively, if the re-oxidation product on the surface of the melt is a liquid oxide, the subsequent behavior of the melt and the properties of the casting are significantly improved. The folding-in of this liquid oxide surface acts to impinge against the opposing liquid oxide surface, assimilating and coalescing to form droplets which mostly float out; no crack-like defect is formed. In fact the larger droplets float out rapidly as a result of the low drag of their favorable spherical shape, thus completely disappearing from the casting. The smaller droplets float more slowly and may be trapped in the casting. However, of course, their small size and rounded shape renders them practically harmless.

The addition of Ca to a final deoxidation addition is helpful in achieving a liquid oxide on the surface of a liquid carbon steel. The limited use of Al followed by Ca in one of its many addition forms, possibly CaFeSi, is suggested to be a more rational technique to achieve alloys with better hot formability and improved properties in service. This happens because CaO and certain other oxides such as Al_2O_3 (and Cr_2O_3 in many stainless steels) form low melting point oxide eutectics. CaO reduces the melting point of Al_2O_3 from approximately 2050 C down to 1350 C and Cr_2O_3 down from 2620 C to 1022 C. It is therefore one of the most effective elements for liquefying oxides.

Metallurgists have known for decades about the benefits of '*inclusion shape control*' and the action of Ca deoxidation to produce clean steels. However, the accompanying mechanism of its action during the pouring of liquid steel to prevent bifilms and substitute the creation of droplets has to be far more important, but so far has been not generally recognised.

(iv) **Elimination of bifilms by chemical reduction.**

In what appears to be a rare example, the bifilm population entrained in an alloy can be eliminated from a metal by a chemical reaction. This seems true for gray cast irons. These irons are thought to be filled with silica-rich bifilms by the inoculation process, which entrains the surface silicon oxide (SiO_2) film as the particles of inoculant penetrate the surface of the melt. The silica bifilms form favored substrates for the formation and growth of graphite, and the growth of the graphite straightens the bifilms, thus lowering properties. Gray cast iron therefore has low tensile strength and is brittle. However, when converted to spheroidal graphite by Mg addition, the Mg in solution immediately reduces the silica to silicon in solution plus MgO as a residual particle. Thus at a stroke, the natural silica-based bifilm population in the melt is eliminated. The Mg treated melt can in principle be free from bifilms [10].

Unfortunately, however, this may last for only a brief moment of its life; although the natural population of bifilms is eliminated, new bifilms rich in magnesium oxide can now be created if the pour into the mold is turbulent, which, unfortunately, is common. These MgO-rich bifilms can now reduce the properties and reliability of ductile irons, on occasions their size and density being sufficient to even cause failure by brittle fracture! Thus Mg-treated irons are highly susceptible to damage by turbulent pouring [1].

Looking into more detail concerning the Mg treatment of iron, some treatments are extremely turbulent as a result of the energetic eruption of Mg vapour bubbles from the melt (the Mg being above its boiling point). The bursting of bubbles at the surface of the melt opens up fresh areas of liquid metal to the oxidizing environment, with the result that on the collapse of the walls of the escaping bubble, a MgO-rich bifilm is formed. Thus the bifilm population in ductile irons may depend critically on what Mg treatment process is adopted.

In summary, if the Mg treatment process can be carried out quiescently, without evolution of bubbles, and can be cast quiescently, without surface turbulence, the ductile iron can in principle be completely free from bifilms. In this condition it can achieve its desired maximum ductility.

Discussion

In the quest for metals and alloys of improved mechanical and corrosion resistant properties, it is necessary to reduce the bifilm population, to zero if possible, but any reduction is valuable. The reasons for reducing the bifilm population are far reaching:

Many bifilms will be expected to intersect the outside surface of the casting, with the result that any corrodant in the environment would be able to enter the casting, penetrating along the easy path between the two halves of the bifilm. The second phase particles precipitated on the outside surface of the bifilm would encourage penetration and dissolution because of the natural corrosion cells which they would provide. In this way both pitting corrosion and grain boundary penetration could be understood. Even the difficult area of stress corrosion cracking might finally be capable of solution [11].

From recent molecular dynamics (MD) studies of atomic processes operating in the crystal lattices of metals, the nucleation of a pore or crack appears to be impossibly difficult except at stresses of the order of the theoretical strength of the metal, i.e. measured in tens of GPa. At these extraordinary high stresses the nucleation of a pore or crack causes the metal to explode apart. It seems that those atomic processes which were assumed to nucleate cracks, such as the dislocation pile up, do not occur in practice; it seems the forces of attraction between atoms are too great to allow atoms to be pulled apart, except at the theoretical strength limit, which, of course, is never normally reached. Similarly, during solidification, as a very low stress inducing process, interfaces cannot be separated to produce incipient cracks; solidification can only create essentially crack-free products. In the absence therefore of crack initiation mechanisms, how do cracks start? In searching the MD literature it is curious that the researchers in this field do not yet appear to have notice their universal finding that the nucleation of pores and cracks in metals are only possible at the theoretical fracture stresses, even though this is massively at variance with everyday experience.

We appear to be left with only bifilms as a crack initiating feature which is capable of opening at modest stresses in line with everyday experience. These are practically universally present, and appear to constitute the Griffith cracks necessary to initiate failure. If this conjecture is correct, it seems that as dislocations are the micro-mechanisms of macroscopic plasticity, bifilms are the microstructural features to initiate macroscopic cracking [2]. Bifilms appear to be a microstructure feature of metals essential to explain otherwise inexplicable aspects of physical metallurgy, particularly failure by fracture and corrosion, possibly including aspects of stress corrosion cracking and hydrogen embrittlement. It will be exciting to test these predictions.

Conclusions

1. If, as seems probable, that bifilms are the only initiators of failure by cracking (i.e. if they *are* the Griffith cracks) when reliable techniques are developed to achieve metals and alloys free from bifilms their properties should achieve near-theoretical strengths and dramatic increases in toughness and elongation.

2. Some of the potential for metals without bifilms is currently illustrated by such examples as
   (i) The Ni superalloy equiaxed polycrystal in comparison to its current single crystal equivalent (but it is predictable that the equiaxed product could have equal, or nearly equal, properties if it also were made free from bifilms).
   (ii) The conversion of flake graphite iron to spheroidal graphite iron (although turbulent pouring techniques can impair or eliminate this natural advantage).

3. In the quest for improved metals and alloys, the achievement of freedom from bifilms is expected to outweigh the potential for the combined effects of all traditional metallurgical pursuits such as alloying and heat treatment. It is seen as the next great potential step forward for metallurgy and engineering.

References

1. J. Campbell; Complete Casting Handbook, 2011, Elsevier, Oxford, UK.
2. J. Campbell; The origin of Griffith cracks *Metall & Mater Trans B* 2011 vol. 42B (No. 6) pp. 1091-1097
3. J. Campbell; "Quality Castings; A Personal History of the Cosworth Casting Process" to be published.
4. J. Campbell; Hoyte Lecture "Stop Pouring; Start Casting" *International J Metalcasting (AFS)* 2012 vol. 6 (No. 3) pp. 7-18
5. J. Campbell; Castings Practice: The 10 Rules for Casting; Elevier, Oxford, UK 2004.
6. J. Campbell and M. Tiryakioglu; Bifilm Defects in Ni-Based Alloy Castings *Metall & Mater Trans B* 2012 Vol. 43B (No. 4) pp. 902-914.
7. J. Erskine, Progress Rail, South Queensferry, Scotland; private communication 2013.
8. C. R. Das, S. K. Albert, J. Swanithansan, S. Raju, A. K. Bhaduri, B. S. Murthy; Transition for crack type IV to type II from B in 9Cr1Mo steel welds. *Metall. & Mater. Trans.* A 2012 vol. 43A, pp. 3724-3741.
9. C. R. Das, S. K. Albert, A. K. Bhaduri, B. S. Murty; B addition and mechanical properties of 9Cr1Mo steels. *Metall. & Mater. Trans.* A 2012 vol. 43A, pp. 2171-2186.
10. J. Campbell; A Hypothesis of Graphite Formation in Cast Irons *Metall & Mater Trans B* 2009 vol 40B (No. 6) pp. 786-801.
11. J. Campbell; "On the Role of β Particles in Stress Corrosion Cracking of Mg-Al Alloys" *Metallurgical and Materials Transactions A*: 2009 40A, (7) 1510-1511.

Shape Casting: 5th International Symposium 2014
Edited by: Murat Tiryakioğlu, John Campbell, and Glenn Byczynski
TMS (The Minerals, Metals & Materials Society), 2014

CORRELATION BETWEEN BIFILM INDEX AND TOUGHNESS OF ALUMINUM ALLOYS

Derya Dispinar*, Cem Kahruman*, John Campbell**

*Istanbul University, Metallurgical and Materials Engineering Department, Istanbul, Turkey
deryad@istanbul.edu.tr, cemcemce@istanbul.edu.tr
**University of Birmingham, Metallurgy and Materials, Birmingham, UK
jc@campbelltech.plus.com

Keywords: melt quality, toughness, bifilm, hydrogen, aluminum, casting

Abstract

The measurement of melt quality of aluminum alloys have been a complex phenomenon. There exists various methods but the general experience show that they were slow, complicated or expensive for use on the foundry floor. In addition, these methods were not quantifiable. For the first time, bifilm index was proposed by the authors which give a numerical indication of the melt quality in millimeters as the total oxide length. Numerous studies were carried out to correlate this index with the mechanical properties. In this work, toughness measurements were made by using MatLab software. Alloy studied was 356 which were cast under various conditions (degassed and upgassed for 0.1, 0.2 and 0.4 cm^3/100g Al) and the results were correlated with bifilm index.

Introduction

The toughness of metals decrease as the population and size of defects increase. Typically, casting defects have been held responsible for low ductility (porosity, inclusions etc). In addition, the secondary phases or intermetallics are blamed to be the cause of low mechanical properties. Many efforts have been spent on quality indices that is aimed to predict the ductility [1-4]. However, the nature of these previously listed defects are actually nucleating on the bifilms that is in the form of folded oxide skin which already exists a crack [5, 6]. It has been shown that high variability of fatigue and tensile properties merely depend on the presence of these bifilms [7-15]. It has also been shown that porosity was formed only in the presence of bifilms regardless of the hydrogen content [8, 16-19]. Dispinar [16] had proposed an index (called bifilm index) as a measure of melt quality prior to casting. Extensive work was carried out to correlate hydrogen content, porosity level and mechanical properties of several Al-Si series alloys.

In this paper, the effect of hydrogen and bifilm content that control the toughness of A356 alloy was investigated. A serious of data collected from the earlier work of authors were re-evaluated by means of area under the curve of the tensile test results using MatLab software.

Materials and Methods

The composition of A356 used in the experiments is given in Table I. 75 kg of charge was melted in a resistance furnace at 740°C. Two melts were prepared: one was degassed down to 0.1 cm^3/100g Al hydrogen levels and Ar-10%H_2 mixture was used to increase the hydrogen level up to 0.2 and 0.4. In the second melt, the opposite hydrogen levels were studied: upgassed first and then gradually degassed. During all the experiments, reduced pressure samples were collected at 100 mbar to measure bifilm index. A sand mould was used where 10 cylindrical bars were cast which were used in the tensile property measurements. MatLab software was used to determine the best fit for each tensile curves and the area under the curve was measured as toughness.

Table I: Chemical analysis of A356

Si	Mg	Mn	Fe	Ti	Na	Sr	P	Al
6.88	0.32	0.002	0.116	0.11	0.0012	0.0005	0.0002	rem.

Results

The raw data collected from the tensile machine was converted to true-stress and true-strain. These were used in MatLab software and "curvefit" function was used to find the best-fit equation. R^2 values of the equations for all the toughness results were found to change between 1 and 0,9996.

The initial approach was to plot toughness versus bifilm index and hydrogen content. These relations are given in Figs 1-3. It can be seen that toughness decreases with increased bifilm index (Figs 1-2). On the other hand, there is not much clear relationship between hydrogen and toughness (Fig 3).

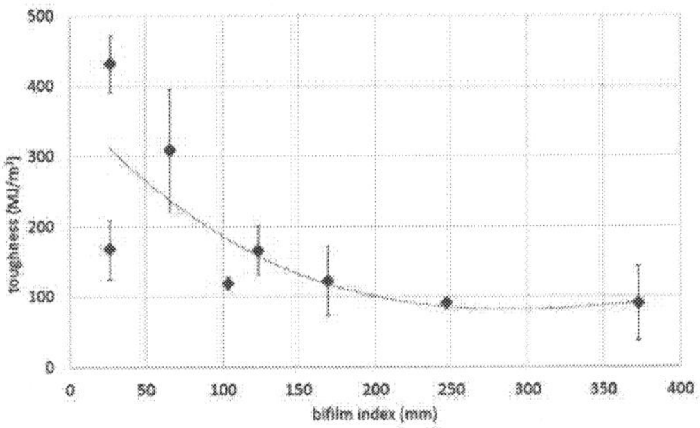

Figure 1: Bifilm index vs toughness for the degassed melt

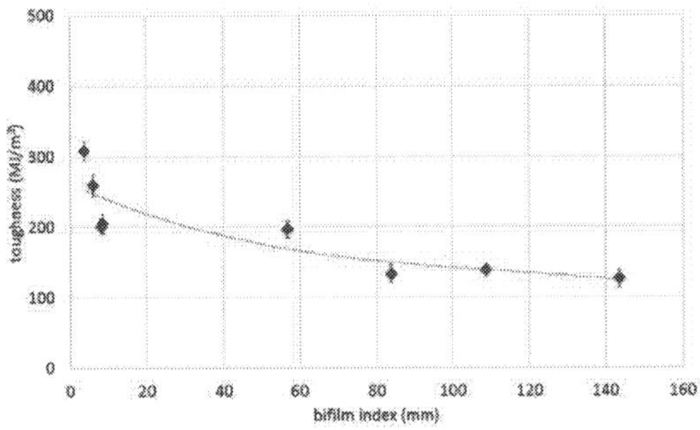

Figure 2: Bifilm index vs toughness for the upgassed melt

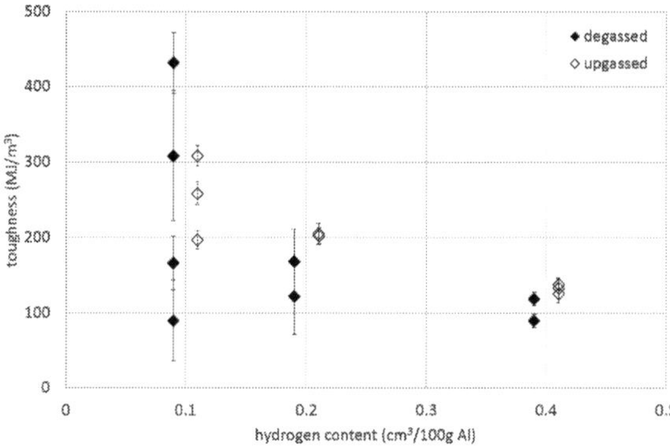

Figure 3: Toughness vs hydrogen content of the melt

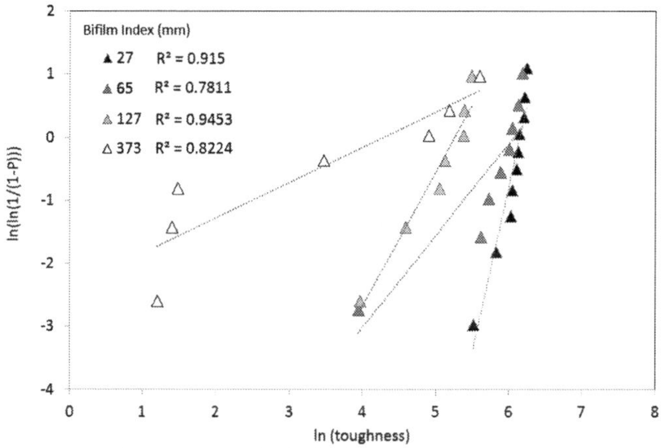

Figure 4: Weibull distribution of toughness values

A Weibull analysis was carried out from the set of toughness values. The values were plotted as ln-ln plot of failure probability versus log property. The resulting slope of the graph is termed "The Weibull Modulus", *m*. This is a measure of the variability of the property being measured. The probability of failure, *F*, as a function of stress, *S*, is given by:

$$F(S) = 1 - (e)^{-\left(\frac{S}{S_o}\right)^{m}}$$

Figure 4 is the Weibull distribution of toughness values with regard to the bifilm index. Although the results do not appear to fit acceptable for Weibull distribution, yet they provide relative information. It can be seen that the melt with the lowest bifilm index (▲ 27 mm) had the highest Weibull modulus; and the casting with the highest bifilm index (△ 373 mm) had the lowest modulus.

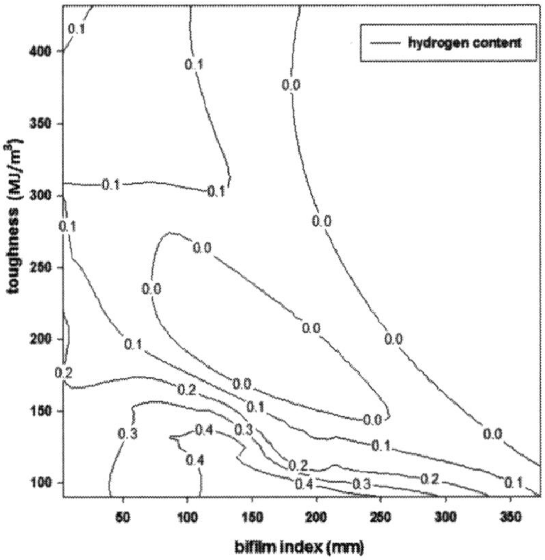

Figure 5: Contour plot of toughness vs bifilm index with varying hydrogen content

Discussion

As seen Figure 3, the interpretation of the hydrogen content and the toughness was not straightforward. It was clear that at high levels of hydrogen, the toughness was significantly low. However, at low levels of hydrogen, the toughness value changes between 100 and 500. This scatter was only explainable by introducing the bifilm index into the relationship. Thus, a counter-map analysis was carried out and variation of toughness versus bifilm index was plotted using hydrogen content that is given in Figure 5. This counter map clearly shows that the highest toughness can be achieved when both hydrogen content and bifilm index is low. It is important to note that even at very low hydrogen levels (i.e. 0 and 0.1 cm^3/100g Al), the toughness can be low depending on the bifilm index. Similar findings were observed when Weibull distribution of toughness was plotted (Fig 4). It can be clearly seen that scatter of tensile test results increases with increased bifilm index.

As seen in Figure 3, even at low hydrogen levels, there is a scatter of toughness values. Thus, the extrapolation of curves drawn in Fig 1 and 2 where y-intercept is zero do not correspond to the same toughness values. For the degassed melt (Fig 1), the y-intercept is around 350 MJ/m^3 and this value is 270 MJ/m^3 for upgassed melt. This relationship can also be seen in Figure 5.

Campbell [5] suggested that the matrix must technically behave as a perfect plastic material; and only the presence of defects, the necking may be hindered to result in a premature fracture or brittle fracture. The presence of folded oxide skins (i.e. bifilms) already act as a stress riser and the earlier works [8-15, 19, 20] have shown that these defects play a significant role on toughness of the cast parts. Thus, it is proposed there is a potential possibility for the inclusion of bifilm index into the existing quality indices [1, 21].

Conclusion

As the bifilm content (i.e. bifilm index) is increased, the toughness decreases regardless of the hydrogen content and the scatter of the result also increase.

Low hydrogen content of the melt does not imply that the mechanical properties need to be high.

In order to achieve high quality castings, bifilm content of the melt has to be as low as possible.

References

1. Caceres, C.H. and J.R. Griffiths, *Damage by the cracking of silicon particles in an Al-7Si-0.4Mg casting alloy.* Acta Materialia, 1996. **44**(1): p. 25-33.
2. Tiryakioğlu, M., J. Campbell, and N. Alexopoulos, *On the Ductility of Cast Al-7 Pct Si-Mg Alloys.* Metallurgical and Materials Transactions A, 2009. **40**(4): p. 1000-1007.
3. Tiryakioğlu, M., J. Campbell, and N. Alexopoulos, *Quality Indices for Aluminum Alloy Castings: A Critical Review.* Metallurgical and Materials Transactions B, 2009. **40**(6): p. 802-811.
4. Tiryakioğlu, M., J. Campbell, and N.D. Alexopoulos, *On the ductility potential of cast Al–Cu–Mg (206) alloys.* Materials Science and Engineering: A, 2009. **506**(1–2): p. 23-26.
5. Campbell, J., *Castings.* 2nd ed. 2003: Buttonworths.
6. Campbell, J., *An overview of the effects of bifilms on the structure and properties of cast alloys.* Metallurgical and Materials Transactions B, 2006. **37**(6): p. 857-863.
7. Dai, X., et al., *Influence of oxide film defects generated in filling on mechanical strength of aluminium alloy castings.* Materials Science and Technology, 2004. **20**: p. 505-513.
8. Dispinar, D., et al., *Degassing, hydrogen and porosity phenomena in A356.* Materials Science and Engineering: A, 2010. **527**(16–17): p. 3719-3725.
9. Green, N.R. and J. Campbell, *Influence of oxide film filling defects on the strength of Al-7Si-Mg alloy castings.* AFS Transactions, 1994. **102**: p. 341-347.
10. Mi, J., R.A. Harding, and J. Campbell, *Effects of the entrained surface film on the reliability of castings.* Metallurgical and Materials Transactions A, 2004. **35**(9): p. 2893-2902.
11. Nyahumwa, C., N.R. Green, and J. Campbell, *Effect of mold filling turbulence on fatigue properties of cast aluminum alloys.* AFS Transactions, 1998. **106**: p. 215-224.
12. Nyahumwa, C., N.R. Green, and J. Campbell, *The concept of the fatigue potential of cast alloys.* Journal of Mechanical Behaviour of Metals, 1998. **9**(2): p. 227-235.
13. Rezvani, M., X. Yang, and J. Campbell, *Effect of ingate design on strength and reliability of Al castings.* Transactions of the American Foundrymen's Society, Vol 107, 1999. **107**: p. 181-188
14. Runyoro, J.J., S.M.A. Boutarabi, and J. Campbell, *Critical gate velocities for film-forming casting alloys: a basis for process specification.* AFS Transactions, 1992. **100**: p. 225-234.
15. Yang, X. and J. Campbell, *Pouring basin desgin.* International Journal of Cast Metals Research, 1998. **10**(5): p. 239-299.
16. Dispinar, D. and J. Campbell, *Critical assessment of reduced pressure test. Part 2: Quantification.* International Journal of Cast Metals Research, 2004. **17**(5): p. 287-294.

17. Dispinar, D. and J. Campbell, *Critical assessment of reduced pressure test. Part 1: Porosity phenomena.* International Journal of Cast Metals Research, 2004. **17**(5): p. 280-286.
18. Dispinar, D. and J. Campbell, *Effect of casting conditions on aluminium metal quality.* Journal of Materials Processing Technology, 2007. **182**(1-3): p. 405-410.
19. Dispinar, D. and J. Campbell, *Porosity, hydrogen and bifilm content in Al alloy castings.* Materials Science and Engineering: A, 2011. **528**(10–11): p. 3860-3865.
20. Liu, L. and F.H. Samuel, *Effect of inclusions on the tensile properties of Al–7% Si–0.35% Mg (A356.2) aluminium casting alloy.* Journal of Materials Science, 1998. **33**(9): p. 2269-2281.
21. Tiryakioğlu, M., J.T. Staley, and J. Campbell, *Evaluating structural integrity of cast Al–7%Si–Mg alloys via work hardening characteristics: II. A new quality index.* Materials Science and Engineering: A, 2004. **368**(1–2): p. 231-238.

Shape Casting: 5th International Symposium 2014
Edited by: Murat Tiryakioğlu, John Campbell, and Glenn Byczynski
TMS (The Minerals, Metals & Materials Society), 2014

IMPACT OF SECTION THICKNESS ON THE MICROSTRUCTURE AND MECHANICAL PROPERTIES OF SEMI-SOLID CASTINGS

Stephen Midson, Youfeng He, Xiaogang Hu, Daquan Li, Fan Zhang, Qiang Zhu
General Research Institute for Non-Ferrous Metals, Beijing, China

Keywords – Semi-solid casting, aluminum alloys, casting thickness, mechanical properties

ABSTRACT

Slower cooling and solidification rates tend to increase both the grain size and dendrite arm spacings (DAS) of conventional castings. Both grain size and DAS impact strength and ductility, and so heavy section castings generally have inferior mechanical properties than thin-walled castings. However, the primary phase of semi-solid castings is not produced during solidification in the mold, but instead is developed in the feed material. Therefore, the size of the primary phase is essentially independent of the solidification rate, and so the mechanical properties of semi-solid castings should be less sensitive to the wall thickness of the casting. This paper describes the results of a study to examine the impact of wall thickness on the microstructure, hardness and mechanical properties of semi-solid castings. Castings were produced at three thicknesses – 7.5 mm, 15 mm and 50 mm. Results are compared to literature data for conventional fully-liquid castings produced at similar section thicknesses.

INTRODUCTION

It has been well documented that cooling and solidification rates have a big impact on the mechanical properties of conventional castings. Faster cooling and solidification rates result in smaller dendrite arm spacings (DAS) and finer grain sizes [1-4], both of which tend to improve mechanical properties [5,6]. The dependence of eutectic spacings on solidification rate has also been established [7,8]. Therefore, heavy section castings, which solidify more slowly, tend to have inferior mechanical properties than thinner-walled castings.

Semi-solid casting is different than all other casting processes, as it uses a feed material that is between 25-50% solid and 50-75% liquid, using high pressure die casting machines to inject the semi-solid slurry into re-usable, hardened steel dies [9-11]. Semi-solid casting requires a special, "globular" microstructure, consisting of globular solid particles surrounded by a matrix of liquid. These globular solid particles are generated in the feed material before the semi-solid material is transferred to the casting die. Therefore, the size of the primary particles is essentially independent of the cooling rate in the die, and so the mechanical properties of semi-solid castings should be much less dependent on section thickness and cooling rate. This can be of considerable significance for components that contain both thin and thick sections.

The objective of this study, therefore, was to characterize the impact of section thickness on the structure and properties of semi-solid castings. Semi-solid castings were produced from aluminum alloy 319S at three thicknesses – 7.5 mm, 15 mm and 50 mm, and microstructure, hardness and mechanical properties were examined at each thickness. The results have been compared to literature data for fully-liquid conventional castings produced from a similar aluminum alloy at comparable section thicknesses.

BACKGROUND

Li et al. [12] performed a study to examine the impact of solidification rate on the mechanical properties of low-iron 319 alloy castings produced from the fully liquid state. The composition of their alloy is listed in Table 1. Castings were produced by pouring the alloy from a temperature of 730°C (superheat of about 125°C) into refractory molds having a water-cooled copper chill employed at the bottom. As the majority of the heat was extracted through the copper chill, the cooling rate reduced at increasing heights above the chill. Li et al. [12] measured the mechanical properties of T6 heat treated samples at four heights above the chill (10, 30, 50 and 100 mm), and so, in effect, their data can be considered to represent the properties of castings produced in chilled metal molds of different section thicknesses (approximately 20 mm, 60 mm, 100 mm and 200 mm thicknesses).

Table 1. Chemical composition (in wt%) of the alloy examined by Li et al. [12], and the nominal composition of the alloy in this study

Source	Si	Fe	Cu	Mg	Ti	Sr
Li et al.	5.95	0.11	3.56	0.043	0.145	0.015-0.020
This study	6.0	<0.15	3.0	0.35	<.20	0.01-0.05

The results from Li et al. [12] are reproduced in Figure 1. Their data indicate that, for castings produced from fully liquid feed material, strength and ductility decreased considerably as the distance from the chill increased. The percentage change in mechanical properties from the 10 mm position to the 100 mm position are summarized below

- YS decreased by 25% from the 10 mm position to the 100 mm position
- UTS decreased by 28% from the 10 mm position to the 100 mm position
- Elongation decreased by 48% from the 10 mm position to the 100 mm position

This confirms the premise that both strength and ductility of castings produced from fully liquid feed material are very dependent upon section thickness and solidification rate, and that properties are lower in thicker section castings.

EXPERIMENTAL PROCEDURES

The castings in this study were produced using the thixocasting (billet) semi-solid casting process. The 89 mm diameter feed material was purchased from SAG in Austria, and was produced using a multi-strand horizontal continuous caster. Electromagnetic stirring generated the globular semi-solid microstructure. Nominal composition of the alloy is listed in Table 1.

The castings of different section thickness were produced using the step die shown schematically in Figure 2. The dimensions of the three steps were as follows: Step 1 – 7.5 mm; Step 2 – 15 mm; Step 3 – 50 mm. Slugs used to produce the castings were cut from the pre-cast feed material and re-heated to the semi-solid casting temperature using a 10-coil carousel-style induction heater (operating at about 1,000 Hz). Final slug temperature was either 580°C or 590°C, corresponding to liquid fractions of 46% and 58%, respectively. Total heating time was about 10 minutes. Table 2 lists the process parameters used to produce the castings.

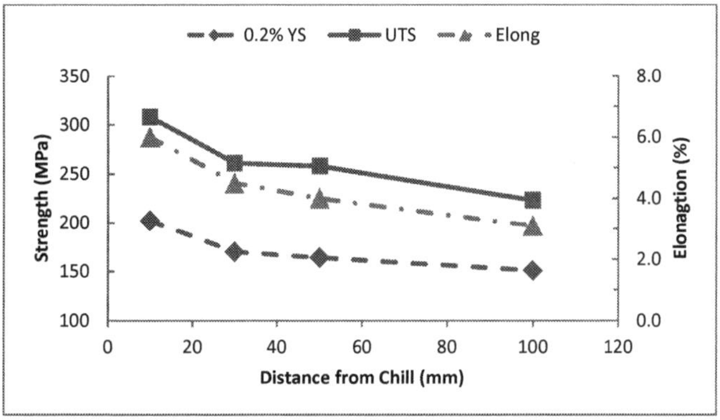

Figure 1. T6 heat treated mechanical properties at different heights above the chill face (data from Li et al. [12])

a)

b)

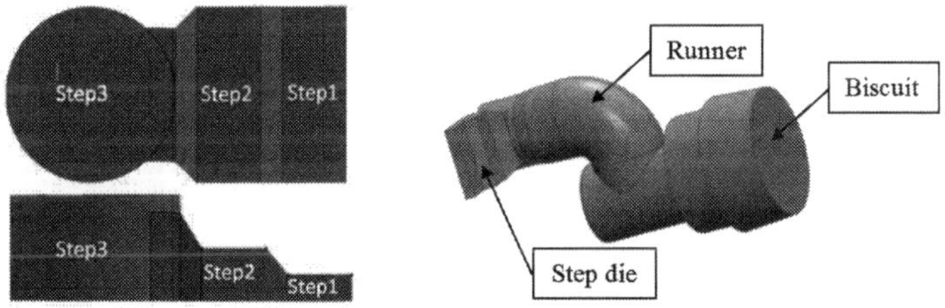

Figure 2. Schematic drawing of the step die.

After casting, the step samples were saw-cut from the runner and T6 heat treated. Solution heat treatment involved 4 hours at 470°C plus 2 hours 40 minutes at 500°C, followed by water quenching. The castings were aged for 10 hours at 170°C. Samples for microstructural analysis were prepared using standard techniques and Figure 3 shows the locations of the samples taken for microstructural analysis. Brinell hardness testing was performed using a 2.5 mm diameter ball and a 62.5 kg load. Sub-size tensile samples with a gauge length of 20 mm and a gauge diameter of 4.0 mm were machined from each of the three steps, ensuring that the tensile samples machined from step 3 were located at least 25 mm from the end of the casting.

Table 2. Semi-solid casting parameters

Condition	Slug Temperature (°C)	Plunger Velocity (m/s)	Intensification Pressure (MPa)
1	580	0.25	112
2	580	0.50	112
3	590	0.05	112

Figure 3. Location of samples for microstructural analysis from each step

RESULTS AND DISCUSSION

As-Cast Microstructures

Figure 4 shows the as-cast microstructure of the samples taken from different locations within the three steps. Figure 5 documents the average size of the primary particles from the nine different locations, showing essentially no differences between the three steps.

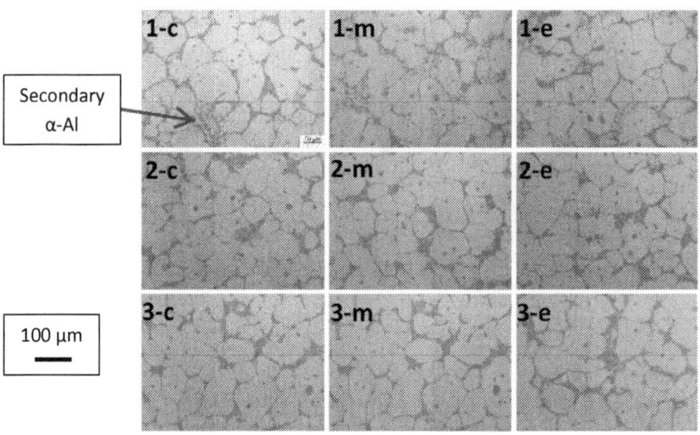

Figure 4. Microstructure in the as-cast condition

As noted earlier, this is to be expected as the primary aluminum particles are formed in the feed material during re-heating to the semi-solid casting temperature. One minor difference observed was the higher amount of secondary α-aluminum observed in step 1 – the secondary α-aluminum forms from the liquid as the semi-solid castings solidify in the die, and is finer-scale due to the fast cooling rate in the un-coated steel die. The reason for the higher amount of secondary α-aluminum in step 1 is not clear.

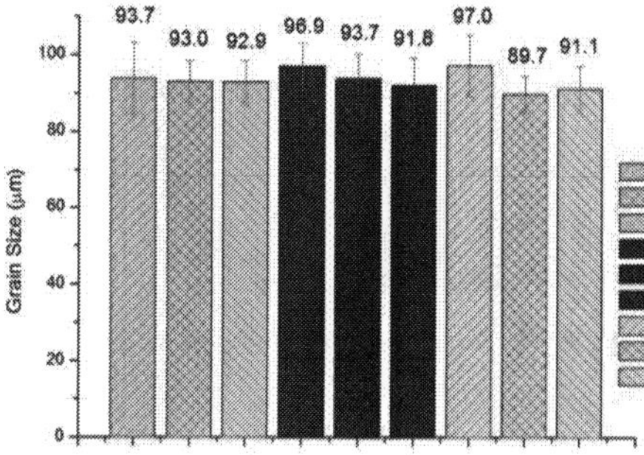

Figure 5. Measurements of the average size of the α-aluminum particles in the as-cast condition

Figure 6 shows measurements of the percent eutectic phase at the center position of each step in the as-cast condition. There is a slightly lower percentage of eutectic (27%) in step 3 than in the other two steps. This phenomenon is common with semi-solid castings, as some segregation of solid and liquid does occur during die filling, with slightly higher eutectic fractions often found towards the extremities of flow.

<u>Heat Treated Microstructures</u>

Figure 7 shows measurements of the average sizes of the primary α-aluminum particles after T6 heat treatment. The α-aluminum particles have grown slightly from the as-cast condition, most likely during the high temperature solution heat treatment.

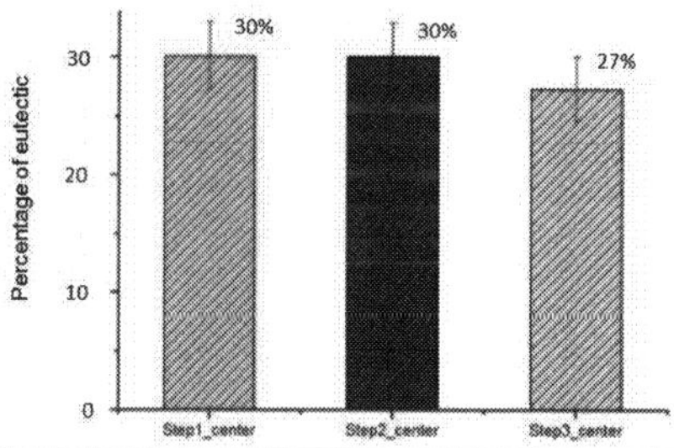

Figure 6. Eutectic percent measured at the center position of each step in the as-cast condition

Figure 8 shows the eutectic microstructure after the T6 heat treatment. As is normal with Al-Si based casting alloys, the eutectic silicon particles have spheroidized during the T6 heat treatment. Most of the silicon eutectic particles are still finer than 5μm, but some of the particles adjacent to the primary α-aluminum particles are closer to 10μm in size. Quantitative measurements were not performed, but the particles do appear to be slightly coarser in the thickest step (step 3).

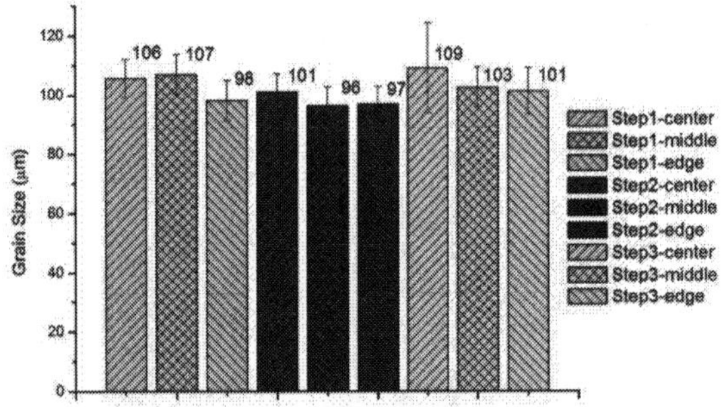

Figure 7. Measurements of the average size of the α-aluminum particles in the T6 heat treated condition

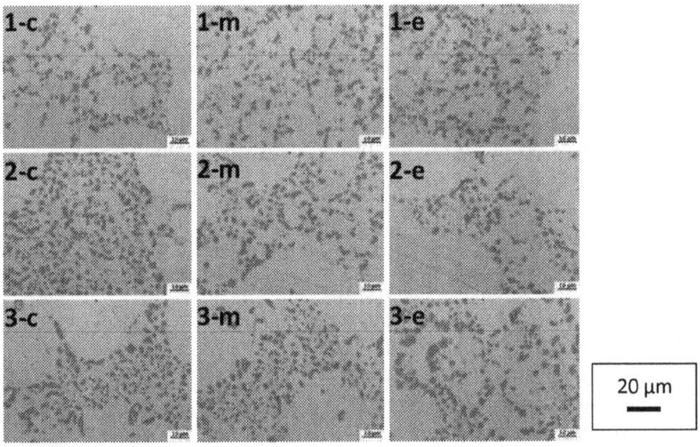

Figure 8. Microstructure of the eutectic after T6 heat treatment

Hardness

Brinell hardness testing was performed at the 18 locations shown in Figure 9. The hardness was very uniform throughout the step castings.

Mechanical Properties

Figure 10 shows the impact of section thickness on the mechanical properties of the semi-solid castings. The first item to note is that both the yield strength (YS) and ultimate tensile strength (UTS) values for the semi-solid castings produced in this study are considerably higher than the strength values reported by Li et al [12]. This is probably partially due to the higher magnesium concentration of the alloy examined in this study, but possibly also due to the higher integrity of the semi-solid castings. Elongation values are similar in both studies.

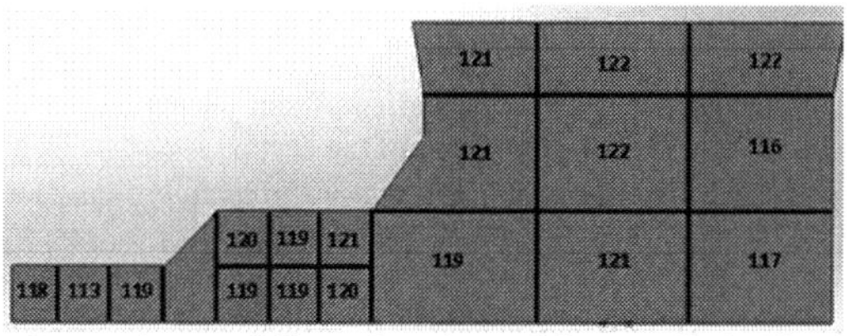

Figure 9. Brinell hardness values at different locations of the step die (T6 heat treated condition)

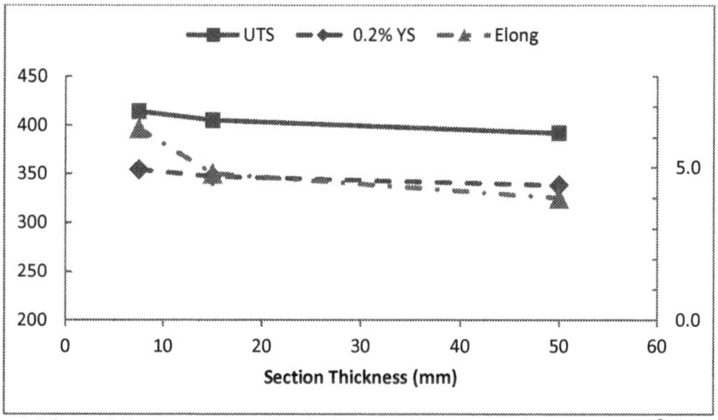

Figure 10: Average mechanical properties of samples cast at 580°C

In addition, the change in strength and ductility values from the thin section (7.5 mm) to the thick section (50 mm) for the semi-solid castings in this study is much smaller than measured by Li et al [12] for fully liquid castings. Based on the data in Figure 10, this is summarized below for the semi-solid castings:

- YS decreased by only 5% from the 7.5 mm thick casting to the 50 mm thick casting
- UTS decreased by only 5% from the 7.5 mm thick casting to the 50 mm thick casting
- Elongation decreased by 37% from the 7.5 mm thick casting to the 50 mm thick casting

The results from this study confirm that hardness and strength are little impacted by section thickness for semi-solid casting. As noted earlier, this is most likely due to the fact that the primary α-aluminum particles are generated in the feed material, and so are insensitive to the slower cooling rate experienced in thick sections. Ductility was observed to be slightly lower in the 50 mm thick section (compared to the 7.5 mm thick section). It is not clear the reason for this, but it is possibly due to the coarser silicon particles observed in the 50 mm section thickness (see Figure 8).

SUMMARY AND CONCLUSIONS

- The objective of this study was to determine the impact of section thickness on the structure and properties of semi-solid castings. As the primary particles are formed in the semi-solid feed material, and so their size should be essentially independent of the solidification rate, the mechanical properties of semi-solid castings are expected to be relatively insensitive to the wall thickness of the casting.
- Analysis of published data for castings produced from fully liquid metal shows that YS, UTS and ductility are all reduced as the section thickness of the casting increases.
- The castings in this study were produced from aluminum alloy 319S using the thixocasting (billet) semi-solid casting process. Castings of different section thickness were produced using a step die, where the cast thickness ranged from 7.5 mm to 50 mm.
- Examination of the microstructure indicated that the size of the primary particles was insensitive to the thickness of the section being cast. After T6 heat treatment, the eutectic silicon particles were slightly coarser in the thick (50 mm) section.
- The hardness of the castings was essentially insensitive to the casting thickness.
- In the T6 condition, the YS decreased by only 5% from the 7.5 mm thick casting to the 50 mm thick casting.
- The UTS decreased by only 5% from the 7.5 mm thick casting to the 50 mm thick casting.
- Elongation decreased by 37% from the 7.5 mm thick casting to the 50 mm thick casting.

ACKNOWLEDGMENTS

The authors would like to acknowledge the financial support by the International Science and technology cooperation projects of China (2012DFA50300), by the National Science and Technology Support Program of China (2012BAE14G01) and by the Innovation Foundation of the General Research Institute for Nonferrous Metals (52611).

REFERENCES

1. B.H. Alexander & F.N. Rhines, Trans. AIME Vol. 188, 1950, 1267
2. R.E. Spear & G.R. Gardner, AFS Trans., Vol. 71, 1963, 209
3. J. A. Horwath & L.F. Mondolfo, Acta Met. , Vol. 10, 1962, 1037
4. H. Matyja, B.C. Geissen & N.J. Grant, J. Inst. Metals, Vol. 96, 1968, 30
5. O. Vorren, J.E. Evensen & T.B. Pedersen, AFS Trans., Vol. 92, 1984, 459
6. G.D. Scott, B.A. Cheney & D.A. Granger, "Fracture Toughness and Tensile Properties of Directionally Solidified Aluminum Foundry Alloys", in: Technology for Premium Quality Castings, E. Dunn, D.R. Durham (Eds.), TMS, Warren dale, PA, 1987, 123
7. M.H. Burden & H. Jones, J. Inst. Metals, Vol. 98, 1970, 249
8. J.D. Livingston, H.E. Cline, E.F. Koch & R.R. Russell, Acta Met., Vol. 18, 1970, 399
9. Merton C. Flemings, Met. Trans. A Vol. 22A (5), 1991, 957
10. Z. Fan, Inter. Mater. Rev., Vol. 47, 2002, 49
11. S.P. Midson, Die Casting Engineer, January 2006, 48
12. Z. Li, A.M. Samuel, F.H. Samuel, C. Ravindram, S. Valtierra & H.W. Doty, Mat. Sci. Eng. A, Vol. 367, 2004, 96

Shape Casting: 5th International Symposium 2014
Edited by: Murat Tiryakioğlu, John Campbell, and Glenn Byczynski
TMS (The Minerals, Metals & Materials Society), 2014

THE RELATIONSHIP BETWEEN ELONGATION AND FATIGUE LIFE IN A206 ALUMINUM CASTINGS

Murat Tiryakioğlu

University of North Florida, School of Engineering, 1 UNF Drive, Jacksonville, FL 32224, USA

Keywords: Quality index, Weibull statistics, fatigue life potential

Abstract

The tensile and fatigue testing results of Staley Jr. et al. [Mater. Sci. Eng. A, v. 460-461 (2007) 324 and Mater. Sci. Eng. A, v. 465 (2007) 136] for A206-T71 castings were reanalyzed. The Weibull distributions for both elongation and fatigue life of castings that were hot isostatically pressed (HIPed) as well as those that received no processing (no-HIP) were correlated. Results indicate that significant improvements in fatigue performance are possible if pores old oxides are eliminated. Extrapolations of results and underlying mechanisms are also discussed.

Introduction

Design concepts are based on the assumption that the maximum stress in an engineered component during its service will be well below the yield strength (σ_Y) of the material selected by the designer. In aerospace castings, in addition to the factor of safety, a casting factor is sometimes used to accommodate the scatter in the structural quality of the castings and to ensure that all stresses within the casting are safely below the yield.

Although castings can be said to be overdesigned so that stresses are well below the yield strength, almost all specifications for cast aluminum alloys require a minimum level of elongation. This requirement can only be interpreted as an expectation for a certain level of structural quality, as evidenced by small number of defects. Therefore elongation specification can be interpreted as a *de facto* fatigue life specification. Although there are a number of publications that link defect size distribution to fatigue life distribution [1,2], to the author's knowledge, there is no study that has addressed the possibility of a link between elongation and fatigue life. This paper is intended to fill that void by reanalyzing data published previously for A206 castings.

Results of Staley, Jr. et al.

Staley, Jr. et al. [3,4] conducted experiments to determine the effectiveness of various hot isostatic pressing (HIP) conditions on the tensile and fatigue properties of A206 aluminum alloy. In this study, specimens were excised from ingots that were either hot-isostatically pressed (HIPed) or that did not receive the HIP treatment (no HIP). Specimens were heat treated to T71 temper. The average yield strength of no-HIP and HIPed specimens were 266 and 290 MPa, respectively. The tensile and fatigue results were analyzed by using the Weibull distribution [5]:

$$P = 1 - \mathbf{exp}\left[-\left(\frac{\sigma - \sigma_T}{\sigma_0}\right)^m\right] \qquad (1)$$

where P is the probability of failure at a given stress (strain, fatigue life, etc.), σ, or lower. The threshold value, σ_T, is the value below which no specimen is expected to fail. The term, σ_0, is the scale parameter, and *m* is the shape parameter, alternatively referred to as the Weibull modulus.

The elongation data for various HIP treatments were combined and reanalyzed in the present study. The Weibull probability plot of elongation for no-HIP and HIPed specimens is presented in Figure 1. Note that the data for no-HIP specimens show a linear trend, which indicates that the threshold is zero [6]. The HIPed specimens, however, have two different trends: (i) at low elongation values, the trend is a curve, rather than a line, which indicates a positive threshold [6], and (ii) at higher elongation values, there is a sudden change in the slope of the fitted curve, which indicates the presence of Weibull mixtures [6] which can be modeled as:

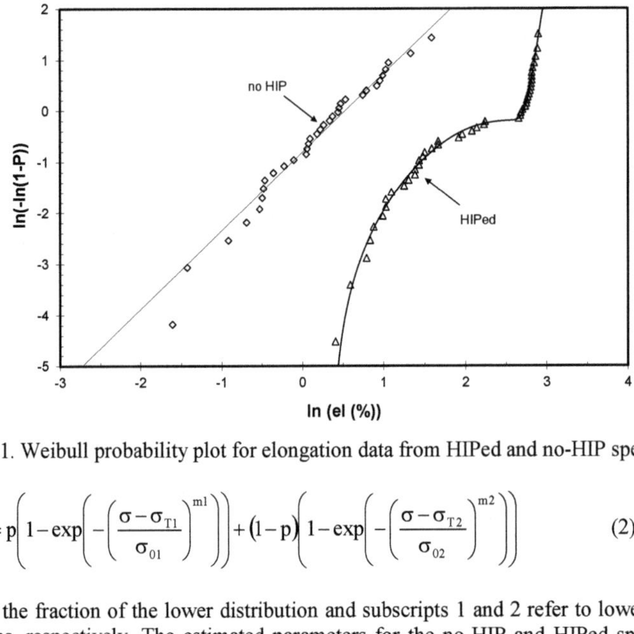

Figure 1. Weibull probability plot for elongation data from HIPed and no-HIP specimens.

$$P = p\left(1-\exp\left(-\left(\frac{\sigma-\sigma_{T1}}{\sigma_{01}}\right)^{m1}\right)\right)+(1-p)\left(1-\exp\left(-\left(\frac{\sigma-\sigma_{T2}}{\sigma_{02}}\right)^{m2}\right)\right) \quad (2)$$

where *p* is the fraction of the lower distribution and subscripts 1 and 2 refer to lower and upper distributions, respectively. The estimated parameters for the no-HIP and HIPed specimens are provided in Table 1.

Table 1. The estimated parameters for the Weibull fits presented in Figures 1 and 2.

	No-HIP			HIPed						
	σ_T	σ_0	m	p	σ_{T1}	σ_{01}	m_1	σ_{T2}	σ_{02}	m_2
e_F	0	1.69	1.54	0.565	1.42	1.36	3.42	12.77	3.75	3.91
N_f	6,063	23,852	0.48	0.633	177,022	171,887	1.14	442,269	7,745,853	2.33

Analysis of fracture surfaces of tensile specimens via scanning electron microscopy (SEM) showed that the defects that led to premature failure in no-HIP specimens were "old" and "young" oxides and associated large pores. In HIPed castings, there were no pores, as can be expected. The "old" oxides were found to be resistant to healing and therefore cracks remained even after the pores around the old oxide bifilms were closed. The lower distribution in HIPed specimens is predominantly due to premature failure from "old" oxide bifilms. The upper distribution represents premature failure due to "young" oxide bifilms that form during pouring and filling of the current casting (as opposed to old oxides representing former damage). The young bifilms, unlike old bifilms, were partially healed during the HIP treatment. Therefore the upper distribution represents the case where an "old" oxide crack either does not exist in the casting or is parallel to the direction of loading. Hence the Weibull mixture model presented in Eq. 2 which is built on the assumption that "no competition" is applicable in this case. In this case, the "old" oxides "win" all the time over partially-healed "young" oxides to lead to premature fracture in HIPed castings. Hence it can be concluded that for elongation to be improved significantly by HIP, the coarse, large "old" oxide films should not be present in the casting.

Fatigue tests were conducted at a maximum stress of 170 MPa, R = 0.1 under load control using a sine waveform at a frequency of 60 Hz in laboratory air. Tests were run to failure or stopped at 10^7 cycles. Weibull probability plot for fatigue life (N_f) results is presented in Figure 2. Note that as in Figure 1, (i) the closure of the pores by HIP results in a significant improvement in fatigue life, by more than an order of magnitude, (ii) results for no-HIP castings can be modeled by a three-parameter Weibull distribution, and (iii) data for HIPed castings come from a Weibull mixture.

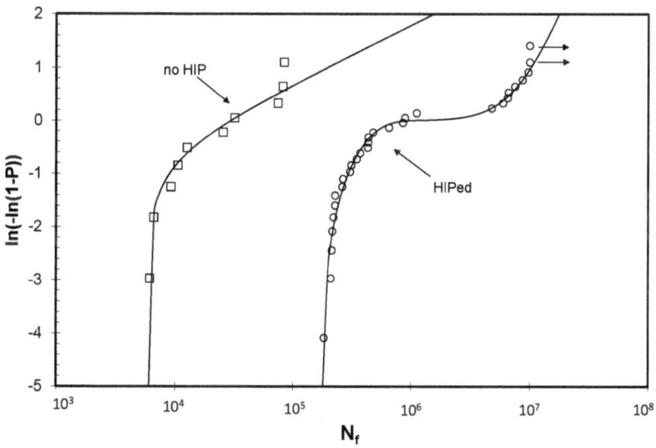

Figure 2. Weibull probability plot for fatigue life data for no-HIP and HIPed specimens.

SEM analysis of fracture surfaces of fatigue specimens showed that all fatigue cracks for no-HIP specimens initiated at surface-connected porosity containing various amounts of alumina and spinel bifilms. All fatigue cracks for HIPed specimens initiated at surface-connected or interior oxides, predominantly "old" oxides. The upper distribution for HIPed specimens in Figure 2 represents failure, to a great extent, from interior defects with no other surface or interior defect visible on the fracture surface. For this to occur, the number of defects per unit volume and their sizes should be small [7]. The probability of having no defects not intersecting the specimen's surface can be written as [8]:

$$P=\left(1-\frac{SD}{V}\right)^{n_V V} \tag{3}$$

where S is the surface area of the specimen, D is the defect size (average diameter), V is specimen volume and n_V is the number of defects per volume. For specimen with a cylindrical gage length, the term S/V is equal to 4/d, where *d* is the specimen diameter. The change in P as a function of D/d for various numbers of expected defects ($n_V V$) values is presented in Figure 3. Note that the number of defects has a much more pronounced effect on the probability than the size of the defect. Therefore, to obtain superior fatigue performance, casting defects should be kept to a minimum in size and number density, if not eliminated.

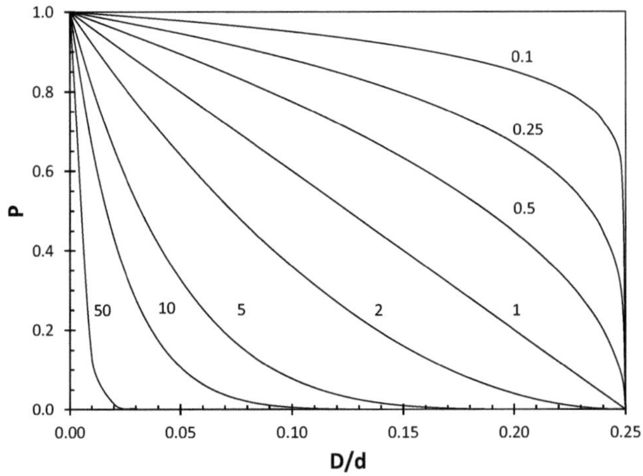

Figure 3. The probability of a defect not intersecting the surface as a function of D/d for various values of number of expected defects in the specimen ($n_V V$).

The Effect of Structural Quality on Fatigue Life

The author and his collaborators [9,10,11] recently analyzed σ_Y-e_F data for various cast Al alloys and found that there is a ductility potential, $e_{F(max)}$, for uniform elongation for these alloys. It can be written as a function of yield strength:

$$e_{F(max)}(\%)=\beta_0-\beta_1\sigma_Y \tag{4}$$

where β_0 and β_1 are coefficients with values of 47.8 and 0.085 MPa^{-1}respectively for A206 [12]. Based on the concept of ductility potential, Tiryakioğlu et al. [12] developed a new quality index, Q_T:

$$Q_T=\frac{e_F}{e_{F(max)}}=\frac{e_F}{\beta_0-\beta_1\sigma_Y} \tag{5}$$

Recently, Tiryakioğlu and Campbell [13] divided the Q_T space into four distinct regions and provided recommendations for quality improvement for each region. When tensile data are in

Region 1 ($0 \leq Q_T < 0.25$), the premature failure is due to "old" oxides from remelted castings and foundry returns and/or ingot. In this region, tensile specimens do not neck and fatigue failure starts from defects on the specimen surface. Region 2 ($0.25 \leq Q_T < 0.70$) represents melts that are free from major "old" oxides but there is still a considerable density of "young" oxides. Tensile specimens may show some necking and there will be occasional fatigue failures initiating from internal defects with facets around them [7]. In Region 3 ($0.70 \leq Q_T \leq 1.0$), tensile specimens are expected to neck and deform significantly beyond ultimate tensile strength. Moreover fatigue fracture is predominantly due to internal defects, exhibiting facets on fracture surfaces.

The tensile and fatigue results shown in Figures 1 and 2 can be interpreted together. To accomplish this, first the quality index Q_T values need to be calculated. The values of $e_{F(max)}$ were calculated as 266 and 290 MPa for no-HIP and HIPed castings, respectively. All elongation values were then converted to QT by using Equation 5. The Weibull distributions for Q_T are presented in Figure 4. The fractions of the three Q_T distributions in each region are given in Table 2. Note that almost the entire distribution for no-HIP castings is in Region 1. In HIPed castings, the lower distribution still has 75% in Region 1. This result indicates how ineffective HIP or even squeeze casting can be when old oxides are present in the casting. In the absence of old oxides, HIP partially heals young bifilms and consequently the size and number of defects is significantly lower, leading to a significant shift of the Q_T distribution towards Region 3. Approximately 54% of the upper Q_T distribution is in Region 3.

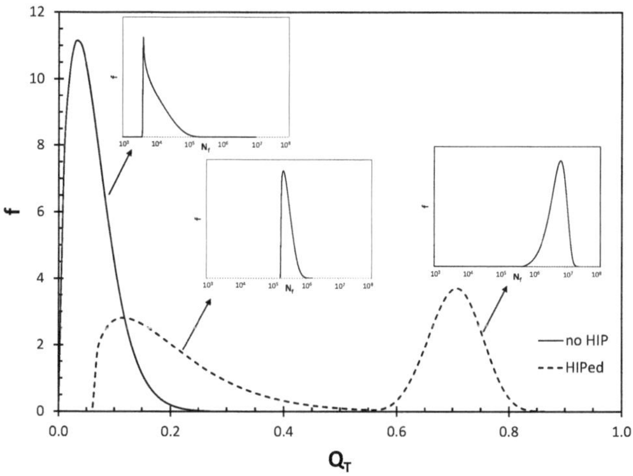

Figure 4. Weibull distributions for Q_T and corresponding Weibull distributions for fatigue life.

The three Weibull distributions for fatigue life are also indicated in Figure 4, showing the corresponding Q_T distribution. The probability of survival after 10^5, 10^6 and 10^7 cycles at R=0.1 and maximum stress of 170 MPa (58% of yield strength) for the three different quality levels of metal if also given in Table 2. When large pores are present (Region 1), the probability of survival after 105 cycles is approximately 15%. When pores are eliminated, all castings survive for at least 10^5 cycles. When old oxides are present in the casting, almost no casting survives after 10^6 cycles. When old oxides are eliminated and bifilms are kept to a minimum (Region 2 and 3), almost 20% of the specimens survive after 10^7 cycles. Although the improvement in fatigue performance with increasing Q_T is significant in the experimental results, it should be noted that there is still more potential because very few castings in this study have a Q_T over 0.80. The fatigue performance can be expected to even higher in castings with Q_T values approaching 1.0.

Table 2. Fraction of distributions for Q_T in each region and probability of survival at three fatigue life limits.

	Fraction in Each Q_T Region			Probability that N_f exceeds limit		
	$Q_T \le 0.25$	$0.25 \le Q_T < 0.70$	$Q_T \ge 0.70$	10^5	10^6	10^7
no-HIP	0.999	0.001	0.000	0.145	0.000	0.000
HIPed-low	0.752	0.247	0.001	1.000	0.003	0.000
HIPed-high	0.000	0.457	0.543	1.000	0.998	0.196

Extrapolations of Results

To correlate the three distributions of Q_T and fatigue life, the (expected) mean values for each Weibull distribution was calculated by using the following equation:

$$\bar{\sigma} = \sigma_T + \sigma_0 \Gamma\left(1 + \frac{1}{m}\right) \qquad (6)$$

where Γ represents the gamma function. The means for Q_T and N_f distributions are cross-plotted in Figure 5. The fit to the data presented in Figure 5 is:

$$\mathrm{Log}_{10}\left(\bar{N}_f\right) = 3.087\ \bar{Q}_T + 4.73 \qquad (7)$$

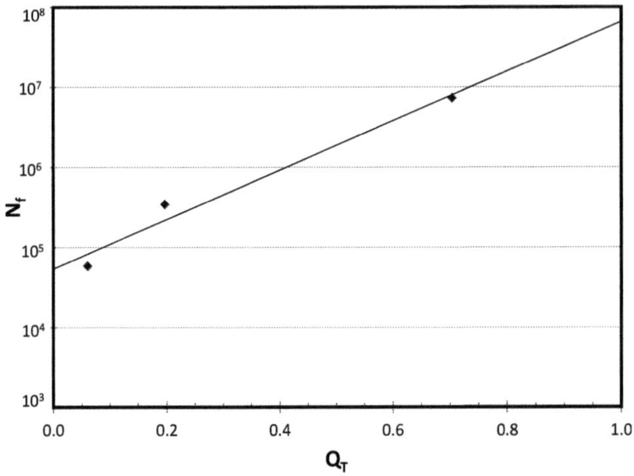

Figure 5. Mean values of three distributions for Q_T and N_f.

When Q_T = 1.0, Equation 7 yields $10^{7.82}$, or 66 million cycles. This fatigue life is quite impressive given the fact that the specimens are tested in pull-pull mode and at a maximum stress that is 58% of yield strength of the alloy. When tested at a maximum stress equal to 60% of its yield strength at R=-1, 6061-T6 was found [14] to have a fatigue life ~10^5.

It is unrealistic that castings will be designed so that the maximum stress during service will be 58% of the yield strength of the alloy but realistic fatigue life expectations at lower maximum stresses can be estimated by [1]:

$$N_{f(eq)} = N_f \left(\frac{\sigma_{max(1)}}{\sigma_{max(2)}} \right)^n \qquad (8)$$

where n is the exponent in the Paris-Erdoğan equation for fatigue crack growth. For cast aluminum alloys, n is approximately 4.2 [1,15]. If we were to estimate the potential fatigue life of A206-T71 when Q_T = 1.0 and maximum stress is 90 MPa (~30% of σ_Y), Equation 8 yields 10^9 cycles. Therefore true fatigue life potential of A206 can be expected to be in gigacycles.

Conclusions

1. The structural quality (Q_T) and fatigue life distributions were matched and correlated. This is the first time that such a correlation is demonstrated.
2. The closure of the pores after HIP leads to significant improvement in fatigue life. However this improvement is well below the fatigue life potential of A206.
3. Fatigue lives in gigacycles are possible for A206 castings at a maximum stress of 30% of yield strength.

References

1. M. Tiryakioğlu: *Metall. Mater. Trans. A,* 40 (2009) 1623-1630.
2. D. Casellas, R. Pérez, J.M. Prado: *Mater. Sci. Eng. A*, A398 (2005) 171-179.
3. J.T. Staley, Jr., M. Tiryakioğlu, J. Campbell: *Mater. Sci. Eng. A*, 465 (2007) 136-145.
4. J.T. Staley, Jr., M. Tiryakioğlu, J. Campbell: *Mater. Sci. Eng. A*, 460-461 (2007) 324-334.
5. W. Weibull: *J. Applied Mechanics*, 13 (1951) 293-297.
6. M. Tiryakioğlu, J. Campbell: *Metall. Mater. Trans. A*, 41 (2010) 3121-3129.
7. M. Tiryakioğlu, J. Campbell, C. Nyahumwa: *Metall. Mater. Trans. B.* 42 (2011), 1098-1103.
8. A. de Bussac, J.C. Lautridou: *Fatigue Fract. Engng. Mater. Struct.* 16 (1993) 861-874.
9. M. Tiryakioğlu, J. Campbell, N.D. Alexopoulos: *Metall. Mater. Trans. A*, 40A (2009), 1000-1007.
10. M. Tiryakioğlu, J. Campbell: *Mater. Sci. Technol.*, 25 (2009) 784-789.
11. M. Tiryakioğlu, J. Campbell, N.D. Alexopoulos: *Mater. Sci. Eng. A*, A506 (2009) 23-26.
12. M. Tiryakioğlu, J. Campbell, N.D. Alexopoulos: *Metall. Mater. Trans. B*, 40 (2009) 802-811.
13. M. Tiryakioğlu, J. Campbell: *AFS Transactions*, (paper no: 13-1525) (2013) 217-222.
14. Q.Y. Wang, N. Kawagoishi, Q. Chen: *Intl. J. Fatigue*, 28 (2006) 1572–1576.
15. Q.G. Wang, C.J. Davidson, J.R. Griffiths, P.N. Crepeau: *Metall. Mater. Trans. B.*, 37B (2006) 887-895.

Shape Casting: 5th International Symposium 2014
Edited by: Murat Tiryakioğlu, John Campbell, and Glenn Byczynski
TMS (The Minerals, Metals & Materials Society), 2014

MAGNESIUM CASTING PROCESSES: THE HIMAC PROJECT RE-ASSESSED

John Campbell[1] and Murat Tiryakioglu[2]

[1]University of Birmingham, Metallurgy and Materials, Edgbaston, B15 2TT, UK
[2]University of North Florida, School of Engineering, 1 UNF Drive, Jacksonville, FL 32224, USA

Abstract

A program within USCAR/USAMP-AMD (United States Car/United States Automotive Materials Partnership – Automotive Materials Division) Project entitled High Integrity Magnesium Automotive Components (HIMAC) was divided into several tasks, some involving the metallurgy of Magnesium casting alloys, and the rest to the production of an automotive control arm casting in AZ91 magnesium alloy by a variety of casting processes. Casting processes included: the Ablation Process, low pressure permanent mold (LPPM), squeeze casting (SC) and T-Mag. LPPM and SC are established processes, whereas the Ablation Process and T-Mag are relatively new processes in the early phases of being productionized. The performance of the castings from the various casting processes was compared. The Ablation Process was found to produce castings with the best combination of strength, ductility, reliability and X-ray soundness. Ablation fatigue resistance showed superior results. However, since ablated specimens did not fail in fatigue, Weibull quantitative results could not be ascertained, whereas, all other processes failure occurred.

Introduction

One task of this research was targeted to evaluate 'Emerging Casting Technologies' for the production of magnesium automotive castings. Established processes such as LPPM and SC were compared with emerging processes including the Ablation Process (an aggregate mold process in which the mold is ablated away allowing the casting to be rapidly cooled with jets of coolant) and the T-Mag process (a variety of tilt casting which effectively delivers a low pressure filling of a permanent mold).

The evaluation was carried out by arranging each casting process to manufacture, so far as possible, 20 identical castings. A control arm, originally a 2.3 kg steel stamping, was redesigned as a 1.8 kg casting in AZ91 alloy and was selected as the standard casting (Figure 1). The 20 castings were sufficient to allow a wide variety of tests to be conducted to compare the processes over a complete range of characteristics that are required of a safety-critical automotive component. However one process, the LPPM with the EM Pump, was unable to provide the full set of 20 samples. Overall, the project achieved a remarkably complete comparison of these different approaches.

The many aspects of the project have been individually documented as part of the final report [1]. Some of these aspects have been commented on and published by previous workers [2]. This evaluation is intended to give a more complete and critical presentation of HIMAC results as pertaining to the manufacture of magnesium alloy castings.

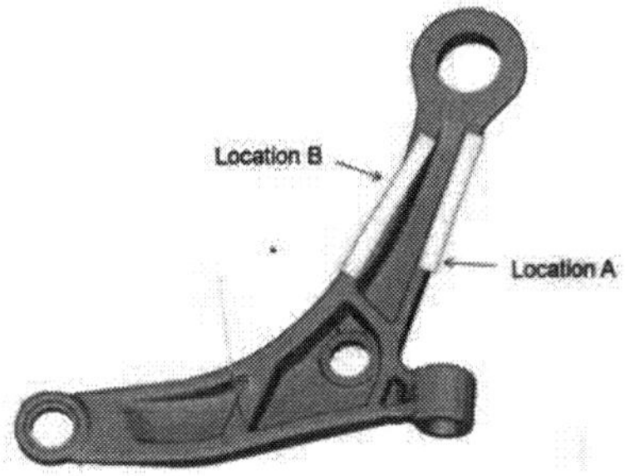

Figure 1: The standard casting and the location of test specimens.

Background to the Casting Processes

The low pressure permanent mold (LPPM) process and the squeeze casting (SC) process are relatively well known conventional processes, and have been critically reviewed by one of the authors [3]. They are therefore not further described here.

Ablation Process

In the Ablation Process, the fundamental limitation to heat flow presented by the air gap is eliminated by removal of the mold so that a coolant can be applied directly onto the surface of the casting. This is achieved with the use of an aggregate mold bonded with a soluble binder. The process consists of filling a mold (assembled as a 'core package', an established precision sand mold technique) with molten alloy. After filling, the ablation coolant is applied immediately to the mold package, which still contains the metal in its liquid state. The direction and velocity of solidification is now under the control of the Ablation Process. The mold is ablated away and the liquid alloy is simultaneously frozen to form the casting.

In common with other sand casting processes (but in contrast to metal die processes) the pattern work (tooling) for the Ablation Process is relatively simple and relatively low cost. In addition the production plant is both simple and inexpensive light engineering. No development or changes relating to the tooling are normally needed since the cooling conditions are under independent software control. Again, in common with other good sand casting techniques, the surface finish is at best approximately equivalent to that produced by permanent mold processes. The Ablation Process seems equally suitable for both prototype and series production, so that the prototype castings exactly simulate eventual production castings, in contrast to die casting and permanent mold processes in which the prototype is simulated to an uncertain degree of accuracy by a sand casting. The Ablation Process had been proven for the production of aluminum alloy castings [4]. Its inclusion in the HIMAC program was to explore its application to the production of magnesium alloy castings.

A gravity filling system was used for this study, but was specifically designed to reduce the entrainment of air and oxides to a minimum. In addition, a woven steel mesh filter was used at the runner gate junction. Molten metal was poured into the pouring basin to give the designed depth required, at which point the alloy depth was sufficient to start the filling of the mold. When filling was complete, ablation was started to solidify the casting.

For the same reasons that the die heating and cooling development was found to be critical for all the die casting processes, the precise conditions for ablation were found to be similarly critical. Several ablation techniques were evaluated before a combination of top and bottom ablation was chosen to produce the simulated production castings. The ablation conditions were quickly and easily changed and tested. No tooling changes for the mold and casting were required. A total of 35 castings were produced and tested from this unadjusted and not fully optimized process.

T-MAG Process

T-Mag presumably stands for 'Tilt Magnesium' process. It is a variant of LPPM, in that it uses a riser tube that delivers the melt into a permanent mold. However, the riser tube is filled by the action of tilting of the whole furnace and die assembly rather than by use of pressure. This mechanical action allows an accurate knowledge of the position of the melt surface, compared to the relative uncertainty of estimating the melt level from the use of pressure measurement. As with all LPPM approaches, however, the great benefits of quiescent filling are threatened from the other conventional sources. At the present time the process melts the charge and delivers liquid metal to the die from the same furnace, involving the necessarily inherent entrainment of the thick oxide skins of the charge materials into the material that is transferred into the die. The scraping of the furnace crucible for routine maintenance will also add to debris that cannot easily be completely removed from the crucible, and may be dislodge during tilting and thus impair the casting (although the developers claim to be working on a furnace re-design to alleviate some of these problems). The provision of a filter at the ingate into the mold certainly helps but does not seem to address these problems completely.

Results and Discussion

Ablation Process and T-Mag castings were subjected to different heat treatments and tensile tests. The mean results for yield strength (σ_Y), tensile strength (S_T) and elongation (e_F) (excluding clearly defective samples) are given in Table 1. Table 2 contains a complete list of tensile properties for the T4 condition). Although the T6 condition gave generally higher strengths as expected, a useful result in terms of all-round performance was found to be a T4 treatment (solution, quench and natural ageing). A T4 treatment was therefore selected as a uniform treatment for all casting processes in this study.

Table 1: Ablation Process and T-Mag Data for AZ91, prior to HIMAC Project.

Process	Alloy	Temper	σ_Y (MPa)	S_T (MPa)	e_F (%)
Ablation	AZ91D	F (As-Cast)	138	172	6.4
Ablation	AZ91E	T4	119	287	15.0
Ablation	AZ91D	T6	143	275	4.9
T-Mag	AZ91E	F	83	156	3.0
T-Mag	AZ91E	T4	75	199	5.5
T-Mag	AZ91E	T6	107	204	3.1

The study has uncovered the fact that the ablation-cooled castings are capable of exhibiting unusual properties which deserve special discussion.

Table 2: Average tensile properties in T4 condition from excised samples (tests by Mississippi State University)

Process	σ_Y (MPa)	S_T (MPa)	e_F (%)
Ablation*	96	308	13.8
LPPM	83	160	3.2
LPPM EMP	97	188	4.5
Squeeze	93	144	2.4
T-Mag	90	208	6.1

*Proprietary T4 type heat treatment by Lite Metals.

The rapid chilling effect of the ablation coolant on the solidifying alloy produced a unique microstructure and high mechanical properties despite clear evidence of high porosity which is not typical of the Ablation Process, but, interestingly, denotes the use of rather poor quality liquid metal for these particular castings. Furthermore, the secondary dendrite arm spacing (SDAS) was in general rather coarse and typical of a sand casting of that section thickness simply because some solidification of dendrites had occurred at the natural rate in a silica sand mold prior to the application of ablation. It is significant therefore that the improved properties are clearly not directly related to the secondary dendrite arm spacing (SDAS) as appears common to all other casting processes.

During ablation, the removal of the mold from around the solidified portions of the cast product ensures that any constraint normally provided by the mold as a result of the contraction of the metal in the mold, is eliminated. This eliminates problems such as hot tearing. It was noteworthy that all of the processes using steel dies in the study had to pay significant attention to reducing hot tearing issues. For instance for LPPM much attention was given to the precise timing of core pulls to reduce any tendency to hot tear. T-Mag process records the tendency for the casting to crack from frictional drag when opening the die.

Table 3 illustrates that the original steel component was itself rather variable in strength, one out of two specimens falling short of its designed yield force target. Even so, the fatigue performance was good, as might be expected for steel compared to aluminum alloys.

LPPM and the Ablation Process were the only two processes that achieved one result that met the yield target of 20 kN. This indicates that in principle the yield target can be met by magnesium alloy castings of this design. However, in the case of LPPM the fact that only one result was this high, and three out of five results were especially low at 9 kN indicates the other three contained serious defects. This is a high percentage of serious strength-reducing defects, and most likely reflects the unfortunate but fundamental issue of cleanness of the melt. [3] The relatively poor results by EM pump are perhaps a surprise, but again, the issue of cleanness to be affecting properties.

When using an EM pump with Al alloys the foundry layout has the benefit of the pump located in the pump well of a large holding furnace. However, for magnesium alloys, a large holding furnace with the benefits of constant melt level is in general not practical. A crucible furnace would be used. There will be enhanced danger in a crucible furnace of picking up the sedimented dense inclusions which will tend to concentrate under the pump. Furthermore, when pumping from a crucible furnace it is unlikely that the melt could be held at the top of the delivery tube and so would be allowed to fall between castings as is normal for LPPM process. This brings the twin severe quality penalties of (i) stirring up the sediment in the base of the crucible back into suspension, and (ii) creates a wash of oxide inside the riser tube that endangers subsequent castings.

Table 3: Destructive Tests on Complete Castings
(Carried out by ADT Engineering & Machine)

Process	Average Yield (Design Target 20 kN)	Average Max Load (kN)	Fatigue kCycles Log-average
Steel stamping	23	25	436
Ablation	18	35	133
LPPM	12	26	135
LPPMEMP	12	25	73
T-Mag	16	31	84

The Ablation Process gave the highest average yield and maximum load, together with outstanding fatigue resistance. Fatigue tests on excised specimens (Table 4) showed that only the Ablation Process met the fatigue specification for this product. Interestingly, ablated casting had variable porosity, but the fatigue results (Table 3) were almost identical. This result indicates that quality measures in magnesium alloys such as area percentage porosity are probably irrelevant as indicators of performance. This conclusion is underlined from the finding that the Ablation Process had, in general, a higher level of porosity than most processes, but properties were in general much higher. The results imply that the Ablation Process might have significantly more potential for improvement. In addition, this observation points to other factors of greater importance than porosity in influencing fatigue; this seems most likely to be the result of the presence of oxide bifilms acting as cracks [3].

Tensile properties of ablated castings were remarkable. Samples excised as seen in Table 2 showed that yield strength at approximately 100 MPa approximately equaled that from other processes but ultimate strength at approximately 300 MPa was 50 % greater than the next largest value, and 100 % better than the worst. Elongation in the Ablation Process at 14 % was over double the next best, T-Mag, and five times better than the worst value.

Table 4: Summary of Fatigue Results on Excised Samples
(Tests carried out at Mississippi State University)

Process	Percentage meeting fatigue endurance to 10^3 cycles at strain amplitude 0.5%	Percentage meeting fatigue endurance to 10^6 cycles at strain amplitude 0.1%
Ablation	100%	100%
LPPM	0%	27%
LPPM-EMP	N/A	0%
SqueezeCast	20%	0%
T-Mag	50%	50%

Some of the most insightful results were those revealed by statistical analysis of ultimate bend strength (S_B) results after four-point bend tests, in which the Weibull distribution was used:

$$P = 1 - \exp\left[-\left(\frac{\sigma - \sigma_T}{\sigma_0} \right)^m \right] \qquad (1)$$

where P is the probability of failure at a given stress, σ, or lower. The threshold value, σ_T, is the value below which no specimen is expected to fail. The term, σ_0, is the scale parameter, and *m* is the shape parameter, alternatively referred to as the Weibull modulus. Note that Equation 1 has three parameters that need to be estimated and is reduced to two parameters when the threshold is zero. Weibull probability plots for S_B data from three processes are presented in Figure 2. The surprise result was that the Ablation Process samples continued to bend to the limits imposed by the bend test rig and showed no signs of cracking. Because failure could not be induced, these results could not be plotted so that a Weibull distribution of failure strengths could not be obtained. After three attempts to break the ablation samples in the standard test rig, tests were transferred to a larger bend test rig in an effort to achieve fracture, but even this did not cause the ablated bars to fracture. The bars simply deformed to maximum extent that could be provided by the apparatus. Thus only three unbroken strength values provided in the standard rig could be determined and compared with the other processes whose samples had been tested in the standard test rig.

For squeeze castings the Weibull analysis shows a broad distribution of defects which could be fitted with a 2-parameter Weibull straight line, indicating that the scatter in strengths could in principle be extrapolated to zero strength (i.e. zero threshold strength). This distribution of strengths would be of serious concern for a safety critical component. It is not easy to speculate why this disadvantageous distribution occurs, but may be associated with turbulent filling conditions for this horizontal squeeze process.

The squeeze cast distribution of strengths contrasts with both of the counter-gravity variants (LPPM and T-Mag) whose distributions of strength indicated a 3-parameter Weibull distribution was appropriate, giving threshold (minimum) strengths of 258.7 and 295.3 MPa respectively, below which no failures are to be expected. This is a reassuring result for these processes.

It is a tribute to the Ablation Process that no Weibull plot could be produced because despite being loaded to much higher bend stresses (values of 455.1, 767.5 and 574.7 MPa were recorded) no specimens could be persuaded to break. However, this is, of course, a frustrating result arising from the selection of bend testing as a measure of strength. If conventional tensile testing had been selected as the test of strength, a Weibull result could also have been obtained for the ablated material. This result would have been of interest, providing a quantification of the reliability of ablated material. Additional information on yield and elongation would also have been of interest of course.

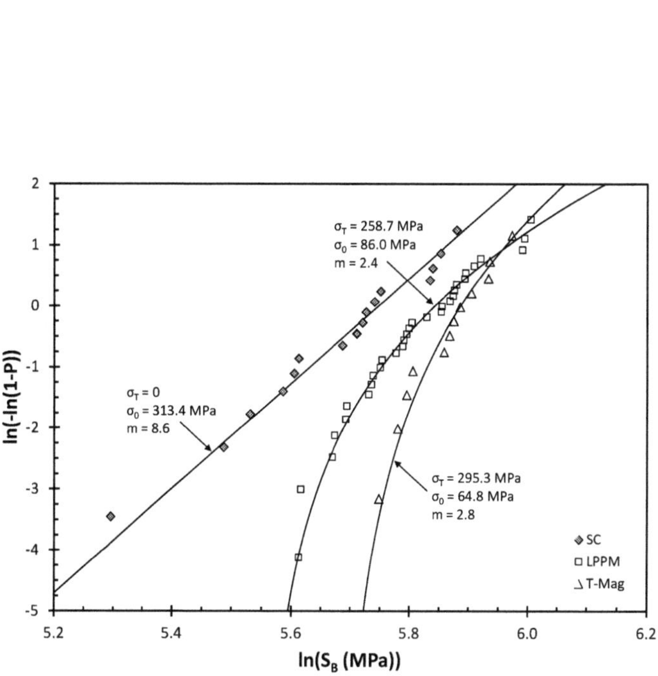

Figure 2: Weibull plots of Ultimate Bend Strength Data.

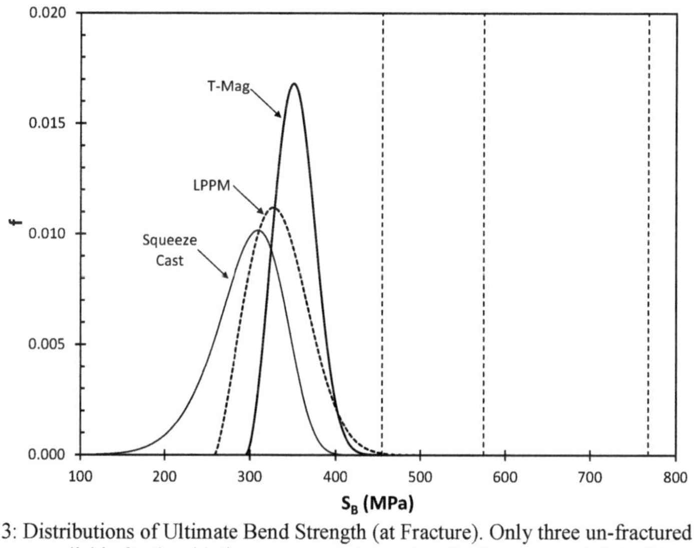

Figure 3: Distributions of Ultimate Bend Strength (at Fracture). Only three un-fractured results were available for the ablation process as denoted as the three vertical dotted lines.

Micrographical analysis showed that most of the processes exhibited a significant scatter of porosity levels, except for squeeze casting, which gave a uniformly low level of residual porosity. Even so, (i) the relatively poor ductility illustrated in the tensile tests (Table 4); (ii) the scatter of strengths in the bend test as revealed by the Weibull analysis (Figure 2) and (iii) the poor radiographic results (Table 5) confirms that the use of pressure for Squeeze Casting may assist a certain amount of pore closure but insufficient to meet radiographic standards and in any case does not appear to assist the welding of pores, particularly as revealed by the Weibull distribution which indicates zero bend strength results appear possible in a sufficiently high population.

Table 5: Radiographic results to E155 (Target was for all castings to meet level 2)

Process	Cracks	Zero Defects	Passing Level 1 %	Passing Level 2 %
Ablation		30	65	90
LPPM		56	67	89
LPPM-EMP	12	0	0	0
Squeeze Cast		0	0	0
T-Mag		85	95	95

Conclusions

All the processes studied in this project were capable of producing a control arm casting in AZ91 magnesium alloy. However, the Ablation Process was found capable of producing AZ91 castings of unprecedented properties and reliability despite the simple fact the process was not fully optimized as evidenced by the high porosity and unrefined microstructure.

Acknowledgements

The authors thank Tessa Lett of Mississippi State University for generous provision and explanation of the details of the bend test studies, and USAMP for kind permission to publish this paper.

References

1. USAMP High Integrity Magnesium Automotive Components (HIMAC) Project; Automotive Metals Division Project AMD 601; Chairmen: R J Osborn (GM), J Zindel (Ford) and S. Logan (Chrysler) 2010 Sept 30 to be made public 2012 Sept 30.
2. L Wang; R. Lett; S D Felicelli; J T Berry; Reliability evaluation of four casting processes for magnesium alloy AZ91. TAFS 2011 paper 11-039 Division 11 pp 9
3. J Campbell; “Complete Casting Handbook” Elsevier, Oxford, England 2011.
4. J. Grassi, J. Campbell, M. Hartlieb and F. Major “Ablation Casting Up-date”, Shape Casting 3rd International Symposium TMS 2008; Editors J Campbell, P N Crepeau and M Tiryakioğlu.
5. M. Tiryakioğlu, J. Campbell: “Weibull Analysis of Mechanical Data for Castings: A Guide to the Interpretation of Probability Plots”, Metall. Mater. Trans. A., v. 41A, p. 3121, 2010.

Shape Casting: 5th International Symposium 2014
Edited by: Murat Tiryakioğlu, John Campbell, and Glenn Byczynski
TMS (The Minerals, Metals & Materials Society), 2014

Manufacturing Cost Modeling of Castings Produced with CRIMSON Process

Binxu Zeng[1], Mark Jolly[1], Konstantinos Salonitis[1]
[1]Cranfield University, Cranfield, Bedfordshire, MK43 0AL, UK

Keywords: CRIMSON, Cost modeling, discrete event simulation

Abstract

The present paper uses the analytical cost estimation technique to estimate the variable cost of the CRIMSON process. In order to assess the performance of the CRIMSON process, the analytically estimated cost is compared with the conventional sand casting process cost. Because of the difficulty of data collection for analytical cost estimation, discrete event simulation method is employed to assist time based data collection. In order to cope with various input of the casting process, a comprehensive cost estimation tool was also developed to assist the cost estimation. It was found that the most expensive variable cost is the raw material cost, which can be 80% of the total variable cost. Furthermore, it is concluded that the CRIMSON process has less variable cost compared with the conventional sand casting process under most of the circumstances.

Introduction

Nowadays, the global market has become increasingly competitive. In the casting foundry industry in particular, such pressure forces the manufacturing organisations to seek continuously for opportunities to improve quality, reliability and productivity with competitive manufacturing cost. For this reason, the novel Constrained Rapid Induction Melting Single mould Up-casting (CRIMSON) process was developed to improve casting quality and reduce energy consumption of the process.

CRIMSON process employs a high powered induction furnace that melts just enough metal for a single mould, and uses a computer controlled up-casting method for optimum filling and solidification. Comparing with the conventional sand casting method, the CRIMSON rapid melting and counter-gravity process has following advantages: short melting time reduces significantly the chance of oxidation and hydrogen absorption. No refining required due to high quality of melt. Simplified running system increases the casting yield as well as the casting quality.

In general CRIMSON process has better quality and energy efficiency compared to the conventional casting process [1]. Despite these advantages, manufacturing cannot consider the process without realistic and accurate cost estimations. Therefore, this paper will present a method to estimate the manufacturing cost of the CRIMSON process. The performance of the process will be assessed by comparison of the cost of the CRIMSON process and the conventional casting process.

Methodology

Cost estimation method

The cost estimation method used in this paper is the analytical cost estimation technique. Using this method, only the production time and hourly rate for the man, machine and resources need to be accurately calculated. The associated manufacturing costs can be calculated by multiplying times and rates together. The main challenge with such an approach is that the accurate and precise collection of production information is time consuming and costly. However, through the exploitation of the advantages of process simulation, it is possible to use the simulation approach to investigate the production time.

Because production time is investigated by using the analytical cost estimation technique, it is easier to assess how cost varies with production quantity. Therefore, the cost estimation method used is the fixed and variable cost method. The fixed cost refers to the capital cost for machines, rental cost for site, development cost for new product and administration cost, etc. These costs are fixed even if there is no product output. Investigating such costs can be relatively simple and easy. Therefore, the cost estimation has been focused mainly on the variable costs of the casting process.

Variable costs refer to process costs such as raw material cost, labor cost, the inventory cost and facility maintenance cost, etc. These costs are a function of the amount of output of the products. Generally, as the output increases, the time for production increases, the raw material requirement increases and the price of material may decrease. The longer the equipment is in operation, the possibility of breakdown increases and the cost of maintenance increases. Therefore, it is easy to see that most variables are inter-connected. Changing one variable may require changes in others. Therefore, to coordinate all the variables and to estimate the cost of the production, a cost estimation model will be developed.

In order to investigate the time of production under different throughput, a process simulation needs to be introduced. The process simulation model is able to represent the variability, interconnectedness and complexity of the system [2]. It is possible to predict the system performance with simulation. Discrete event simulation is ideal for such an analysis, and for this reason WITNESS simulation package was employed, which can be used for simulating full production runs over an arbitrary period [3]. WITNESS allows material flow to be modelled and tracked through each production process, which is a good way to discover problems and suggest improvements.

Assumptions for model development

In order to estimate the cost of the CRIMSON casting process, a complete casting model needs to be developed. Therefore, a casting foundry model was developed for this purpose. A survey of foundries was attempted to investigate parameters such as cycle time, casting yield, Operational material Efficiency[1] (OME) and recovery ratio[2]. The survey results are combined with well thought reasonable estimates to build the foundry model.

Casting weight

Because the CRIMSON furnace can melt up to 30 kg of aluminum, it is sensible to investigate the influence of casting weight at the limit of production performance. Because of metal loss

1 Represents how many raw materials have passed through the process and been shipped.

$$OME = \frac{Mmt-Mws-Mwr}{Mmt} \times 100\% \quad (1)$$

Where Mmt stands for metal melted, Mws stands for metal waste sold and Mwr stands for metal waste recycled in-house (fettling loss, machining loss, and scrap). By using the OME, the efficiency of good casting per unit mass can be calculated.

2 It considers the scrap that is recycled from the process as a percentage of the material put in.

during the casting operation, the actual weight of the casting is less than 30 kg. OME is used to calculate the casting weight and associated melting weight. The OME for the CRIMSON and conventional casting sand processes is 34% and 27%, respectively. Under 34% OME, the CRIMSON process can produce a maximum 10 kg good casting. Therefore, it makes sense to investigate the influence of casting weight on productivity from 1 to 10 kg for both casting processes (Table I).

Table I Depending on the capacity of the CRIMSON furnace, a maximum 10 kg aluminium casting can be produced. From 1 to 10 kg, the corresponding weights for the conventional casting sand process are also shown

Goot Casting Weight(kg)		1	2	3	4	5	6	7	8	9	10
OME (%)	CRIMSON	34	34	34	34	34	34	34	34	34	34
	Conventional	27	27	27	27	27	27	27	27	27	27
Charge Weight (kg)	CRIMSON	2.94	5.88	8.82	11.76	14.71	17.65	20.59	23.53	26.47	29.41
	Conventional	3.70	7.41	11.11	14.81	18.52	22.22	25.93	29.63	33.33	37.04

Cycle time for Melting

Melting is the most time-consuming operation of any casting process. The cycle time determines the maximum number of operations that can be performed each day. For the CRIMSON process, a 300 kW induction furnace is used. Table below is an example of using 40 KW power output. Currently, there is no lid to the furnace and by assuming 50% efficiency, the time for melting is calculated in Table II.

Table II Time required to melt different weights of metal to make one casting under the CRIMSON process

Good casting (kg)	1	2	3	4	5	6	7	8	9	10
Charge Metal (kg)	2.94	5.88	8.82	11.76	14.71	17.65	20.59	23.53	26.47	29.41
Energy (kJ)	3908	7815	11723	15631	19538	23446	27354	31261	35169	39076
Time (Min)	1.62	3.26	4.89	6.51	8.14	9.77	11.40	13.03	14.65	16.28

For the conventional sand casting process, a batch melting method is adopted. In this case, a 500 kg gas furnace is used for the melting operation. If it is assumed that the furnace is completely empty after each cycle of melting, then according to the survey result, two hours will be used to melt such amount of metal.

Table III Time required melting different weights of metal to make one casting under the conventional process

Good Casting (kg)	1	2	3	4	5	6	7	8	9	10
Charge Metal (kg)	3.704	7.407	11.11	14.81	18.52	22.22	25.93	29.63	33.33	37.04
time per casting (min)	0.89	1.78	2.67	3.56	4.44	5.33	6.22	7.11	8.00	8.89
Number of casting can be produced	135.00	67.50	45.00	33.75	27.00	22.50	19.29	16.88	15.00	13.50

Customer requirements

Customer requirement for casting products depends on their demand. As an assumption, following table shows the customer demand from low volume to high volume. These demands will be fed into the WITNESS simulation to work out time for production

Table IV Quantity of the shipment by customer requirement

shipping	100	200	300	400	500	700	1000	1500	2000	2500
unit	3000	4000	5000	7000	10000	15000	20000	25000	35000	50000

Supplier information

Supplier delivers raw materials twice a week to ensure sufficient supply for production

Information flow

1. All communication between customer and supplier is electronic.
2. Production control receives 30-day forecasts and daily orders from the customer.
3. Production control transmits monthly forecasts and weekly orders to supplier.
4. There is a daily schedule released to the shop floor.
5. The combination of push and pull signal is used in the production flow.

Special assumptions for shop floor operation

1. Every month has 30 working days.
2. Foundry operates three shifts daily. Ignoring break times, the working time is 1440 minutes.
3. The conventional melting furnace can supply 500 kg of aluminium every 120 minutes. Depending on the weight, the CRIMSON melting furnace can supply up to 30 kg of aluminium every 16 minutes.
4. One-piece flow manufacturing method[3] is adopted for both casting processes to eliminate the work in process. Therefore, there is no inventory during the casting process. There is no batch production required. Because of the speciality of the casting process, the one-piece flow starts from the shakeout operation at the cold end[4] of the casting process. For the conventional casting process, the raw material enters into production flow every 120 minutes. For the CRIMSON process, the raw material enters into production flow at a rate depending on the melting time providing that the mold production rate is adequate.
5. Based on reasonable assumptions, each operation requires one operator (9 operations for conventional and 6 for the CRIMSON), except for preheating, melting, refining, holding and casting operations in the conventional casting process. Preheating, melting and refining processes can share an operator in the conventional casting process and the holding and casting operation can share one operator as well. Therefore, total 7 operators for conventional casting process.
6. The production period for both casting processes is set as one year
7. Setup time is ignored due to the long period of production.
8. Due to the uncertain shapes of the casting products, there is no point in investigating casting solidification for a particular casting shape. Therefore, an average solidification time 30 minutes is used for all castings.

Development of casting cost estimation model

According to the assumptions, the variable costs modeled are raw material cost, energy cost for melting and labor cost. The total variable cost can be calculated as below:

[3] One-piece flow production is also called the Cellular Manufacturing Method. It aims to move the products through the production process one piece at a time, at a rate determined by customer demand.

[4] Casting can be divided into hot end and cold end processes. The hot end refers to the liquid state of the casting operation, in which all of the operations have to be continuous. The cold end refers to operations dealing with solidified metal. The processes at the cold end can be discrete.

$$C_{Total\ varible} = C_{raw\ material} + C_{energy} + C_{labour} \quad (2)$$

Raw material cost

Because casting production is a continuous process, it is possible to use recycled material for the casting production. The cost of raw material with recycle and reuse can be presented as below:

$$c_{metal} = \frac{total\ shipment}{n} \times [c_{unit\ cost\ of\ scrap} \times (\sum_1^n W_{mn} \times RR + W_{m1}) + c_{unit\ of\ new\ Al} \times \sum_1^n (W_{m1} - W_{mn} \times RR) \quad (3)$$

$$W_{n+1} = W_n \times RR \quad (4)$$

Where c_{metal} is the cost of metal, $c_{unit\ cost\ of\ scrap}$ is the unit value of in-house scrap, $c_{unit\ of\ new\ Al}$ is the unit cost of new aluminum alloy, W_{mn} is the weight of the metal melted at a particular cycle, RR is the recovery ratio for the casting process; it represents how much metal can be recycled in one cycle, n is the number of operation cycles; it represents how many cycles before the recycled metal is fully replaced by new metal Similarly, the sand cost can be presented as below:

$$c_{sand} = \frac{total\ shipment}{n} \times [c_{unit\ cost\ of\ reclamation} \times (\sum_1^n W_{sn} \times RR + W_{s1}) + c_{unit\ of\ sand} \times \sum_1^n (W_{s1} - W_{sn} \times RR) \quad (5)$$

Energy cost

Regarding the melting, the conventional sand casting process uses a gas furnace to melt the aluminum to its melting point, after which it is transferred to a holding furnace to be superheated to 700–750 °C. By contrast, the CRIMSON process only uses electricity to melt metal up to750 °C.

$$C_{CRIMSON} = \{[c_{ps} \times (t_{melt} - t_{room}) + c_{pl} \times (t_{super} - t_{melt}) + E_f] \times m_{melt}\} \div e \times C_{unit\ energy} \quad (6)$$

$$C_{conventional} = 2.2 \times \{[c_{ps} \times (t_{melt} - t_{room}) + c_{pl} \times (t_{super} - t_{melt}) + E_f] \times m_{melt} \div e \times C_{unit\ energy}\} \quad (7)$$

Where $C_{CRIMSON}$ is the energy cost for the CRIMSON process, $C_{conventional}$ is the energy cost for the conventional process, c_{ps} is the specific heat of metal in the solid phase, c_{pl} is the specific heat of metal in the liquid phase, t_{room} is room temperature, t_{melt} is melting temperature, t_{super} is the superheated temperature, m_{melt} is the mass of molten metal, e is the melting efficiency , $C_{unit\ energy}$ is the unit energy cost (electricity or gas), $\boldsymbol{E_f}$ is the heat of fusion, 389 kJ/kg (°C)

Labor cost

Labor cost is a function of the equipment, the labor and the time required to produce a certain amount of products. As stated previously in the assumptions, the CRIMSON process has six operators and the conventional process has seven operators. Assuming the national minimum wage is used, the labor costs will be:

$$C_{labour} = \sum_1^n (N_n \times Wage_n) \times t \quad (8)$$

Where

C_{labour} is the cost of the labour, $Wage_n$ is the wage for particular job *n*, N_n is the number of the operators for job *n*, t is calculated by process simulation

Development of the cost calculation tool

Figure 1 shows the relation between the variable cost and the variables. Most of the variables are connected with more than one variable cost. Any change of a parameter might cause a different cost estimation result. Moreover, not only might the total cost of the production change but something like the time distribution and cost contribution might change as well. Therefore, in order to gain a complete view of the cost model, it was necessary to develop a tool (implemented on a spreadsheet) that can integrate all the variables

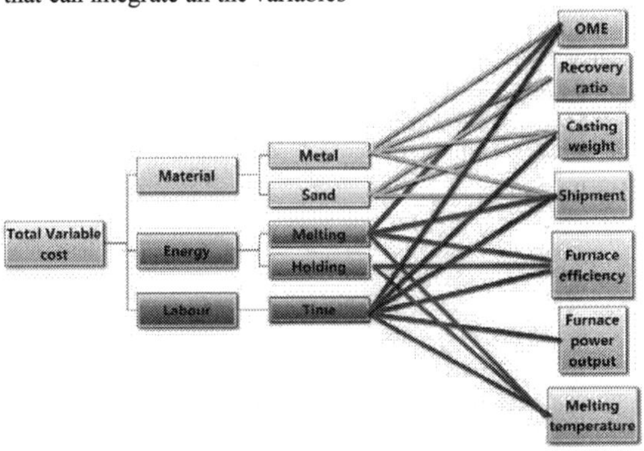

Figure 1 Schematic of the relation between variable cost and variables

Results and Discussion

In this section, different combinations of the variables are fed into the discrete event simulation model and the cost estimation tool. The manufacturing cost of the same casting made with the CRIMSON and conventional sand casting process are investigated.

Table V the variable investigated

type of influence	variables
OME	low / high
sand recovery method	Secondary/thermal
material value	low/middle/high
casting size	1kg --10kg
batch size	100--30000
furnace power output (kW)	40--300

Table V presents all the scenarios that were investigated. A box plot has been used to illustrate the distribution of all the cost comparison of the CRIMSON and conventional sand casting processes.

The CRIMSON costs are divided by the conventional casting costs for the box plot calculation. The CRIMSON process is considered an expensive choice when the result is greater than 1, whereas the conventional casting process is more expensive when the ratio is less than 1. From Figure 2, it is easy to see that most data lie to the left-hand side of the base line. This means that the conventional casting process has higher total variable costs in most circumstances. However, the CRIMSON process can be expensive when a low furnace power output is adopted, because low power output prolongs the production time, which increases the labour cost.

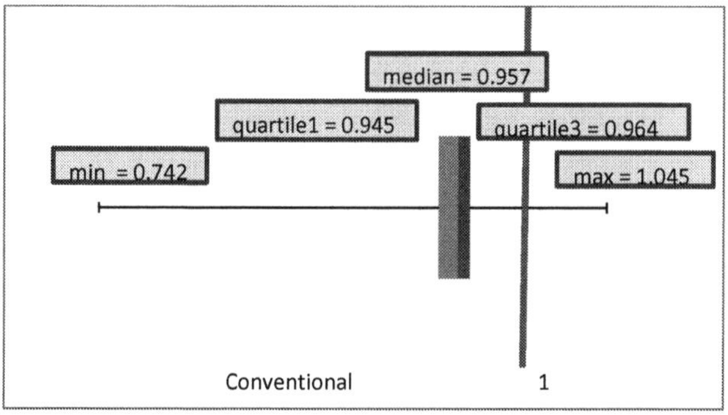

Figure 2 Comparison of the CRIMSON and conventional sand casting processes. The red line is the base line of the comparison. Left-hand side of the red line means that conventional casting is expensive, the right-hand side of the red line means that the CRIMSON process is expensive

Based on the results, the average cost contribution for each variable cost can also be plotted (Figure 3). Irrespective of whether the CRIMSON or the conventional sand casting process is used, the energy costs only contribute about 1% of the total variable costs. In contrast, the metal costs contribute the greatest effect on the variable costs. Assuming a worst case scenario in which 1 kg of metal is heated to its melting point by gas furnace. The energy consumption is 14444 kJ / kg (50 % energy efficiency). As the energy price shows on Europe's Energy Portal website [4], natural gas in the UK currently costs £0.024 / kWh. Therefore, the conventional casting process costs £0.1 to melt metal. After melting, the metal will be transferred to holding furnace. As the literature shows, the holding operation costs on average 1.2 times more than the melting process [5]. The holding cost is £0.12. Therefore, the total cost is £0.22 / kg. On the other hand, the aluminium alloy is about £1.6 / kg to £2.0 / kg, approaching 1% of the material cost.

In addition to the average results of the variable costs, the figure also indicates the distribution of each variable cost. The sand cost has the widest distribution because of the sand price. It can be as low as 10% when silica sand is used and as high as 70% when zircon sand is used. Because the sand has such a wide distribution, it affects the overall contribution of the metal cost. In contrast to the sand cost, the metal contribution is high when the sand cost is low and the metal contribution is low when the sand cost is high. Therefore, the metal has a similar distribution to the cost contribution.

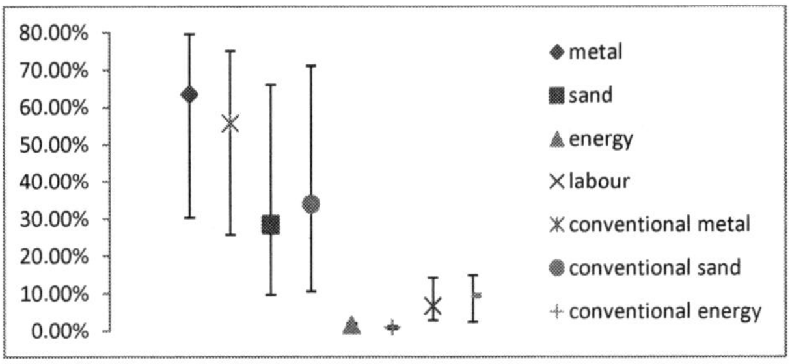

Figure 3 Average cost contribution and distribution of each variable cost

Conclusions

The manufacturing variable cost of the CRIMSON process was investigated in this paper. In order to serve this purpose, a comprehensive foundry model was developed and analytical cost estimation method was used. Because of the diverse process variables driving the casting process, a cost estimation spreadsheet was developed to assess the manufacturing cost. By using the spreadsheet, various input combinations were investigated. From the spreadsheet result, both conventional and CRIMSON processes indicate that the raw metal is major cost contributor in the variable cost. Its contribution can vary from 30% to 80% of the total variable cost. Followed by raw metal cost, the sand cost is the second cost contributor in variable cost, which can contribute 10% to 70%. However, the energy cost only contributes 1 % of the variable cost. Because the CRIMSON has higher material utilization compared to the conventional casting process, it uses less material to make casting products. This is the main reason why the CRIMSON process has less manufacturing cost compared with conventional casting process under most circumstances. Such results indicate that cost reduction should focus on the utilization of raw material. It is also a clear reminder that the primary goal of energy saving is not necessary cost reduction, but it is also related to environmental impact reduction and sustainability, to indicate few. The natural ability of CRIMSON, as a counter-gravity filling process, also enjoys the massive benefit of the superb quality of the product, and the hugely important benefit of the reduction of scrap, with clear advantage of avoiding the necessity to re-make the product more than once to achieve a saleable quality.

Reference

1. B. Zeng, M. Jolly, and X. Dai, "Designing Novel CRIMSON running system through numerical simulation method for the purpose of reducing the energy content of aluminium investment casting," *TMS Annual meeting 2013*, pp. 39 - 46.
2. S. Robinson, *Simulation: The proactive of Model Development and Use* (Chichester : John Wiley & Sons Ltd, 2004).
3. P.L. Markt, and M.H. Mayer, "Witness simulation software a flexible suite of simulation tools", *Simulation Conference*, (1997), 711-717
4. Portal, Europe's Energy. Fuel Prices. Europe's Energy Portal. [Online] nd. http://www.energy.eu/.
5. Department of the Environment, Transport and the Regions, "Energy Consumption Guide 38, Non-Ferrous Foundries," Best Practice Programme, 1997.

Shape Casting: 5th International Symposium 2014
Edited by: Murat Tiryakioğlu, John Campbell, and Glenn Byczynski
TMS (The Minerals, Metals & Materials Society), 2014

ON WEIBULL MIXTURES IN MECHANICAL PROPERTIES OF CASTINGS

Murat Tiryakioğlu

University of North Florida, School of Engineering, 1 UNF Drive, Jacksonville, FL 32224

Keywords: Fatigue life, defect distributions, failure mechanisms

Abstract

A review of Weibull mixtures in the mechanical properties of castings is provided. Two distinct scenarios which result in Weibull mixtures in the mechanical properties of castings are introduced: (i) multiple defect distributions that lead to failure by the same mechanism are present in the casting, and (ii) there are multiple failure mechanisms in effect regardless of the type and size of defects. These two scenarios are discussed in detail with examples from the literature.

Introduction

Statistical models for fracture are all built on the concept by Griffith [1] that the difference between the ideal and actual performances can be attributed to the presence of defects (flaws) weakening the structure and resulting in premature failure. Hence the statistical distributions of fracture-related mechanical properties, i.e., fracture stress, elongation, fracture toughness, fatigue life, etc., can all be linked to the underlying defect distribution. Statistically, this implies that the worst (largest) defect is the one that determines the fracture-related mechanical properties [2], which constitutes the "weakest link", based on the theory developed by Pierce [3]. The "weakest link" theory applies in situations that are analogous to the failure of a chain when its weakest link has failed [4]. Based on the "weakest link" theory, Weibull [5] introduced an empirical distribution, for which the cumulative probability function is expressed as:

$$P = 1 - \mathbf{exp}\left[-\left(\frac{\sigma - \sigma_T}{\sigma_0}\right)^m\right] \qquad (1)$$

where P is the probability of failure at a given stress (strain, fatigue life, etc.), σ, or lower. The threshold value, σ_T, is the value below which no specimen is expected to fail. The term, σ_0, is the scale parameter, and *m* is the shape parameter, alternatively referred to as the Weibull modulus.

One of the most commonly used methods of presenting the Weibull fits to data is the Weibull probability plot. After rearranging, Equation 1 can be written as

$$\ln[-\ln(1-P)] = m\ln(\sigma - \sigma_T) - m\ln(\sigma_0) \qquad (2)$$

Note that Equation 2 has a linear form when the left-hand side of the Equation is plotted versus $\ln(\sigma\text{-}\sigma_T)$ with a slope of m and an intercept of -m $\ln(\sigma_0)$. Alternatively, the Weibull probability plot can be obtained when the left-hand side of the equation is plotted versus $\ln(\sigma)$. This method of presentation gives a straight line relationship only when $\sigma_T=0$. This is demonstrated in Figure 1, in which Weibull distributions for three datasets for elongation (e_F) of sand cast 319 [6], permanent mold- and ablation cast A356 [7] are plotted. Note that the trend of the curve at low values of ln(σ) is influenced by the value of the threshold, σ_T; a positive threshold results in higher slope at low values of ln(σ). When the threshold is negative however, the slope decreases with decreasing ln(σ). For mechanical properties, a negative threshold is, of course, meaningless. A negative threshold is indicative of a Weibull mixture [8]. The treatment of Weibull mixtures is complicated and will be discussed in detail below.

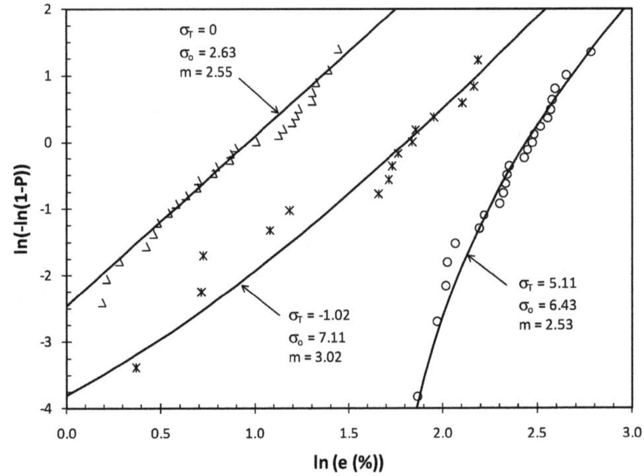

Figure 1. The Weibull probability plots for the elongation of three aluminum casting datasets: sand cast 319 (Δ) as well as A356 cast in permanent molds (*) and by the ablation process (O) [8].

Weibull Mixture Models

Evidence for the presence of a Weibull mixture is that the slope in a probability plot (i) decreases significantly as observed for the top-filled castings in Figure 3, or (ii) approaches zero and then increases again [9], like an inflection point. Coincidentally, it was Weibull [10] who analyzed the tensile strength of malleable cast iron reported by Pearson [11] after noticing that 15 of the 75 castings were poured at a different location. For the mixture of distributions, Weibull [12] suggested the following form of the probability P of failure:

$$P = pP_1 + (1-p)P_2 \tag{3}$$

where p is fraction of the Weibull distribution number 1 in the mixture and subscripts 1 and 2 refer to the two Weibull distributions. The author previously [8] observed that Weibull mixtures are not uncommon in castings. However, there are two scenarios in which a Weibull mixture can be the result: (i) when multiple types of defects are present in castings, and (ii) when there are two failure mechanisms. These instances are discussed below.

Scenario 1: Multiple Defect Distributions

Prior studies [13,14,15,16,] have shown that there are multiple defect distributions in castings, including bifilms (both thick, coarse "old" and thin and fine "young" films) and pores. The treatment of the link between defect size distributions and the resultant Weibull mixture was discussed in detail by Johnson [17]. Equation 3 applies when defect distributions are mutually exclusive, i.e., when one type of defects always "wins" over the other type. When there is some competition between the defect types, the mixture methods are as follows [18];

$$P = 1-(1-P_1)(1-P_2) \tag{4}$$

$$P = (1-p)P_1 + p(1-(1-P_1)(1-P_2)) \tag{5}$$

Equations 4 and 5 are for concurrent and partially-concurrent defect distributions, respectively. According to Johnson, to determine which kind of Weibull mixture is appropriate, (i) the types of defects causing the premature fracture have to be identified on the fracture surfaces, and (ii) it should be determined whether the two defect distributions are competing with each other. From a process viewpoint, most Weibull plots for mechanical data for castings can be expected to reveal two populations of defects: the original rather fine scattering of defects remaining in suspension in the original poured liquid (prior damage), and the large new bifilms (new damage) that would have been produced during the pour if the filling system was not designed properly. Because the filling systems of many castings are poorly designed, the Weibull mixtures can be expected to be the norm, rather than the exception. A single Weibull distribution can normally be expected in castings whose filling systems were designed so that no new damage to liquid metal takes place during filling. Therefore one should take certain process information, such as melt quality and the design of the filling system into account during the Weibull analysis of casting data.

An example for Scenario 1 was provided by Eisaabadi et al. [19] who investigated the effect of stirring (and therefore melt quality getting progressively poorer] and the presence (or absence) of a foam filter in the filling system on the tensile properties of Al-7%Si-Mg alloy castings. They had three casting conditions: (A) unstirred, filter, (B) stirred, no filter, and (C) stirred, filter. The resultant elongation values are presented in Figure 2. Note that the linear trends for conditions A and B in their Weibull probability plots indicate two-parameter Weibull distributions. Data from Condition C have two distinct regions with different slopes. Analysis of fracture surfaces via scanning electron microscopy showed that all castings indicated with full triangles failed due to the presence of coarse, "old" oxides, which managed to go through the filter. Those indicated with hollow triangles, however, had "young" oxides on their fracture surfaces and there were no "old" oxides. Hence, the filter was effective for some castings, although the initial melt quality was poor due to stirring. Therefore, the Weibull mixture is due to different defect distributions. It can be argued that the two types of oxide bifilms exist concurrently in the castings that failed at lower elongation (full triangles), but only "young" oxide bifilms are in the casting with higher elongations. Hence the partially-concurrent Weibull mixture model presented in Equation 5 should be applicable. However, there is a significant gap between the lower and upper distributions for elongation. When there is an "old" bifilm in the specimen, there is no competition and "old" bifilm "wins" all the time over "young" oxide bifilms. Therefore the mutually exclusive version (Equation 3) can be used when data from two distributions do not overlap although the defects are partially concurrent in the specimens.

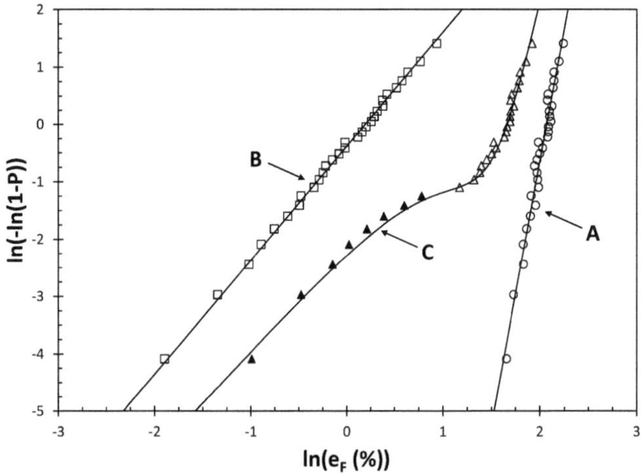

Figure 2. The Weibull probability plots for three conditions of Al-7%Si-Mg alloy castings [19].

Scenario 2: Multiple Failure Mechanisms

There are several studies [13,14,16,20] in which the type of fatigue crack initiators were identified for all specimens. Wang et al. [20] investigated the fatigue life (N_f) variability in A356 alloy castings and attributed fatigue failures, pores, oxides and slip planes. Wang et al. presented separate Weibull probability plots due to these three fatigue crack initiators. The data of Wang et al. are presented in Figure 3.a, along with three-parameter Weibull fits (Wang et al. assume σ_T = 0 although trends in all three datasets strongly suggest $\sigma_T > 0$). It should be noted that pores and oxides, as reported by Wang et al., were connected to the surface of fatigue specimens. Moreover, slip planes occurred when the fatigue initiating defects were not connected to the surface. These internal defects were surrounded by the slip planes (alternatively referred to as facets) and therefore the fatigue initiators were still defects, such as oxides, which were subsurface. When all data are plotted on the same Weibull probability plot, Figure 3.b., it becomes evident that fatigue life data with pores and oxides on the surface come from the same Weibull distribution. Fatigue life of specimens with internal fatigue crack initiators, however, has a distinctly different Weibull distribution. Because there is no overlap between the two distributions, Equation 3 was used for the Weibull mixture.

That fatigue life has different distributions when fatigue initiating defects are on the surface or subsurface is noteworthy. Reanalyses of data from the literature [13,21] by the author [22,23] for the fatigue life of A206 and A356 cast aluminium alloys yielded similar results, which are presented in Figure 4. Note that fatigue life is consistently higher when fatigue initiators are subsurface and there is no overlap between the data from the lower and upper distributions. Subsurface crack initiation in fatigue has received considerable attention recently [24,25,26,27,28,29]. The Wöhler curve for the subsurface defects is different from the one for those on the surface, as shown schematically in Figure 5. The location of the S-N curve for internal defects will depend on the size of the internal defect. [30]. Separate S-N curves for surface and subsurface defects were found [28] for a cast Al-10%Si-4%Cu-0.6%Mg alloy when specimens were subjected to a surface treatment. In several studies, cracks were observed to grow from structural defects at or shortly after the first stress cycle [31,32,33]. However in cases where there is no defect on the surface large enough to initiate a propagating crack, fatigue crack initiation from a subsurface defect takes significantly longer, resulting in increased fatigue life. When results in a fatigue experiment may come from castings, some of which have defects on the surface while the rest have subsurface defects, we can expect to have two fatigue life distributions, as presented in Figures 3.b and 4.

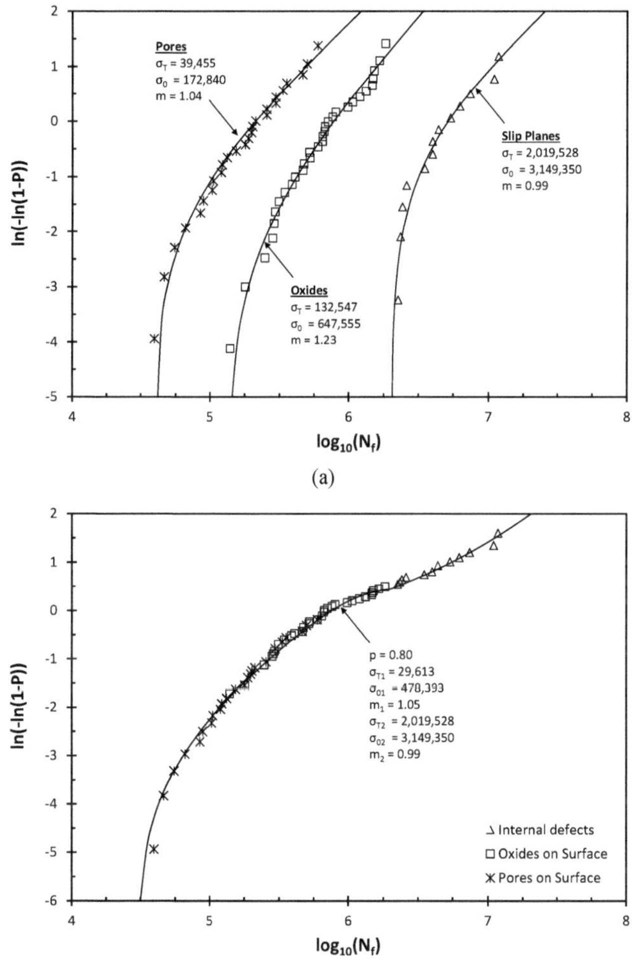

(b)

Figure 3. Reanalysis of data by Wang et al. [20]. Weibull probability plots for fatigue life (a) for three different fatigue crack initiators based on original classifications by Wang et al., and (b) after all data are combined and slip planes are interpreted as internal defects.

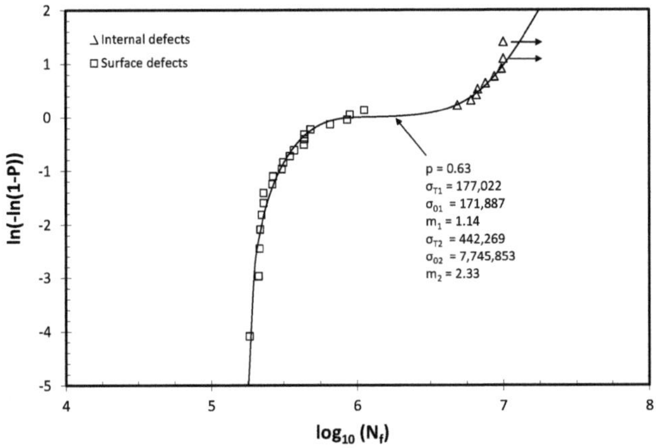

(a)

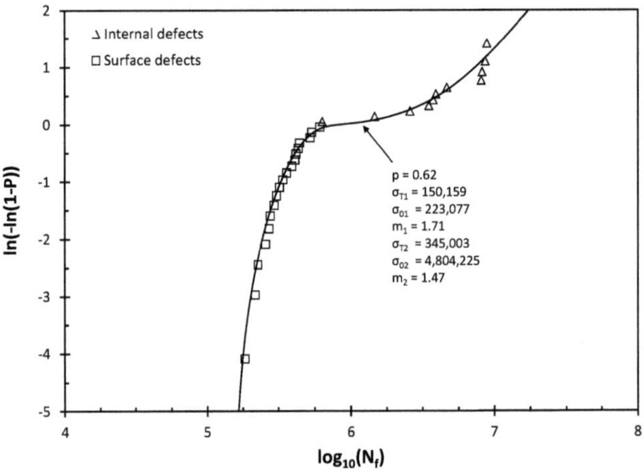

(b)

Figure 4. Weibull probability plots for Weibull mixtures in fatigue life data of (a) A206-T71 castings [22] and (b) A356 alloy castings [23].

It has been the author's experience that Weibull mixtures (i) in tensile data usually are due to two distinct defect distributions and are a result of different failure mechanisms in fatigue, and (ii) can be represented by using Equation 3 only. The author has not analyzed a dataset for which Equation 4 or 5 was necessary.

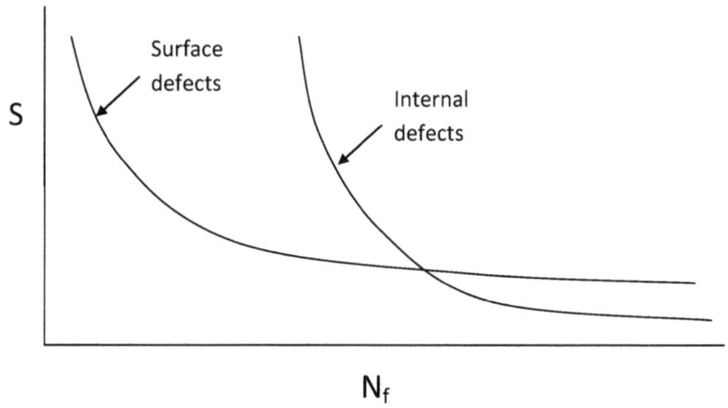

Figure 5. Schematic illustration of S-N curves of fatigue tested specimens failed from cracks initiated at surface and internal (subsurface) defects.

Conclusions

- Weibull mixtures are common in fracture data of castings.
- There are two scenarios which lead to Weibull mixtures: (i) presence of two distinct defect distributions (applicable usually in tensile data) but the same failure mechanism, and (ii) different failure mechanisms despite the presence of different types of defects (applicable in fatigue data).
- The mutually exclusive version of the Weibull mixtures, as originally suggested by Weibull, can be applied to all Weibull distributions where there is no overlap between the lower and upper distributions, which, to the author's experience, has always been the case.

Acknowledgements

The author would like to thank Dr. John Campbell for his most helpful comments.

References

1. A.A. Griffith: *Phil. Trans. Royal Soc. London. Ser. A*, 221 (1921) 163-198.
2. B. Epstein: *J. American Stat. Assoc.* 43 (1948) 403-412.
3. F. T. Pierce: *J. Textile Inst.* 17 (1926) T355-T368.
4. T.T. Shih: *Eng. Frac. Mech.*, 13 (1980) 257-271.
5. W. Weibull: *J. Applied Mechanics*, 13 (1951) 293-297.
6. H. Zahedi, M. Emamy, A. Razaghian, M. Mahta, J. Campbell, M. Tiryakioğlu: *Metall. Mater. Trans. A.*, 2007, vol. 38A, pp. 659-670.
7. M. Tiryakioğlu: unpublished work, Alotech, 2006.
8. M. Tiryakioğlu, J. Campbell: *Metall. Mater. Trans. A*, 41 (2010) 3121-3129.
9. S. Jiang, D. Keçecioğlu, *IEEE Trans on Reliability*, 1992, vol. 41, pp. 241-247.

10. W. Weibull: A statistical theory of the strength of materials. Proc. The Royal- Swedish Institute for Engineering Research. Nr. 151, 1939.

11. E. S. Pearson: "The Application of Statistical Methods to Industrial Standardisation and Quality Control", *British Standard No. 600*—1935

12. W.Weibull, The phenomenon of rupture in solids, Royal Swedish Institute of Engineering Research (Ingenioersvetenskaps Akad. Handl.), Stockholm, Vol. 153, 1–55, 1939.

13. C. Nyahumwa, N. R. Green, J. Campbell; *Metall. Mater. Trans. A*, 32A (2001) 349–358.

14. Q. G. Wang, C. Davidson, J. Griffiths, P. Crepeau; *Metall. Mater. Trans. B*, 37B (2006) 887–895.

15. C. Nyahumwa, N. R. Green, J. Campbell, *AFS Trans.*, 106 (1998) 215-223.

16. B. Zhang, D. R. Poirier, W. Chen; *Metall. Mater. Trans. A*, 30 (1999) 2659-2666.

17. C. A. Johnson: *Frac. Mech. Ceram.* 5 (1983) 365-386.

18. C. D. Tarum: SAE Technical Paper, 1999-01-0055, 1999.

19. G. Eisaabadi B. P.Davami, S.K.Kim, M.Tiryakioğlu: *Materials Science and Engineering A*, 579 (2013) 64-70.

20. Q.G. Wang, D. Apelian, D.A. Lados: *J. Light Metals*, 1 (2001) 73-84.

21. J. T. Staley, Jr., M. Tiryakioğlu, J. Campbell: *Mater. Sci. Eng. A*, 465 (2007) 136-145.

22. M. Tiryakioğlu, in this volume.

23. M. Tiryakioğlu, J. Campbell, C. Nyahumwa: *Metall. Mater. Trans. B*, 42 (2011) 1098-1103.

24. O. Umezawa, K. Nagai and K. Ishikawa: *Tetsu to Hagane*, 75 (1989) 159-166.

25. H. Mughrabi: *Fatigue and Fracture of Eng. Mater. Structures*, 25 (2002) 755-764.

26. K. Sadananda, A.K. Vasudevan, N. Phan: *Intl. J. Fatigue*, 29 (2007) 2060–2071.

27. C. Przybyla, R. Prasannavenkatesan, N. Salajegheh, D. L. McDowell: *Intl. J. Fatigue*, 32 (2010) 512–525.

28. Y. Nakamura, T. Sakai, H. Hirano, K.S. Ravi Chandran: *Intl. J. Fatigue*, 32 (2010) 621–626

29. C. Bathias: *Fatigue Fract Engng Mater Struct.*, 22 (1999) 559–565.

30. O. Umezawa, K. Nagai: *ISIJ International*, 37 (1997) 1170-1179.

31. B. Skallerud, T. Iveland, G. Härkegård: *Eng. Fracture Mech.*, 44 (1993) 857-874.

32. S.A. Barter, L. Molent, N. Goldsmith, R. Jones: *J. Eng. Failure Analysis*, 12 (2005) 99-128.

33. B.R. Crawford, C. Loader, A.R. Ward, C. Urbani, M.R. Bache, S.H. Spence, D.G. Hay, W.J. Evans, G. Clark, A.J. Stonham: *Fatigue Fract. of Eng. Mater. Struc.*, 28 (2005) 795-808.

Shape Casting: 5th International Symposium 2014
Edited by: Murat Tiryakioğlu, John Campbell, and Glenn Byczynski
TMS (The Minerals, Metals & Materials Society), 2014

MICROSTRUCTURE AND MECHANICAL PROPERTIES OF AUTOMOTIVE COMPONENTS DIE CAST WITH SECONDARY ALUMINUM ALLOYS BY SEED SEMI-SOLID PROCESS

Giulio Timelli[1], Stefano Capuzzi[1], Stefano Ferraro[1], Alberto Fabrizi[1], Leonardo Capra[2]

[1]University of Padova – Dep. of Management and Engineering;
Strad. San Nicola 3; Vicenza, 36100, Italy
[2]Raffineria Metalli Capra SpA;
Via Quinzano 96; Castel Mella (BS), 25030, Italy

Keywords: Aluminum alloys; Semi-solid; SEED; Microstructure; Mechanical properties

Abstract

The microstructure and mechanical properties of an automotive component die-cast by Swirled Enthalpy Equilibration Device process are analyzed. Two secondary aluminum alloys have been used, AlSi7Cu3Mg and AlSi9Cu3(Fe) alloys. The castings have been investigated by means of computer tomography to evaluate the concentration and distribution of casting defects, especially in the regions of thick wall thickness. Microstructural investigation has been carried out by means of optical microscope and image analysis technique. The results reveal a not uniform distribution of primary α-Al globules throughout the casting, with a solid fraction ranging from 48 to 58%, and eutectic segregation phenomena. The castings have been mapped in terms of mechanical properties and quality index. The variation of the mechanical properties reflects the presence of local casting defects and alloying segregation.

Introduction

The use of Al-Si foundry alloys is continuously growing in the automotive sector being driven by the need of decreasing vehicle's weight. At the same time, the development of safety and structural components with increasing quality at low cost and reduced cycle time is a key requirement for automotive companies as a consequence of enhanced competitiveness and complex market. Several efforts have been paid to successfully develop high integrity die-casting processes to stretch the capabilities of conventional high-pressure die-casting (HPDC) while preserving short cycle times and providing dimensional stability and other beneficial characteristics of castings. The objectives, considered to extend the capabilities of HPDC, involved the elimination or reduction of casting defects, such as gas entrapment and solidification shrinkage, and the alteration of as-cast microstructure [1]. Among these technologies, semi-solid metal (SSM) processing has established its great potential as commercially viable technology [2].
The most of scientific works and industrial applications on SSM processing has been directed to hypoeutectic primary Al-7Si alloys [2], such as A356 and 357, with low impurity content. Even if the selection of the alloy composition is restricted to the region of practical interest for semi-solid metalworking (fraction solid ~0.5), it is however important to find out a compromise between mechanical performance and reasonable material cost.
Secondary Al foundry alloys, which are the most used Al die-casting alloys, show lower cost than primary ones, but present small amount of impurities and trace elements, with wider composition tolerance limits compared to primary alloys. These characteristics depend on

recycling process and have generally negative effects on the mechanical properties and final quality of conventional die-cast components [3].

The aim of this work is to study the SSM processing of secondary Al casting alloys over the microstructural and mechanical properties. Two different hypoeutectic Al-Si-Cu alloys, such as AlSi7Cu3Mg and AlSi9Cu3(Fe), were investigated. On the other hand, the Swirl Enthalpy Equilibration Device (SEED) process, recently developed by Rio Tinto Alcan [4], was applied to produce a lower arm automobile suspension used as demonstrator.

Experimental procedure

Secondary AlSi7Cu3Mg and AlSi9Cu3(Fe) (EN AB-46300 and EN AB-46000 respectively, equivalent to the US designation A320 and A380) foundry alloys were supplied by Raffineria Metalli Capra as commercial ingots and separately used during the experimental trials. The ingots were pre-heated at 150°C in a tower furnace with a capacity of 800 kg, and subsequently brought to the melting temperature at 670°C. The molten metal was then transferred into an electric resistance furnace and hold at 640°C. The chemical composition of the alloys, measured on separately poured samples, is shown in Table I.

Table I. Chemical composition of the experimental alloys (wt%).

Alloy	Si	Fe	Cu	Mg	Mn	Zn	Cr	Ni	Ti	Al
AlSi7Cu3Mg	6.521	0.718	2.921	0.416	0.407	0.729	0.083	0.053	0.135	bal.
AlSi9Cu3(Fe)	8.823	0.695	2.137	0.318	0.200	0.904	0.051	0.065	0.043	bal.

The preparation of the slurry was carried out by means of the rheocasting SEED process. A detailed description of the semi-solid forming and process parameters used in the SEED process is given elsewhere [5]. Briefly, the melt was carefully poured into a boron-nitride coated steel vessel (diameter 9 mm, height 350 mm and wall thickness of 2.5 mm) preheated at 130°C. Further, the whole system was swirled for 115 seconds at a rate of 150 rpm and then transferred to the shoot sleeve of a cold chamber die-casting machine and injected into the die cavity. In this work, all tests were performed without drainage of the excess eutectic liquid at the end of the swirling period.

The plunger velocity of the HPDC machine was lower than 1 ms^{-1} for the filling phase; a pressure of about 115 MPa was applied once the die cavity was full to guarantee high-integrity die-castings. By means of a dynamic shot control system in the HPDC machine, every casting was documented with its shot profile, to monitor the final quality and repeatability.

The analysed component is a lower arm automobile suspension (Fig. 1). The weight of the Al alloy casting was 4.2 kg, including the runner, gating and overflow system, while the weight of the final component is 1.3 kg. Typical cycle time, from die cavity filling to solidified casting extraction, was 65 seconds. More than 130 castings were produced with each alloy.

In order to assure an acceptable level of soundness and to localise the porosity distribution, the castings were analysed by means of a Wälischmiller Rayscan 200 computer tomography (CT) equipped with a microfocus X-ray.

Samples for microstructural investigations were drawn from several regions of the castings and mechanically prepared according to standard metallographic practice. Microstructural analysis was carried out by means of an optical microscope equipped with image analyzer software for quantitative investigations.

Tensile specimens with rectangular cross section were machined from different locations of the castings in order to map the mechanical behavior of the component. The specimens had a total length of 90 mm and a thickness of 4 mm, a gauge length and width of 30 and 10 mm respectively. The tensile tests were done with a crosshead speed of the tensile testing machine of 2 mm/min. Experimental data were collected and processed to provide yield stress (YS, actually 0.2% proof stress), ultimate tensile strength (UTS), elongation to fracture (s_f) and quality index Q, which takes into account the UTS and s_f [6].

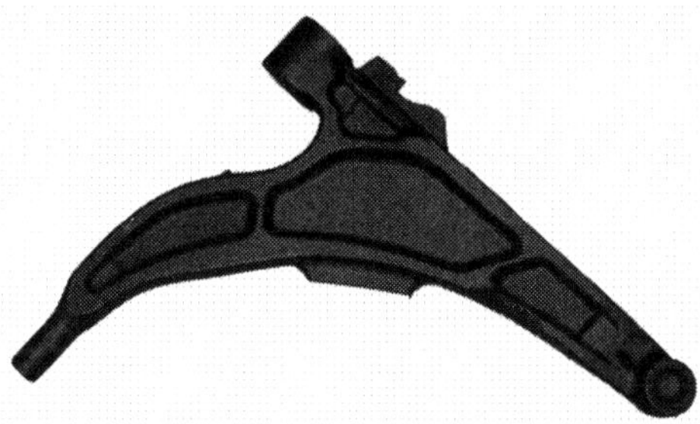

Figure 1. Lower arm automobile suspension produced by rheocasting SEED process.

Results and discussion

The variation of solid fraction as function of the temperature was preliminary evaluated for both the experimental alloys by means of Fourier thermal analysis (Fig. 2). A detailed description of the experimental procedure is given elsewhere [7]. The AlSi7Cu3Mg alloy presents higher *liquidus* temperature than the AlSi9Cu3(Fe) and greater amount of primary α-Al phase, which mainly controls the *apparent viscosity* of the slurry during the filling stage. The AlSi7Cu3Mg alloy exhibits a lower eutectic temperature and greater amount of Cu-bearing phases, which precipitate at ~495°C. In this alloy the high Mg and Cu content depresses the eutectic temperature and increases the fraction of brittle intermetallic compounds [8].

The CT scans evidenced the presence of casting defects in the regions with thicker wall thickness, independently of the casting alloy (Fig. 3). Detailed investigations revealed how these defects are mainly ascribed to coarse solidification shrinkage, which is more referred to casting design and process parameters than to the experimental Al alloys used in the present work. Fig. 3 shows how porosity are located at the end attaches of the suspension arm, i.e. the connection regions with the space frame and the steering knuckle, respectively.

Furthermore, local variations of material density, which appear as *flow lines*, were revealed by CT investigations in the regions of thin wall thickness. Flow lines, indicated in Fig. 3 by dashed arrow in the central part of the component, can be produced mainly during the filling of the die cavity. Liquid segregation, separating the solid phase from the liquid phase, can eventually occur during the process. Generally, a very fast injection velocity causes the influx of oxide skin, and a very slow injection velocity causes liquid segregation due to the velocity difference between the liquid and solid phases [9]. In order to prevent liquid segregation, it is a key requirement to consider injection velocity, gate section and number, location and size of overflows during die design.

Fig. 4 shows the microstructure at different regions of AlSi7Cu3Mg alloy casting. The material consists on largely of globular/degenerated dendrites of primary α-Al dispersed in the residual Al-Si eutectic and secondary intermetallic phases. The microstructure is not uniform throughout the casting; regions enriched of primary α-Al phase contrast zones with higher eutectic content. It is proposed that eutectic segregation forms during the filling stage of the die cavity through exudation mechanism and liquid separation from the slurry.

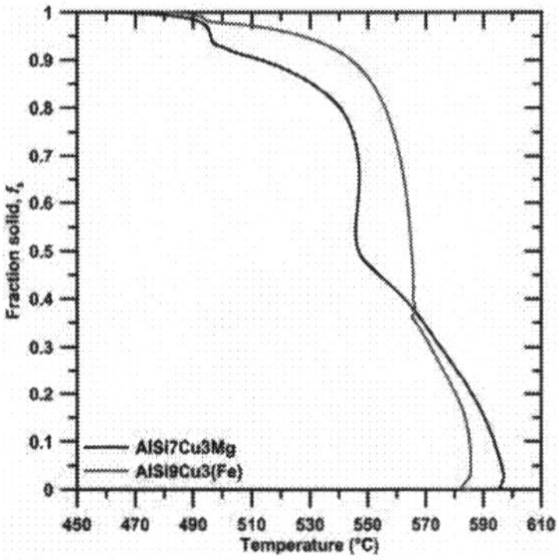

Figure 2. Evolution of solid fraction as a function of the temperature for the analyzed alloys.

Generally, exudation occurs when the solidifying surface region has sufficient strength to pull away from the die wall drawing enriched liquid to the casting surface [10]. On the other hand, due to the local variation of the filling speed, liquid separation occurs, for instance, when the thickness of the casting varies or where obstacles impede the flow of the melt [9]. By varying the filling speed, the rapid changes in the direction of the melt flow cause the front of the molten metal contacts the wall of the die and the speed of flow decreases. Consequently, liquid segregation may take place.

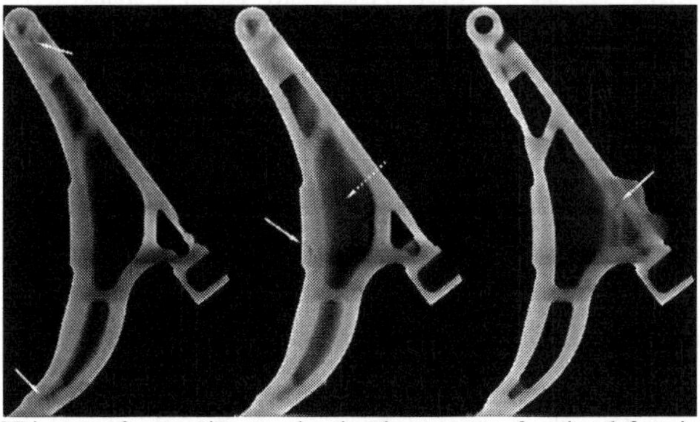

Figure 3. CT images of suspension arm showing the presence of casting defects indicated by arrows. The pictures refer to AlSiCu3Mg component. Dashed arrow evidences the flow lines of the material.

Fig. 4 highlights how the region with the greatest eutectic segregation is located in front of the ingate, where the CT revealed density variations in the form of flow lines, as reported in Fig. 3. In this zone, as well as where the content of primary α-Al globules is reduced, secondary α-Al dendrites form, nucleating on existing α-Al particles or within the residual liquid (Fig. 5).

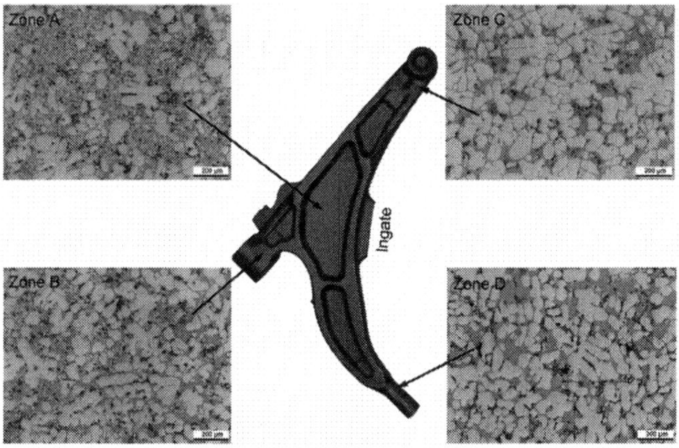

Figure 4. Microstructure of the AlSi7Cu3Mg alloy at different locations of the casting.

Cu-bearing particles were also detected and identified as θ-Al_2Cu intermetallics. Al_2Cu phase occurs in the form of both pockets of fine eutectic (Al + Al_2Cu) and block-like Al_2Cu particles. The former is due to high cooling rate while the latter is consequence of high fraction of Fe-rich intermetallics, nucleating site for Al_2Cu and resulting from relatively lower cooling rate.
Fe-rich phases were observed mainly in the form of blocky α-$Al_x(Fe,Mn,Cr)_ySi_z$ particles and few needle-like β-Al_5FeSi compounds.

Figure 5. Microstructure of the semi-solid AlSi7Cu3Mg alloy.

The quantitative results of average area fraction of Cu- and Fe-rich phases, and primary α-Al globules, estimated at different casting positions in the AlSi7Cu3Mg alloy casting, are summarizes in Table II.
Fig. 6 shows the microstructure at different regions of AlSi9Cu3(Fe) alloy casting. Compared to AlSi7Cu3Mg alloy, lower content of α-Al phase was observed throughout the casting due to higher Si content. Eutectic segregation phenomena similar to AlSi7Cu3Mg components occur, confirming how the reasons are mainly referred to the casting process and component design than to the experimental material. Further, Table II summarizes the quantitative microstructural results referred to different casting positions.

Table II. Area fraction of Cu- and Fe-rich compounds, and of primary α-Al globules estimated at different regions of the AlSi7Cu3Mg and AlSi9Cu3(Fe) castings; st. deviation in parentheses.

Alloy	Location*	Cu-rich phases (%)	Fe-rich phases (%)	primary α-Al phase (%)
AlSi7Cu3Mg	A	2.7 (0.5)	0.3 (0.1)	47.8 (5.2)
	B	2.8 (0.5)	0.4 (0.2)	56.5 (4.1)
	C	2.6 (0.6)	0.4 (0.3)	54.9 (6.4)
	D	2.7 (0.3)	0.4 (0.3)	57.3 (2.7)
AlSi9Cu3(Fe)	A	2.4 (1.1)	2.3 (0.8)	45.3 (5.4)
	B	2.5 (1.0)	2.1 (0.9)	51.0 (6.0)
	C	1.4 (0.4)	1.2 (0.4)	51.2 (4.7)

* as indicated in Fig. 4 and 6

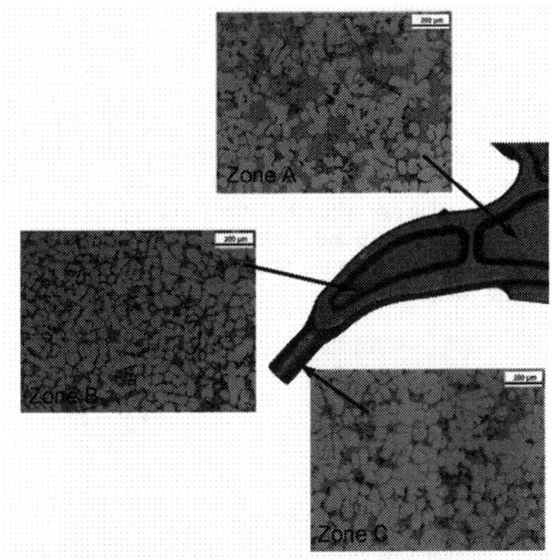

Figure 6. Microstructure of the AlSi9Cu3(Fe) alloy at different locations of the casting.

The AlSi9Cu3(Fe) alloy reveals higher amount of Fe-bearing compounds than the AlSi7Cu3Mg alloy. This can be explained considering the initial Fe, Mn and Cr content, and the holding temperature before SSM forming. These alloying elements can form primary intermetallic compounds that have a high specific gravity and can settle to the floor of the furnace as *sludge* [11]. Sludge formation has been shown to be dependent on the alloy's chemistry, melting and holding temperatures, and time. It has been defined a sludge factor for Al-Si-Cu alloys [12]. This factor is calculated from the formula:

$$\text{Sludge Factor} = (1\times \text{wt.\%Fe}) + (2\times \text{wt.\%Mn}) + (3\times \text{wt.\%Cr}) \quad (1)$$

where Cr and Mn are the most deleterious elements for sludge formation.

In the present work, the AlSi7Cu3Mg alloy contains more Mn and Cr than the AlSi9Cu3(Fe) alloy, while the Fe content is similar. Furthermore, the casting procedure was constant during the entire experimental campaign, especially the holding temperature for slurry forming set up at 640°C. In Al-Si-Cu alloys, Flores *et al.* [13] revealed that the sludge particles form most readily at about 640°C, reaching an average size of 40 μm after holding for just one to two minutes.
In order to map the mechanical properties throughout the casting, uniaxial tensile specimens were machined as shown in Fig. 7. The results of the mechanical properties for both the alloys are listened in Table III. In general, it is observed how the AlSi7Cu3Mg alloy exhibits higher ductility and *Q* values than the AlSi9Cu3(Fe) alloy, while the UTS is almost similar.
On the other hand, the mechanical results evidence great data dispersion, for elastic and plastic properties. This behaviour can be ascribed to the presence of casting defects and macrosegregations in the castings. For instance, the region in front of the ingate (zone B) shows the lowest mechanical performance due to the presence of porosity. However the variability of YS reflects a different distribution of primary and secondary α-Al phase, with a local content variations of atoms (Mg, Cu, Zn and Si) in solid solution in the α-Al matrix.

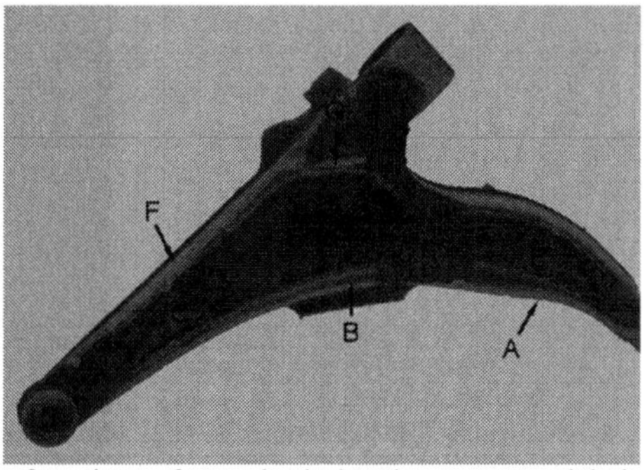

Figure 7. Position of specimens for mechanical testing; arrows and dashed frames indicate transversal and longitudinal sectioning, respectively.

Table III. Average mechanical properties estimated at different regions of semi-solid AlSi7Cu3Mg and AlSi9Cu3(Fe) castings; st. deviation in parentheses.

Alloy	Position	YS (MPa)	UTS (MPa)	s_f (%)	*Q (MPa)*
	A	150 (18)	227 (28)	1.9 (0.4)	272 (51)
	B	174 (14)	244 (14)	1.8 (0.9)	267 (76)
	C	161 (14)	297 (18)	4.6 (1.0)	456 (28)
AlSi7Cu3Mg	D	170 (17)	289 (7)	3.5 (0.8)	413 (26)
	E	163 (9)	262 (21)	2.3 (1.1)	322 (75)
	F	165 (15)	290 (18)	4.3 (0.9)	432 (29)
	G	165 (8)	250 (27)	3.0 (0.2)	319 (69)
	A	169 (13)	250 (8)	1.5 (0.3)	268 (31)
	B	158 (1)	194 (17)	0.8 (0.2)	135 (35)
	C	187 (3)	261 (22)	1.2 (0.4)	232 (74)
AlSi9Cu3(Fe)	D	187 (5)	294 (19)	2.3 (0.8)	366 (77)
	E	180 (11)	241 (32)	1.2 (0.4)	212 (83)
	F	167 (17)	236 (30)	1.5 (0.9)	238 (78)
	G	175 (6)	216 (10)	0.9 (0.1)	149 (24)

Conclusions

In the present work, detailed investigations have been carried out to document the semi-solid metal processing of secondary Al casting alloys over the microstructural and mechanical properties. The AlSi7Cu3Mg and AlSi9Cu3(Fe) alloys have been used to produce a lower arm automobile suspension as demonstrator. The following conclusions can be drawn from the experimental results.

- Casting defects, mainly solidification macro-shrinkage, are revealed in the regions with thicker wall thickness, independently of casting alloy.
- Local variations of material density appear as *flow lines* in the regions of thin wall thickness.
- In both the experimental alloys, the microstructure is not uniform throughout the casting and this is related to phase separation between solid and liquid during material filling.
- Lower amount of Fe-bearing compounds is revealed in the AlSi7Cu3Mg alloy due to the formation and sedimentation of primary sludge particles in the holding furnace.
- The AlSi7Cu3Mg alloy exhibits higher ductility and *Q* values than the AlSi9Cu3(Fe) alloy, while the UTS is almost similar.
- The mechanical results evidence great data dispersion, which is referred to the presence of casting defects and macrosegregations in the castings.

References

[1] E.J. Vinarcik, *High Integrity Die Casting Processes* (New York, NY: John Wiley & Sons, 2003), 3-12.
[2] D.H. Kirkwood et al., *Semi-solid Processing of Alloys* (Berlin Heidelberg: Springer-Verlag, 2010), 109-168.
[3] M.E. Schlesinger, *Aluminum Recycling* (Boca Raton: CRC Press; 2007), 176-192.
[4] D. Doutre, G. Hay and P. Wales, *Semi-solid concentration processing of metallic alloys* (US Patent No. 6,428,636, August 6 2002).
[5] J. Langlais and A. Lemieux, "The SEED Technology for Semi-Solid Processing of Aluminum Alloys: A Metallurgical and Process Overview", *Solid State Phenomena*, 116-117 (2006), 472-477.
[6] C.H. Cáceres, A Rationale for the Quality Index of Al-Si-Mg Casting Alloys, *Int. J. Cast Met. Res.*, 10 (1998), 293-299.
[7] F. Piasentini, F. Bonollo and A. Tiziani, Fourier thermal analysis applied to sodium eutectic modification of an Al-Si7 alloy, *Metal. Sci. Technol.*, 23 (2005), 11-20.
[8] L. Heusler and W. Schneider, Influence of alloying elements on the thermal analysis results of Al-Si cast alloys, *J. Light Met.* 2 (2002), 17-26.
[9] P.K. Seo, D.U. Kim and C.G. Kang, Effects of die shape and injection conditions proposed with numerical integration design on liquid segregation and mechanical properties in semi-solid die casting process, *J. Mater. Process. Technol.* 176 (2006) 45-54.
[10] C.M. Gourlay, H.I. Laukli and A.K. Dahle, Defect Band Characteristics in Mg-Al and Al-Si High-Pressure Die Castings, *Metall. Mater. Trans. A* 38 (2007) 1833-1844.
[11] M. Makhlouf and D. Apelian, "Casting characteristics of aluminum die casting alloys" (Report DEFC07-99ID13716, WPI, 2002).
[12] J.L. Jorstad, Understanding sludge, *Die Cast. Eng.* 11/12 (1986) 30-36.
[13] A. Flores et al., "Kinetics of the formation, growth and sedimentation of the phase $Al_{15}Fe,Mn_3Si_2$ in Al-Si-Cu alloys" (Paper presented at the Int. Die Casting Congress and Exposition, Detroit, Michigan, USA, 30 Sept. - 3 Oct. 1991), 293-297.

Shape Casting: 5th International Symposium 2014
Edited by: Murat Tiryakioğlu, John Campbell, and Glenn Byczynski
TMS (The Minerals, Metals & Materials Society), 2014

The Effect of Grain Structure on Casting Durability Assessment in Al-Si Alloys.

Glenn Byczynski & Robert Mackay

Nemak of Canada Corporation

Abstract

Casting durability may be assessed by several methods. In industrial powertrain castings engineering design is linked to the assessment of casting properties in particular high cycle fatigue performance. In the past porosity has been shown to be the most deleterious microstructural constituent in Al-Si cast alloys. Porosity is nucleated by oxide biflims and evolves during solidification due to segregated hydrogen gas and/or liquid feeding difficulties in the mushy zone during solidification. Porosity and oxide films have been reported to control casting durability as assessed through the fatigue staircase and calculated -3σ plots.

The authors will show in this work that grain structure can also play a major role in terms of controlling fatigue life. Specifically the presence of columnar grains in non grain refined casting structures can lead to low and unpredictable fatigue lives. The mechanisms of fatigue failure due to grain structure are reviewed and the use of grain refiner (in-mold process) to improve fatigue performance and counteract undesirable structure are discussed.

Introduction

Engine blocks cast with Al-Si-Cu alloys in sand or semi-permanent molds often have demanding service requirements and push the design limits of these components. Cast component characteristics, pressure tightness and durability, are typically controlled by the presence of porosity and thus the need to understand and/or control dispersed porosity is paramount[1-7]. Dispersed porosity in thick cast sections can be reduced by employing melt treatments such as hydrogen degassing practices and filtration, particularly when changing the local casting component geometry to quicken the solidification time is not an option. Typically in sand casting operations in order to achieve higher fatigue performance in thicker casting sections a metallic chill is placed in the mold and in direct contact with the liquid metal that has filled the mold cavity.

Casting processes using the aluminum alloy 319 in precision sand or semi-permanent mold commonly use additions of Si modifiers (normally master alloys containing Sr or Na[4-6]) or grain refiners (e.g. Tibor or Al-5Ti-1B[3]). The use of these Si modifiers and/or grain refiners have profound effects on pore distribution within the cast structure as they will increase dispersed porosity throughout the casting at the expense of gross localized shrinkage while also modifying the Si structure, and/or reduce grain size[1].

Experimental Methodology

Casting Method

In precision sand casting a room temperature mold or core package is employed. The mold is composed of individual interlocking sand cores that can be produced in a variety of core making methods. One of the most popular is the phenolic urethane cold box method employing silica or zircon sand and a two part amine cured binder system. The assembled sand mold may be filled using gravity or various counter gravity methods e.g. Low Pressure or Electromagnetic or Mechanical Pump. Both counter gravity casting processes are capable of producing a controlled quiescent filling of the mold that reduces the formation of bifilms in the casting[9-11]. Bifilms are known to impair casting quality and fatigue and tensile performance. Metal front velocities are controlled to be well below 0.5 m/s , and the typical fill times can range from 17 to 25 seconds depending on the casting size. One of the key components of this casting process has been the use of in-mold grain refinement; meaning placement of the master alloy rod into the runner system of the casting. While unusual in application the precision sand rollover process (where the ingates become the feeding system after rotation of the mold) allows for adequate dissolution of the master alloy and sufficient grain refinement of the targeted areas (Figure 1.).

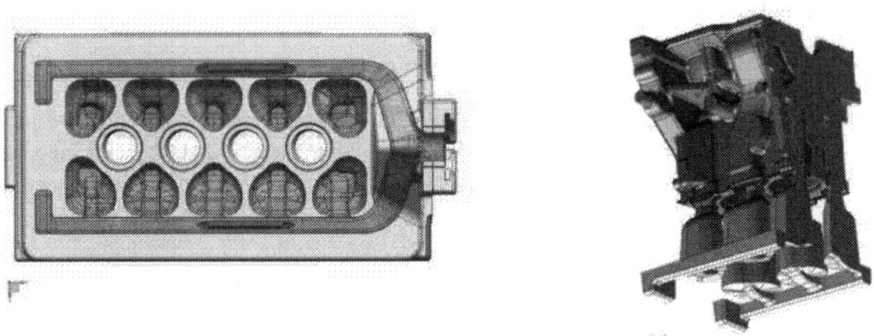

a) b)

Figure 1: a) Layout of runner system with In-mold placement of Tibor, b) location of bulkhead with respect to the rigging system.

Elevated Temperature Fatigue Staircase Method

Elevated temperature fatigue testing was conducted for this research to simulate cyclical stress conditions in an engine block of an operating engine. Fatigue test sample preparation and testing were carried out as per the ASTM E 466-96 protocol. TestStar™ IIs fatigue software by MTS was used to monitor the sinusoidal axial fatigue at a frequency of 98Hz. The stress ratio was R = -1, (R is defined as $\sigma_{min}/\sigma_{max}$). Maintaining a test temperature of 150°C was done using resistance heater tape material (Omega) in direct contact with the test specimen. Temperature is monitored by a K-type thermocouple that is held in place on the reduced (gauge) section using 3M glass fiber electrical tape. It typically takes 3 to 5 minutes for the reduced gauge section to reach target temperature.

Each fatigue test was used to plot the fatigue staircase which can be described as follows: If a life goal of 10^7 cycles was achieved without a failure the test is considered a "run out" at a given stress level. A subsequent test would be conducted on a fresh fatigue test sample at an incrementally higher stress level (incremental step used for this research was 5 MPa). A sample not achieving the 10^7 cycle life goal is considered a failed test, and results in the next sample being tested at an incrementally lower stress.

The mean stress of the staircase study would be determined by taking the average stress of all the fatigue test samples which failed (higher value of stress) or all the fatigue test samples which passed (lower value of stress), whichever is lower in number. This was provided along with a standard deviation of the results. The logic of using the lower number of sample test results is to negate the effect of a poor estimate of starting stress[11-12].

The alignment process of the fatigue test frame performed is meant to addresses two types of possible misalignment—concentric and angular. The fatigue test frames are equipped with an alignment collar (MTS Model 609) that allows alignment in conjunction with a specimen fitted with 12 strain gauges that is connected with a computerized alignment data acquisition and analysis system (MTS 709). This is the system that is used to align the fatigue frames in accordance with ASTM E1012-05

Light Optical & Scanning Electron Microscopy

Microstructural assessment for porosity levels (area fraction and maximum pore diameter), and Secondary Dendrite Arm Spacing (SDAS or λ_2) within the engine block bulkhead sections were measured using the Clemex Image Analysis System (JS-2000). The λ_2 was measured by taking the line intercepts of at least 10 dendrites and then dividing by the number of secondary arms in the same intercept. Image Analysis (IA) systems were calibrated using a Clemex calibration scale before any cast structure features were measured. AFS Si Modification Rating was also performed using Light Optical Microscopy and comparing to the AFS Si Modification Chart. Grain size was assessed using stereomicroscopy to assess the presence of columnar grains.

Scanning Electron Microscope (SEM) analysis of the cast polished structure and the fracture surfaces of fatigue test samples were conducted using a JEOL JSM 5800 system. The observation of the fatigue test sample fracture surface was done in the Secondary Electron (SE) mode under magnifications ranging from 20X to 250X.

Results

Figure 2 shows the results of the fatigue staircase plot for chill bulkhead sections specimens with and without Tibor additions. The first difference that can be identified is the fact that the mean value of the staircase is marginally higher with Tibor is added. Secondly, the variation in performance (gauged here by the values of -3σ and 95% CI) is lower for the case when Tibor is added. The authors will set forth an argument that the reason for the above observations are influenced not by grain size, but the rather the suppression of columnar grains that straighten oxide bifilms in the gauge area of the test specimen.

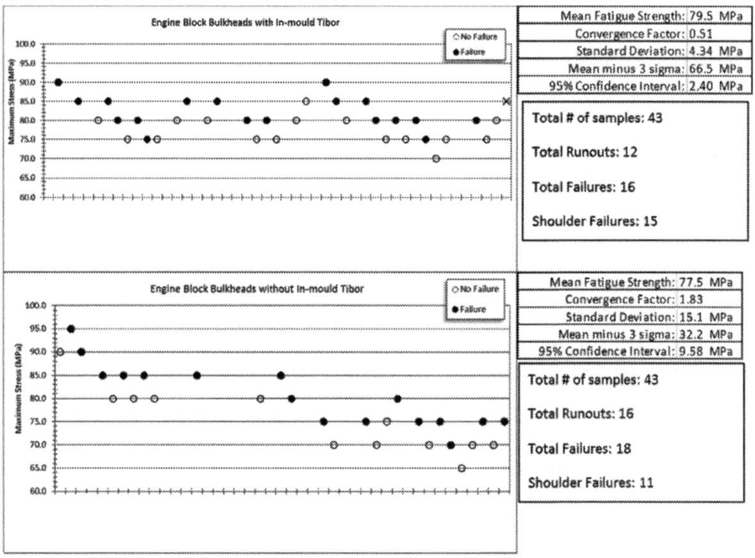

Figure 2: Summary of Fatigue Performance along with Statistical Results

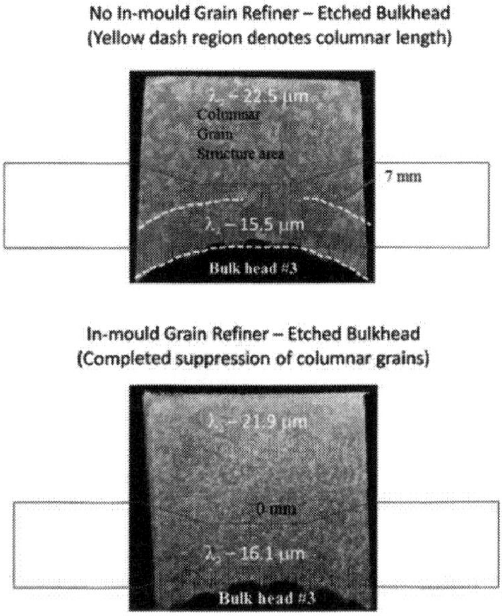

Figure 3: Microstructure of bulkhead with and without in-mold Tibor

In order to understand any difference in fatigue performance, both the bulkhead microstructure and fatigue fractography need to be assessed. Figure 3 shows both λ_2 and grain size of the bulkhead section where the reduced gauge section of the fatigue test bar is extracted. The λ_2 is not affected by the presence of in-mold Tibor, however the columnar grains are completely suppressed.

Fractographic analysis of the fatigue fracture surfaces reveal that all failures were initiated by an oxide film defect or an oxide induced shrinkage pore. In non grain refined castings consistent evidence of columnar grains were seen on the fracture surface (See Figure 2a). These appeared as "feathery grains" of almost featureless dendritic planes. Wrinkled remnants of oxide films were seen on the edges of these planar features suggesting that the columnar grains possibly impinged on and straightened existing bifilms as some authors suggest[13,14]. Specimens with in-mold Tibor additions no columnar grains were seen.

An additional factor that needs to be addressed is that both staircase plots have nearly the same number of samples that failed outside the gauge section and are to be considered a null test under the ASTM E466 protocol. Both regimes outlined in this study had to run nearly the same number of total test samples in order to achieve the 30 fatigue test samples to calculate the statistically relevant parameter for fatigue durability (e.g. -3σ). The main reason for this inconsistency is the λ_2 profile along the fatigue test samples. As shown in Figure 1, the fatigue test sample is extracted from just above the "half-round". The local geometry, wall thickness and position of the metallic chill result in the grip ends solidifying with coarser λ_2 than at the gauge section. As reported widely in the metallurgical literature fatigue performance drops with increasing λ_2. Refinement in grain structure due to the in-mold Tibor additions had no influence on the number of null fatigue tests that had to be performed.

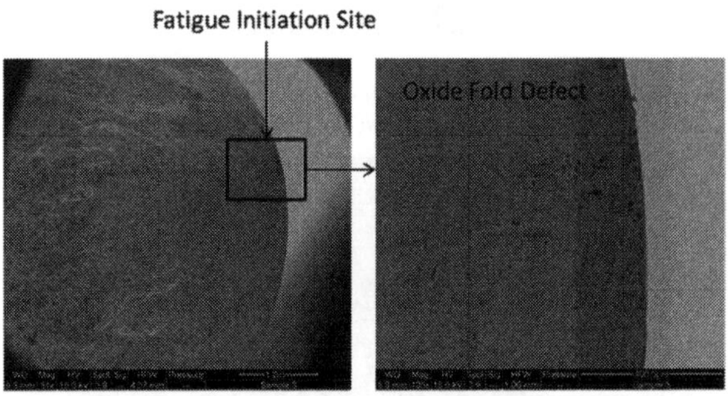

Figure 4a: SEM Image of Typical Fatigue Failure with In-mould Tibor

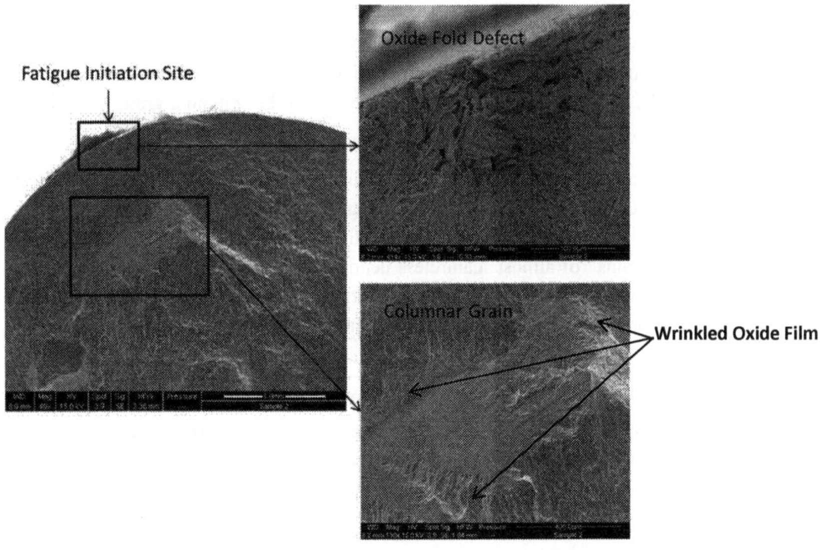

Figure 4b: SEM Image of Typical Fatigue Failure with no In-mould Tibor

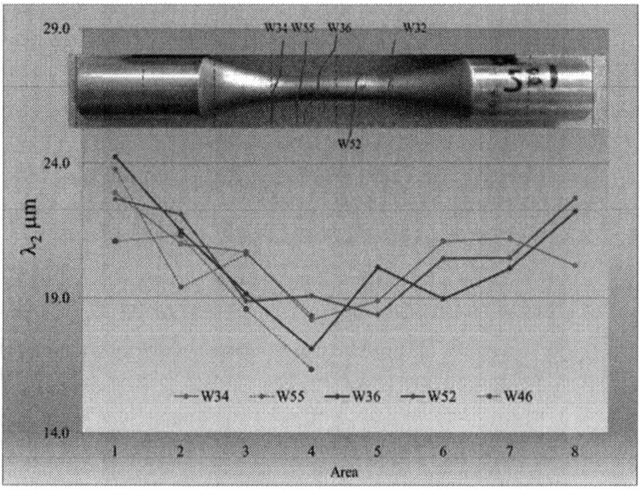

Figure 5: λ_2 profile along fatigue test bar.

Conclusions

This study set out to understand the role of in-mold grain refiner on fatigue staircase performance in industrial castings. The following conclusions can be drawn from this work:

1) Due to the presence of strong thermal gradients during solidification and particular sample orientation and sectioning (in this case axial fatigue sample perpendicular to the main direction of heat extraction) resultant columnar grains in the casting structure can negatively influence the fatigue life of aluminum castings

2) Columnar grains and oxide films were found to be collocated on the fatigue fracture surface of specimens with short fatigue lives. This leads to the suspicion that the two are inextricably linked and play a major role in the fatigue performance.

3) The addition of in-mold grain refinement reduced the degree of scatter in the fatigue test result, resulting in an improved -3σ value. This is thought to be related to (a) the provision of grain nucleation sites resulting in fine grain size and (b) the limitation of the exaggerating effect that columnar grains have on the size of oxide films (through their straightening and inflation) and consequently the limitation of the cumulative negative effects on fatigue performance

References

1) J. E. Gruzleski & B. M. Closset, The Treatment of Liquid Aluminum-Silicon Alloys, *American Foundry Society, Inc.*, 1990, 256pp, 1990.

2) Aluminum Casting Technology, Revised by J. L. Jorstad, W. M. Rasmussen, Edited by D. L. Zalensas, 2nd Ed., *American Foundry Society, Inc.*, 201pp, 1993.

3) Q. S. Hamed, M. Dogan & R. Elliot, "The Dependence of Secondary Dendrite Arm Spacing on Solidification Conditions of Al-7Si-0.5Mg Alloys Treated with TiBAl and TiBAl/Sr Additions", *Cast Metals*, Vol. 6, Number 6, pp.47-53, 1993.

4) F. Paray & J. E. Gruzleski, "Microstructural-Mechanical Property Relationships in a 356 Alloy, Part 1: Microstructure" *Cast Metals*, Vol. 7, Number 3, pp. 29-40, 1994.

5) F. Paray & J. E. Gruzleski, "Microstructural-Mechanical Property Relationships in a 356 Alloy, Part 2: Mechanical Properties" *Cast Metals*, Vol. 7, Number 3, pp. 153-163, 1994.

6) F. Paray & J. E. Gruzleski, "Factors to Consider in Modification", *AFS Transactions*, Vol. 102, pp. 833-842, 1994.

7) E. L. Rooy, "Mechanisms of Porosity Formation in Aluminum", *Modern Casting*, pp. 34-36, Sept. 1992.

8) G.Byczynski and J.Campbell, "A study of crack initiation sites in high cycle fatigue of 319 aluminum alloy castings", Shape Casting: The John Campbell Symposium, TMS (The Minerals, Metals and Materials Society), 2005, p. 235- 244.

9) Couper, M. J., A. E. Neeson, et al. (1990). "Casting defects and the fatigue behaviour of an aluminium casting alloy." Fatigue and Fracture of Engineering Materials and Structures 13(3): 213-227.

10) Skallerud, B., T. Iveland, et al. (1993). "Fatigue life assessment of aluminum alloys with casting defects." Engineering Fracture Mechanics 44(6): 857.

11) Dixon, W. J. and A. M. Mood (1948). "A method for obtaining and analyzing sensitivity data." Journal of the American Statistical Association 43(241): 109-126.

12) Brownlee, K. A., J. L. Hodges, et al. (1953). "The up-and-down method with small samples." Journal of the American Statistical Association 48(262): 262-277.

13) J. Campbell, "Entrainment defects", Materials Science and Technology, vol. 22, pp.127-145, Feb 2006.

14) J. Campbell, "An Overview of the Effects of Bifilms on the structure and Properties of Cast Alloys", Metallurgical and Materials Transactions B-Physical Metallurgy and Materials Science, vol. 37B, pp. 857-863, 2006.

Shape Casting: 5th International Symposium 2014
Edited by: Murat Tiryakioğlu, John Campbell, and Glenn Byczynski
TMS (The Minerals, Metals & Materials Society), 2014

NEAR-NET-SHAPE PROCESSING OF 2024 ALUMINIUM ALLOY BY SIMA METHOD

Hüseyin ÖZDEŞ, İlker ERDENİZ, Eray ERZI, Derya DIŞPINAR
İstanbul University, Metallurgical and Materials Engineering Department
Avcilar, 34320, Istanbul, Turkey

Keywords: SIMA, semi-solid, filling, microstructure, tensile property

Abstract

Semi-solid forming methods are commercially used for both cast and wrought aluminium alloys. The main advantages are elimination of runners and feeders compare to the casting methods. In addition, near-net-shape parts can be produced. Non-dendritic structures are also known to have homogeneous microstructure. One of the methods of obtaining spherical microstructure is called SIMA (Strain Induced Melt Activated). In this work; this method was applied to AA2024 alloy. In order to obtain optimum spherical structure heat treatment was carried out 590, 610, and 630°C at certain time intervals. Then, moulds in the shape of a spanner with different thicknesses were used. Mould filling ability was characterized by means of microstructural analysis and using different mould temperatures. Tensile test samples were produced from the castings followed by T6 heat treatment and mechanical properties were investigated.

Introduction

Wrought aluminium alloys have common applications for aerospace and automobile industry due to superior features compare to cast aluminium alloys [1]. Different aluminium alloys are used for parts with complex geometry. In addition, the industry is looking for the ways to reduce the cost with less processing time and less forming operations [2].

Semi-solid metal processing technology has been developed by Spencer et al at M.I.T during the 1970s, led by Flemings [3]. Semi-solid metal forming method has many advantages such as prolonged mould lifetime, low effort of forming, obtaining near net shape. This technology can be divided into reo-forming and thixo-forming [4]. Technically, thin shaped, non-dendritic and spherical microstructure are needed before the obtaining semi-solid state (typical dendritic and spherical microstructure shown in Fig 1). For this purpose, there exists variety of methods like magnetic stirring, electromagnetic stirring, however thixo-forming process is being widely used [5]. The process includes the preparation and reheating of deformed billet. Semi-solid forming process offers good surface finish, improved mechanical properties and low energy cost compare to the other semi-solid technologies. The technic involves applying force when the material temperature is between the solidus and liquidus temperatures.

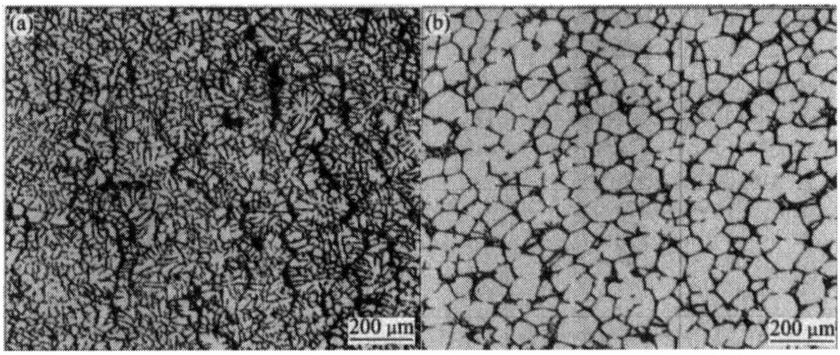

Figure 1. Typical (a) dendritic and (b) spherical microstructure of AA2024

The most efficient technological approach was introduced in 1981 [7-8]. Researchers named the technic as SIMA (Strain Induced Melt Activated). SIMA process is based on the development of a series of process steps. Basically hot extruded or rolled bars are subjected to a cold treatment. Then work piece is heated to semi-solid temperature. Then, a non-dendritic and spherical microstructure is obtained [7]. SIMA process were tested for Al, Mg, Cu and Fe alloys and have been used for the production of aluminium alloys since 1983. Technically it can be applied to larger work piece however it is not economically compare to MHD method [7].

SIMA process has been improved for fixing the difficulties of parts production in various conditions. As seen in Fig.2, after casting, parts are subjected to severe deformation above or below the recrystallization temperature. Thus dendritic structure is broken and thin, long, deformed structure is obtained [7-8].

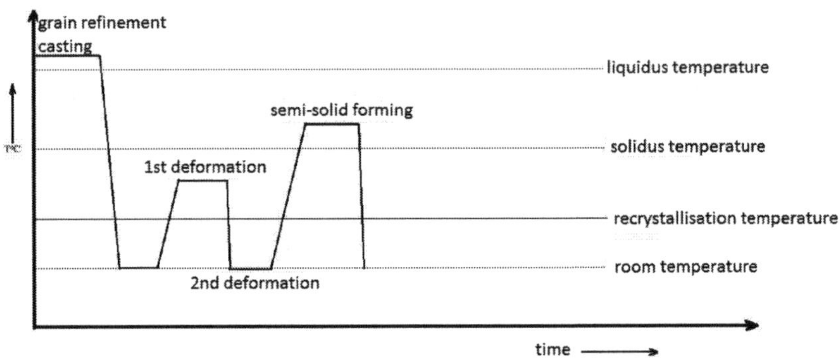

Figure 2. Schematic representation of semisolid forming [6]

Then a cold deformation is applied at low-temperatures. The aim of the cold deformation to increase the dislocation density. The dislocations which are observed as planes create lower grain boundary [7] which will form large angle grain boundary during the heat treatment as shown in Fig.3.

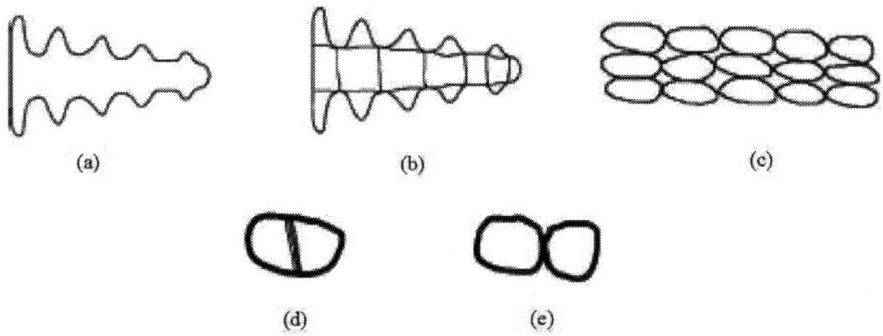

Figure 3. Microstructural change during semi solid process [9]: (a) dendrite arms, (b) deformation on dendrite arm, (c) grain elongation, (d) grain deformation, (e) grain separation

Parameters such as heating time, temperature, deformation rate are critical factors that determine the semi-solid microstructure in SIMA method [3]. Therefore, this paper focuses on the mould filling ability of AA2024 alloy after the optimization of the heating time and temperature by using SIMA process.

Experimental Work

The composition of AA2024 which is used in this work shown in Table 1.

Table I: Composition of alloy used in the work

	Zn	Cu	Mg	Mn	Si	Fe	Cr	Ti	Al
2024	0,35	3,8	1,2	0,51	0,27	0,31	0,028	0,033	kalan

Alloys had been supplied as cold deformed cylindrical bar (diameter 22 mm).

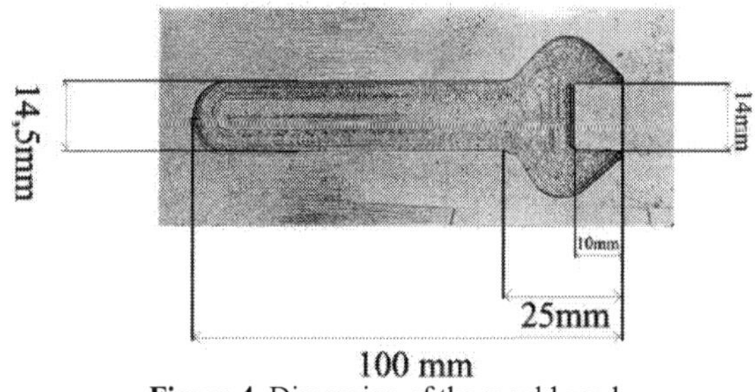

Figure 4. Dimension of the mould used

Figure 5. Experimental procedure for the forging step

Sample were sectioned into 20Øx10 mm dimension and subjected to heat treatment at 590°C, 610°C and 630°C for 5 to 30 minutes with 5 minute increments. These samples were subjected to metallographical examinations in order to find the optimum spherical microstructure. After the achievement of the optimum structure, bars were cut in the dimension of the mould cavity (Figure 4) and SIMA process was carried out to fill the moulds by using 1 ton force to produce part that is shown Figure 5.

During the production of the parts, different mould temperatures were investigated (room temperature, 200°C and 350°C – 600°C; with and without lubrication). After the samples were taken out off the mould, the samples were sectioned and metallographical examination was carried out at selected locations (edges, mid, centre; longitudinal and transvers). Grain size and sphericity were measured by using image analysis software.

The samples were then machined into the dimension given in Figure 6 and tensile testing was conducted.

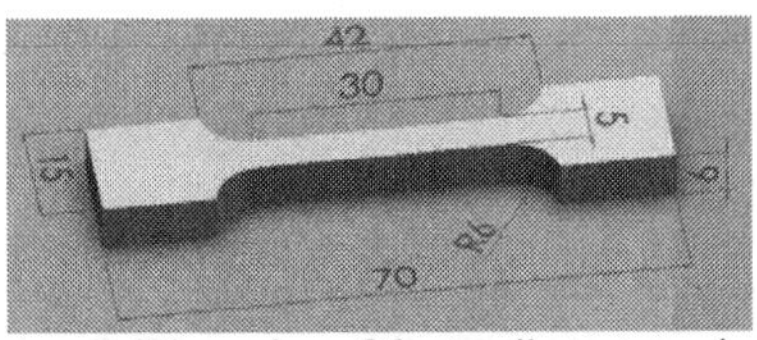

Figure 6: Dimension of the tensile test specimen

Results and Discussion

As a results of observation of cold deformed AA2024 which is kept at 590, 610, 630°C temperature at certain time intervals, optimum spherical microstructure is obtained at keeping 15 minutes at 630°C temperature (Fig 7). Akbaş [6] observed that the temperature of obtaining optimum spherical microstructure for 7075 alloy is at 575°C. This shows that this is depended upon the pre-processes that the material has been through and varies for different alloys.

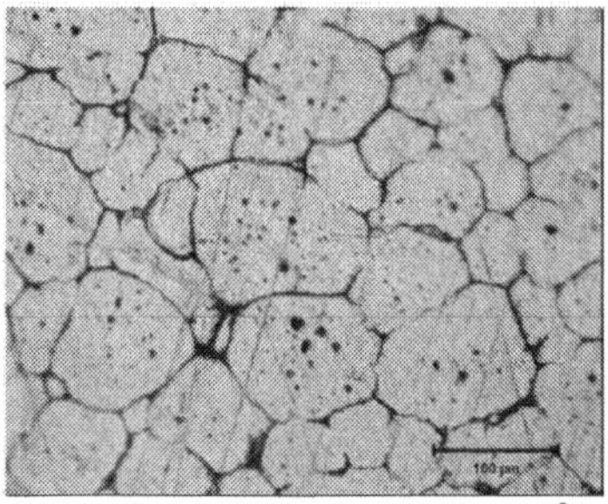

Figure 7: Optimum spherical microstructure obtained at 630°C – 15 minutes of holding

After optimum spherical microstructure was obtained, 1 ton force was applied to hot, cold and lubricated mould for observing bar's mould filling ability. The results are shown in Figure 8-9. As seen in Figure 9, bars which are pressed to hot mould showed better filling properties. The correspondent microstructures are given in Figs 10-11. When pre-heated mould is used, it can be seen that the grains begun to deform longitudinally (Fig 11a).

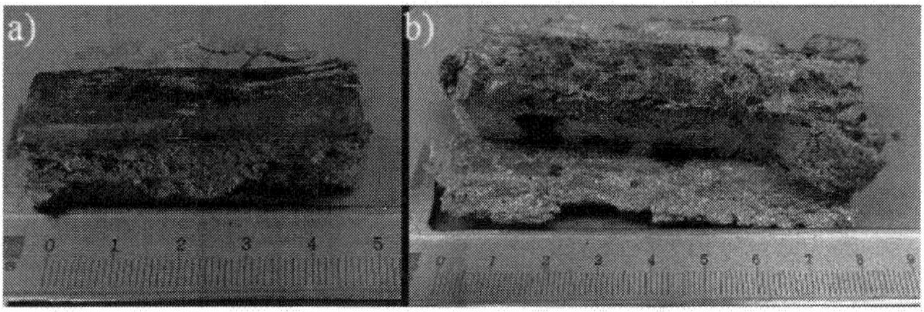

Figure 8. Samples forged at (a) cold and (b) pre-heated moulds

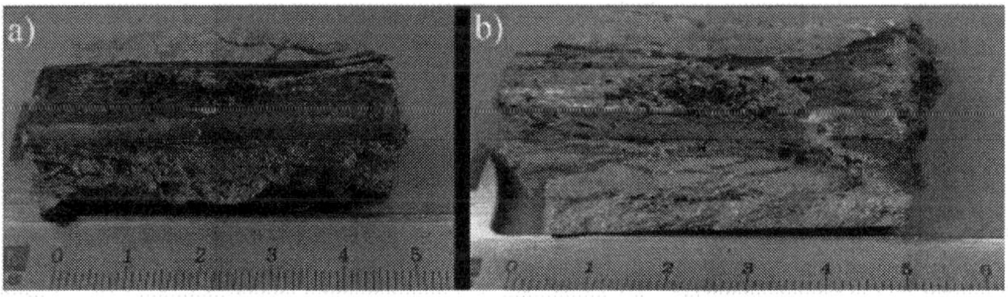

Figure 9. Samples forged at (a) non-lubricated and (b) lubricated moulds

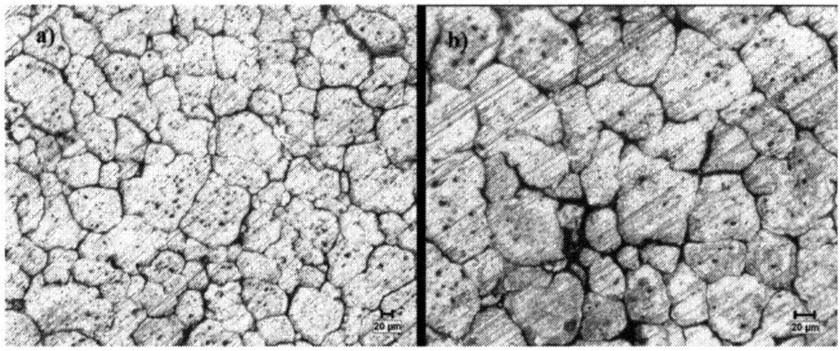

Figure 10: Microstructure forged in the mould at room temperature (a) cross-sectional, (b) longitudinal

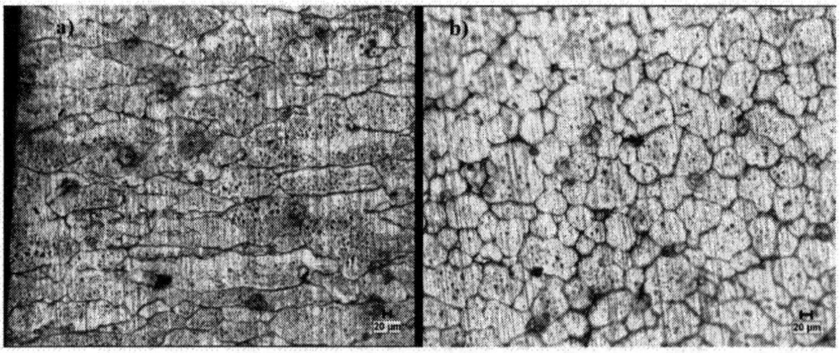

Figure 11: Microstructure forged in the mould at 350°C (a) cross-sectional, (b) longitudinal

After the tensile test samples were machined into the dimension given in Fig 6, Archimedes principle was used to measure the density of the samples. A graph was plotted against density of the samples and the tensile test results. As seen in Fig 12, for ultimate tensile strength, there appears to be trend such that as density increases the value increases. However, it is important to note that there is a significant amount of scatter (Fig 12a). This situation is more complex for the elongation values (Fig 12b). During forging operations, it is important that a dense and non-porous part is obtained. It is concluded that the variations in the results do not only represent the porosity levels, but it is also related with the microstructure, as well as the quality of the alloy (i.e. extrinsic defects that may be present). Figure 12 also appears to have two separate populations. The lower population seems to be close to linear. On the other hand, the top population is logarithmic. As seen Figure 13, there are heterogeneously solidified regions are present and in some cases, there are even rapidly solidified fine eutectics between the grains. In addition, as seen in Figure 11, there are various different grain size and morphologies. These are the reasons for two separate populations in Figure 12.

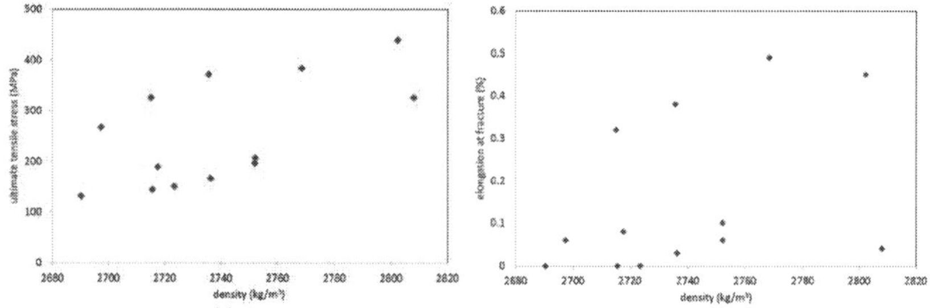

(a) (b)

Figure 12: Comparison of density of tensile test samples with (a) UTS and (b) elongation

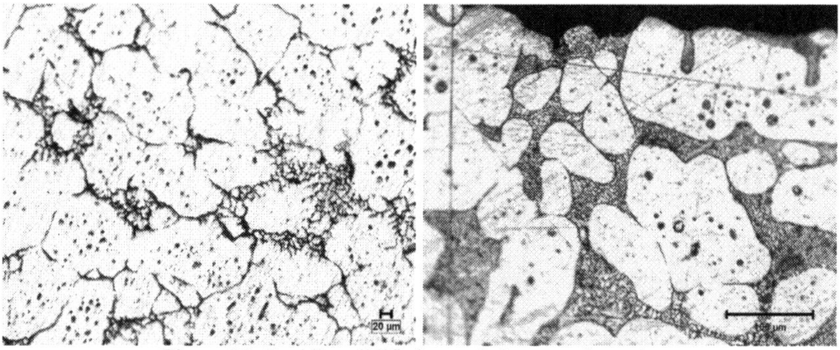

Figure 13: Rapidly solidified eutectic between the grains

Conclusion

The following results were obtained about thixotropic microstructure and part producing for semi-solid forming with SIMA method for AA2024.

Optimum spherical microstructure were obtained at 630°C for 15 minutes of holding.

Mould's temperature has a significant effect on near-net-shape forming.

There was no significant effect of lubrication for the filling properties.

Tensile properties scatter in a wide range, possibly due to heterogeneous solidification of the liquid eutectic phase.

References

1. G. Hong-min, Y. Xiang-jie, "Preparation of Semi-Solid Slurry Containing Fine and Globular Particles for Wrought Aluminum Alloy 2024", Trans. Nonferrous Met. Soc., China 17(2007) 799-804.

2. R. Kopp, D. Neudenberger, G. Winning, "Different Concepts of Thixoforging and Experiments for Rheological Data", Journal of Materials Processing Technology, 111 (2001) 48-52.

3. S. Nurşen, S. İ. Etem, T. Metin, Ö. Özgü, Ç. Önder, "Mechanical Properties and Microstructural Evaluation of AA5013 Aluminum Alloy Treated in the Semi-Solid State by SIMA Process", Journal of Materials Processing Technology, 148 (2004) 103-107.

4. W. Shun-cheng, L. Yuan-yuan, C. Wei-ping, Z. Xiao-ping, "Microstructure Evolution of Semi-Solid 2024 Alloy During Two-Step Reheating Process", Trans. Nonferrous Met. Soc., China 18(2008) 784-788.

5. T. Adnan, S. Nurşen, "Effect of Ag addition on microstructure of semi-solid formed aluminium alloys and their statistical evaluation", C. B. Ü. Technical Bulletin, 16, 2, 2011.
6. A. Nurşen, "Non-dendritic Aluminum alloys production by SIMA Method and heat treatment", Istanbul University, Metallurgical and Materials Eng Dpt, Master Thesis, August1995.
7. D. Dispinar, "Solution heat treatment of spherical grain aluminum alloys ", Istanbul University, Metallurgical and Materials Eng Dpt, Master Thesis, February 1999.
8. A. Turkeli, "Thixotropic material production of 7001 Aluminum Alloy by SIMA method", 5th Materials Symposium, Denizli, Turkey, 5, 1, 52-62, 1993.
9. E. Yiğit, S. Nurşen, S. İ.Etem, "Spherical grain formation and mechanical properties of A360 by SIMA method", 3rd National Design Production and Analysis Conference, Balıkesir October 2012.

SHAPE CASTING:
5th International Symposium 2014

Solidification and Microstructure II

Shape Casting: 5th International Symposium 2014
Edited by: Murat Tiryakioğlu, John Campbell, and Glenn Byczynski
TMS (The Minerals, Metals & Materials Society), 2014

REDUCED PRESSURE TEST (RPT) FOR BIFILM ASSESSMENT

Derya Dispinar*, John Campbell**

*Istanbul University, Metallurgical and Materials Engineering Department, Istanbul, Turkey
deryad@istanbul.edu.tr
**University of Birmingham, Metallurgy and Materials, Birmingham, UK
jc@campbelltech.plus.com

Keywords: bifilms, reduced pressure test, melt quality, quantification, aluminum

Abstract

The assessment of the cleanness of liquid aluminum alloys has traditionally been viewed as difficult, but has been tackled by a number of techniques, all of which have been comparatively expensive and complicated. In the past the presence of entrained defects known as bifilms in metals has not been realised because they are often only a few molecules thick, and so remain invisible, or at best, difficult to detect. The reduced pressure test (RPT) is sensitive to the size and number of entrained oxides which appear to be the major defects affecting metal quality in terms of workability, mechanical properties and corrosion. In addition, the test is a confirmation of the fact that such oxides are double, and thus conveniently called 'bifilms'. The RPT is a simple, low cost, but fundamentally appropriate technique for the assessment of (a) the number and (b) the size of bifilms in a melt. This paper outlines an initial draft procedure. Future users will be welcome to suggest improvements.

Introduction

The assessment of the cleanness of liquid aluminum alloys has traditionally been viewed as difficult, but has been tackled by a number of techniques, all of which have been comparatively expensive and complicated. Furthermore, the reliability of all of the various methods has been open to fundamental questions. In contrast, the reduced pressure test (RPT) is a low cost, simple procedure which has in the past been viewed mainly as a qualitative test of hydrogen in solution in the melt. In the past the presence of bifilm cracks in metals has not been realised because they are often only a few molecules thick, and so remain invisible, or at best, difficult to detect.

The RPT is a valuable technique for the assessment of (a) the number and (b) the size of bifilms in a melt.

In an attempt to quantify this aspect, the test has been carried out by measuring the density of the sample. Such a use of the test is not criticized, but rather seen as only one complementary aspect of the important information that the test can reveal to the operator. The further aspect of the RPT which is described in this draft procedure is the assessment of the double oxide films, known as bifilms.

Bifilms are entrainment defects. These features are introduced (i.e. entrained) during pouring or other turbulence of the melt surface, in which the oxide on the liquid surface is folded over on itself, to form a double oxide film, with a central dry, unbonded interface. These unbonded interfaces act as cracks that can remain in suspension in the liquid. For many liquid metals the bifilms float out within minutes because they are much less dense than the melt. However, for liquid aluminium alloys the bifilm has almost neutral buoyancy, and so will remain in suspension for hours. Thus oxide bifilms are a special problem in the casting of Al alloys. When poured with the metal into the mold they become frozen into the casting where they act to reduce the ductility, giving low elongation and ultimate strength values and poor fatigue performance. In the past the presence of bifilm cracks in metals has not been realised because they are often only a few molecules thick, and so remain invisible, or at best, difficult to detect.

The RPT is a valuable technique for the assessment of (a) the number and (b) the size of bifilms in a melt.

The reduction of pressure on the metal during freezing of the sample causes the bifilm to expand its residual layer of air or gas trapped in the dry space between the films. The expansion of the residual air is a valuable feature of the test because

(i) As the films are forced apart the bifilm becomes visible as an open crack (when the films are close together it can be virtually invisible).

(ii) Immediately after being poured into the sampling crucible, the bifilm is ravelled by turbulence into a compact, convoluted form, as a tiny crunched-up ball. The expansion of its internal gas promotes the unfolding outwards of the bifilm, as the bud of a flower opens its petals, or the unfurling of the sails of a boat. Thus the full size of the bifilm is more clearly seen by the application of reduced pressure. The length of the crack observed on the sample surface can then be taken as the length of the bifilm. This unfurling of the bifilm to its full area takes several minutes. Thus the sample must not be frozen too quickly otherwise the full size of the bifilms will be underestimated. On the other hand of course, to be convenient as a 'quick test', the freezing used for the RPT must not be inconveniently long. A freezing time of the order of 2 minutes is about right for 356 type alloys. Other alloys may be found to require a different time.

(iii) If there is some hydrogen gas in solution in the metal, the gas may expand the crack into a spherical bubble or pore. From the point of view of operating the test in a simple way, the diameter of the pore (or its greatest dimension if not perfectly round) is taken as roughly equivalent to the size of the originating bifilm. This is clearly an approximation that, if anything, will *overestimate* the size of the original bifilm because with very high gas content the diameter of the gas bubble may outgrow the size of the initiating bifilm. However, in general, this admitted limitation of the test tends to have only a minor effect for most working levels of hydrogen content.

Equipment

A steel spoon is required for sampling from the furnace or ladle. It requires a bowl diameter of approximately 100 mm and should be coated with a suitable protective wash such as that based on boron nitride (BN). The wash requires to be thoroughly dried.

If a ceramic fiber ladle is used instead of a steel spoon (with a view to delay freezing in the ladle as much as possible) the significant increased hydrogen content of the sample tends to aid bifilm assessment but will make invalid assessment of hydrogen using the density check.

A solid ceramic ladle from Pyrotek hold promise as use for a sampling ladle for sample collection from the melt but whose contribution to hydrogen contamination of the sample

appears to be significantly less than that of the fiber material, while retaining the temperature of the sample for longer. However, hydrogen contamination is probably still higher than that of a steel ladle. This material is still undergoing evaluation.
Stainless steel cups are required, approximately 35 mm diameter x 40 mm high, of a tapering conical shape to allow the sample to be easily turned out of the cup after solidification. The thickness of the metal should be approximately 1 mm. The cups require to be treated with care since after some use they become annealed, becoming weak and delicate. If the cup is damaged by denting or distorting the cast sample is not easily removed. A small pair of steel tongs is required to handle the hot cups and cast samples. A vacuum bell jar or cover is required that can be pumped down to 100 mbar (one tenth of an atmosphere, corresponding to a residual pressure of 76 mm (3 in) mercury){recalling that 1 atm = 760 mm (30 in) mercury} within a time of 10 to 15 seconds. A saw (preferably a band saw for speed and convenience) is needed to cut the sample down its center. A belt linisher is convenient and fast to carry out the major part of the smoothing the surface. Silicon carbide paper on a glass plate, washed by water to flush away grinding debris, is required for the final finishing of the surface. Alternatively, after sawing, the surface could be finished in a lathe to a near mirror finish, using a diamond tipped tool.

Procedure

The spoon is heated by floating the bowl on the surface of the melt for approximately 10 seconds. The spoon is then used to scrape the local surface clean of oxide so far as possible, and is then dipped through the surface, with a sliding or cutting action designed to disturb the melt surface as little as possible, dipping to a depth of approximately 100 to 200 mm. The spoon is then oriented upwards to bring out a sample from this depth. (The aim is to avoid sampling the melt just under the melt surface which often contains a higher accumulation of floating bifilms; some bifilms contain more air and so tend to float, whereas some with less air will tend to sink. However, both flotation and sedimentation are slow because of the very small density difference between the bifilm and the melt).

The sample is transferred swiftly to the RPT station which should be located conveniently nearby, and the sample cup filled with metal, to create a sample with dimensions approximately 35 diameter x 35 mm tall, weighing 75 to 100 g.

The sample cup may be placed on a piece of 3 mm thickness ceramic fiber sheet to insulate the cup from the base of the vacuum container. The target is to achieve a solidification time in the region of 4 minutes for 356 type alloys, noting that other alloys may require a slightly different time.

When the sample is seen to be solid, easily checked by a slight tap on the bell jar to verify the absence of ripples on the sample surface, the vacuum is released, the bell jar removed, and the cup lifted out with tongs. The cast sample is turned out of the cup. Two routes are possible:

(i) The sample can be checked for density by weighing it in air (while still hot and dry) and in water, after which it can be sectioned and prepared for bifilm assessment.
(ii) The sample is quenched in water and immediately cut and prepared for bifilm assessment.

Clearly, both the density and bifilm quantification are valuable, and can be performed in sequence on the same sample. However, this takes time. Time can saved by having two sets of equipment, or one vacuum chamber capable of accepting two samples, and carrying out both density and bifilm tests concurrently.

In addition, the density technique can be supplemented by the casting of a third cup from the same spoon sample. The third cup sample is allowed to solidify in air to provide a nearly fully dense sample as a baseline against which the density of the RPT sample can be compared to obtain a porosity value. These procedures for the density and possible porosity determination are commonly available and not presented here.

After quenching the sample into water to allow it to be conveniently handled, the sample is cut down its center by sawing, then ground on successively finer SiC papers to achieve a uniform, fine, satin, surface finish.

There are now two approaches to quantifying the defects that are revealed on the prepared section. Assessment can be carried out (i) manually by the operator or (ii) automatically by scanner and appropriate software. These two techniques are detailed separately below.

(i) Direct Operator Measurement

Although counting by unaided eye might be viewed as having significant potential for operator bias or error, a single operator should be able to achieve a sufficiently consistent performance. Occasional checks of performance can ensure that this relatively quick and easy method does give sufficiently useful and sufficiently accurate results.

For the measuring the sizes of defects it is adequate, with care, to simply use a millimeter rule. This can be read to within 0.5 mm without significant effort, and is adequate for the purpose of summing the total maximum lengths of defects. The small errors involved tend to cancel. Even so, some quick and easy training on a range of a few standard test pieces would help an operator to develop sufficient accuracy.

A typical surface as observed by the operator may have a fairly widely spread of sizes and numbers of round pores, thin cracks or other irregular shaped voids. Task (i) is to count the total number of defects. This is easy and straightforward, but perhaps neglecting the very fine features of less than 0.5 mm maximum length. Task (ii) is to sum the total of the maximum lengths of all the defects. This is best explained by an example:

For instance, on an imaginary sample, there may be 2 defects 3.5 mm long, but 15 defects with a maximum length approximately 2 mm. These can be literally ticked off with a felt pen as they are counted. In addition there may be 32 defects with maximum length approximately 1 mm. Finally, there may be 22 defects with largest dimension roughly 0.5 mm. All smaller defects are neglected. The total number of defects is therefore 2 + 15 + 32 + 22 = 71.

We can also find the total length of bifilms, sometimes referred to as the Bifilm Index, as follows

2 x 3.5 mm	= 7 mm
15 x 2 mm	= 30 mm
32 x 1 mm	= 32 mm
22 x 0.5 mm	= 11 mm
Total	= <u>80 mm</u>

Thus for this sample the 71 defects have a total length of 80 mm indicating that the average size of defect in this case is 1.13 mm. As it happens, it seems a size of the order of 1 mm is common for many Al alloys although this can often vary by a factor of 2 from up to 2 mm and down to 0.5 mm. However, bifilms of length 10 to 15 mm have been measured in some holding furnaces where it seems that the bifilms appear to grow with time, possibly as a result of a 'panting' behavior as they circulate in the convection currents from the top to the bottom of the furnace, cycling in pressure, and therefore in size, but with each expansion being irreversible, the bifilm growing by a ratcheting action, since with each cycle the newly created oxide area of the bifilm cannot shrink.

Following example is taken from a well-degassed holding furnace. It was not clear whether the apparently clear rim was in fact clear of bifilms, and they had been pushed into the center of the casting by the advancing solidification front, or whether the outer bifilms had simply not opened because of rapid freezing and insufficient segregated hydrogen.

(ii) Instrumented Measurement

Good surface finish is required for automatic measurement of the number and maximum length of pores on the cross-section of the RPT samples. The best surface condition is achieved by grinding on successively finer silicon carbide papers and polishing with alumina or magnesia polishing powder to a mirror finish. However, with some practice, polishing might be avoided if the grinding with the finest grit paper is sufficiently even and fine, and if it is found that the measurement software is tolerant.

The freshly saw cut surface seen below is unsatisfactory for bifilm assessment for 2 reasons:

(i) the surface is plastically distorted by the cutting action of the saw so that pores are to some unknown extent smeared over, resulting in a low count.

(ii) automatic scanning because of the surface will be unsatisfactory as a result of the 'chatter' or 'resonance' of the saw creating light and dark areas of the surface.

Machining of the sample to produce a satisfactory surface without further preparation can be achieved by in a lathe or milling machine using diamond-tipped cutting tools. Ceramic cutting tools are not satisfactory in general because they tend to smear the surface, concealing smaller features.

Once the surface is prepared to an adequate condition, a scanner is used to scan the samples at 600 dpi. A typical scanning result is seen below:

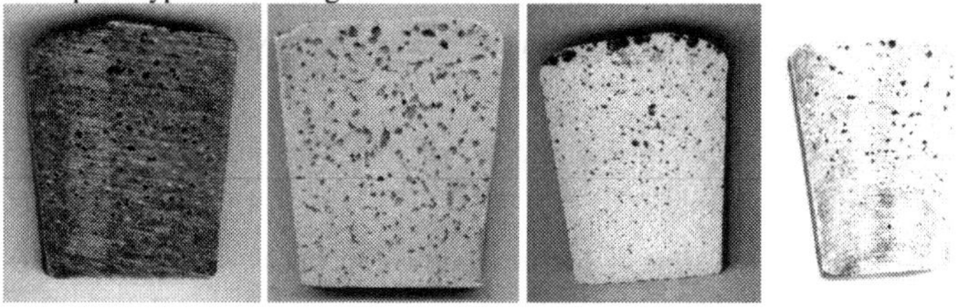

These scans are likely to generate a problem during measurement because there are too many dark areas around the surface such the shadow of the sample. It is recommended that these dark regions should be cleaned during scanning by simply adjusting the brightness and contrast (>70% brightness and >70% contrast) to eliminate shadows and enhance (darken) the pores. An example is given above (right).

The digitized RPT images are then subjected to image analysis. There are several software packages available. The one used in this report as an example is SigmaScan Pro 5.

The image is first required to be calibrated. For 600 dpi scanned images, 815 pixels correspond to 35 mm. A threshold is then applied to select pores:

The area, shape factor and maximum length of pores (major axis length) can be selected in the software. The software measures the parameters and the results are reported as spreadsheet.

The results need to be corrected as there might be noise in the image which may affect the results. Thus, the data is sorted in the Excel and any pore that has a shape factor greater than 1 is deleted as meaningless. (Shape factor is the measure of sphericity. This dimensionless measure is defined as 4π times the object's area divided by its perimeter squared. A perfect circle will have a shape factor of 1. The shape factor for a line will approach zero.) As a result of experience over thousands of samples, pores with area smaller than 0.1 mm^2 can also be deleted from the results, because such a tiny dot has a negligible effect on the result and because such signals can be trivial imperfections that do not represent pores or might even be simply noise.

If in doubt, it is also possible to use the software to measure each pore individually. This is, of course, somewhat time consuming but the result should be the same to within approximately ±10 mm with the automatic procedure outlined above.

The sum of "major axis length" is calculated in Excel to give the bifilm index (total length of bifilm) in millimetres.

The total number and total length counts can be used without further processing, since this raw data is very easily understood, and such clarity is a definite advantage. The variability of the size of the area of the reduced pressure sample does, of course, directly affect the values, but this effect is limited to approximately a factor of 2, which is probably within the margin of error for these measurements. A factor of 2 might be considered embarrassingly large, but is far better than no value. Furthermore, most other measurements of metal quality are either not quantified at all, or have aspects which are fundamentally wrong.

There are times, however, when it is helpful to take account of the observed area of the reduced pressure sample. Thus the number and length values can be easily converted into values per unit millimeter. Thus for values of 20 pores of total length 200 mm measured on a surface area 20 mm x 20 mm = 400 mm^2, the corresponding values are 0.05 bifilms per mm^2 and 0.5 $mm/mm^2 = 0.5\ mm^{-1}$.

Such values per unit area are of doubtful general value. A more generally understood and useful value is obtained by converting the raw data to values per unit volume.

A conversion to values per unit volume can be approximated by assuming that the total length value is similar in a plane at right angles to the measured plane, and by assuming that the value applies approximately uniformly to a cubic volume of depth similar to the outline of the area of measurement. Thus if the number of bifilms is N and total length L, and the area of measurement A^2 the number per unit volume is N^2/A^3 and the total area per unit volume is L^2/A^3. For the example above the values would work out to be $20^2/20x20x20$ = 0.05 bifilms per cubic millimeter, and 2002/20x20x20 = 5 mm length per cubic millimeter = 5 mm^{-2}. The latter unit is, of course, curious and somewhat counter-intuitive, and thus may require to be quoted with care. In the interests of clarity, it might be better to retain the rather redundant form mm/mm^3.

Production of a nearly parallel-sided RPT sample designed for X-ray radiography assessment.

It is sometimes interesting and valuable to view the unfurled bifilms in an RPT sample using radiography. The cylindrical RPT sample is not convenient for viewing by radiography because of the several exposures that are needed to view the whole sample. To achieve a useful clear result the cylindrical RPT specimen requires to have a parallel-sided slab 10 to 15 mm thick excised out of its center.

It is much more convenient to cast a parallel-sided RPT sample for studying the unfurled bifilms by radiography. A thin shell stainless steel mold is not recommended because its slim thickness would mean that freezing would be too fast to allow the bifilms to unfurl to their maximum extent, so that their effect might be underestimated.

A sand mold offering a slower rate of freezing but delivering the convenient (adequately) parallel slab shape for radiography is recommended.

The dimensions (mm) of the mold and the RPT sample are given below:

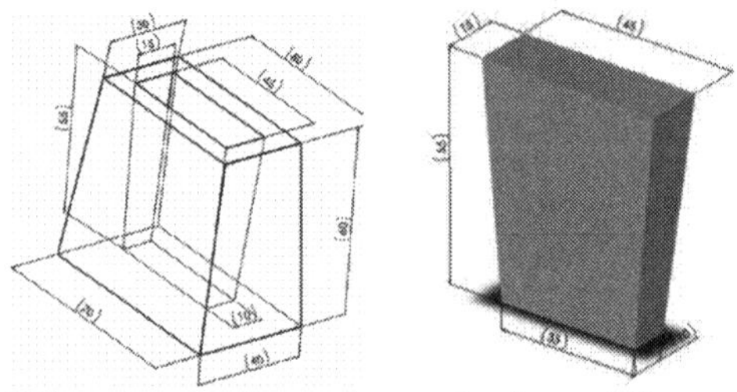

RPT sand cast slab sample can also be used for quantitative bifilm assessment. As for the cylindrical sample, it is cut vertically down its mid-plane, ground and polished as usual. The sectioned RPT sample is shown below:

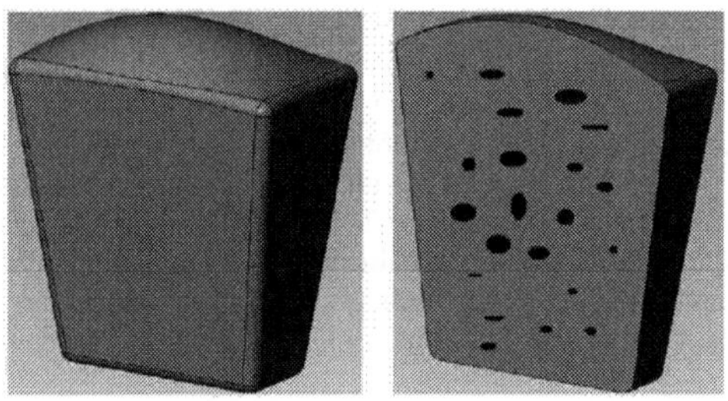

Procedure to Assess Bifilms in a Solid Sample

It is possible to use the RPT technique to sample the bifilms contained in a solid metal sample such as an alloy ingot or casting.

A sample should be cut from the ingot or casting at a point that is believed to be representative of the whole piece. The sample can be as small as 10 mm cube, or possibly as large as 35 mm cube (N.B. Although this 35 mm cube size seems practical and reasonable, such simulations of a full-size RPT sample have not so far been tested).

The cube sample can probably be melted without further preparation in one of the stainless steel cups. The cup is placed in a muffle furnace set to a temperature above the melting point of the alloy. The small deformation of the sample on melting, to fill the lower part of the cup will probably not introduce nor lose any bifilms during this small movement of the melt.

If it is suspected that the small movement on melting may affect the result the cube-shaped sample can be placed in a melting crucible immersed in sand or other ceramic powder. On melting the sample retains its shape exactly. Even stamped identification numbers and letters on the sample will be reproduced without distortion.

When the sample has melted, the crucible with its molten sample is withdrawn from the furnace, taking care to disturb the molten sample as little as possible. The crucible and its undisturbed contents are then placed under the bell jar of the RPT equipment and the vacuum applied as usual. After the sample has solidified the vacuum is released, the sample removed and studied using density and sectioning to measure bifilm data as usual.

Tests have confirmed that such sampling of solidified ingots does reliably reflect the true bifilm content of the melt from which the ingots were cast. Thus the use of RPT to assess solidified material seems sound.

Shape Casting: 5th International Symposium 2014
Edited by: Murat Tiryakioğlu, John Campbell, and Glenn Byczynski
TMS (The Minerals, Metals & Materials Society), 2014

Prediction of Misruns in Thin Wall Castings Using Computational Simulation

J. Jakumeit, E. Subasic, M. Bünck

Access e.V., Intzestr. 5, 52072 Aachen, Germany
j.jakumeit@access-technology.de, e.subasic@access-technology.de, m.buenck@access-technology.de

Keywords: casting simulation, defect prediction, misrun, thin wall casting, sand casting, centrifugal investment casting

Abstract

The prediction of misruns is challenging because flow and solidification have to be computed in a strongly coupled manner. Effects of surface tension, wetting angle and reduced melt flow due to the solidification must be modeled with high precision.
To accomplish these requirements a finite-volume method with arbitrary polyhedral control volumes is used to solve flow and solidification in a strongly coupled manner. The Volume-of-Fluid approach is used to capture the phase separation between gas, melt and solid in connection with a High-Resolution Interface-Capturing scheme to gain sharp interfaces between phases. An additional source term in the momentum equation was implemented to model the resistance of the dendrite network to the melt flow.
This methodology was applied to predict misruns in thin walled aluminum sand and TiAl centrifugal investment castings. Validation against casting trials using simplified geometries is followed by successful prediction of misruns in complex industrial applications.

Introduction

For the production of sound, thin walled castings the fluidity of the melt is an important property. It is a measure of the ability of a metal melt to flow into a mold. In industrial practice, fluidity is measured as the distance metal flows after pouring into a standard fluidity test channel. During the fluidity test design development, there were different proposals regarding the channel shape and its cross-section dimensions. The first record on the molten metal flow characteristics investigations originates by T.D. West in 1902, and the first trials with fluidity spiral geometry were done by Saito and Hayashi in 1919 [1]. Molten metal fluidity is affected by many factors: chemical composition of the melt, melt superheat, grain refinement, melt quality, mold coatings [2]. In recent years, numerical simulation was used to predict the fluidity of Al and Mg alloys [3, 4].
A drawback of modern, light weight alloys as TiAl is its low fluidity, which easily leads to misruns during the casting process. In centrifugal casting it is possible to overcome problems related to misruns by means of the centrifugal force, which is used to pressurize the melt and thus force its flow into the narrow areas of the mold cavity. The combination of thin geometries like turbine blades and centrifugal casting makes the simulation especially demanding: effects of surface tension and wetting angle must be modeled with high precision together with the acting gravitational, centrifugal and Coriolis forces. To the knowledge of the authors, there are just a few unsuccessful attempts to predict misruns in such kind of centrifugal casting processes reported in the literature. Sung and Kim [5] attempted to model γ−TiAl turbo charger wheel

castings, but were unable to reproduce misruns in simulation as observed in real casting. Though Wang et. al. [6] could successfully model gas entrapments during tilt casting, but they did not explicitly show a stop of melt flow due to solidification. Instead, a temperature decrease or fraction solid increase at the melt free surface was used as indicator for a misrun, a common practice in misrun prediction [7].

In this work, STAR-Cast [8, 9] is used to address this challenging simulation task. The underlying methodology is shortly described in the next chapter. Special attention is paid to those features, which are crucial for a prediction of misruns. The third chapter presents the validation by comparing simulation results experimental findings for simple casting geometries. The fourth chapter describes two industrial application examples: an Al sand casting and a TiAl centrifugal investment casting. The work closes with a short summary.

Theory

The simulation of the complete casting process is based on a Finite-Volume method using control volumes (CVs) of an arbitrary polyhedral shape. The transport equations for mass, momentum, energy, and phases in integral form are applied to each CV, whereby the surface and volume integrals are approximated using the midpoint rule. Linear equation systems for each variable are solved using either conjugate-gradient or algebraic-multi-grid iterative solvers. The equations describing mass (1), momentum (2), energy (3) and phase conservation (4) in a volume V bounded by the surface S are given below (for details see [10]):

$$\frac{\partial}{\partial t}\int_V \rho dV + \oint_S \rho\vec{v}\, ds = 0, \tag{1}$$

$$\frac{\partial}{\partial t}\int_V \rho\vec{v} dV + \oint_S \rho\vec{v}\vec{v}\, ds = \oint_S T\, ds + \int_V (\rho\vec{f}_b + S_\sigma + S_{rot}) dV, \tag{2}$$

$$\frac{\partial}{\partial t}\int_V \rho h dV + \oint_S \rho h\vec{v}\, ds = \oint_S \vec{q}\, ds + \int_V S_Q\, dV, \tag{3}$$

$$\frac{\partial}{\partial t}\int_V \rho C_i dV + \oint_S C_i\vec{v}\, ds = \int_V S_{Ci} dV, \tag{4}$$

where t is the time, ρ is the density, $\boldsymbol{v}$ is the velocity vector, $\boldsymbol{T}$ is the Cauchy stress tensor, $\boldsymbol{f_b}$ is the body force, h is the thermal enthalpy, $\boldsymbol{q}$ is the heat flux vector, C_i is the volume concentration of the phase i and $S_{(\rho/Q/rot/Ci)}$ are source terms described below.

Mass conservation (1), pressure and velocity conservation (2) are coupled via SIMPLE-algorithm [11, 12]. The transient term is discretized based on implicit Euler segregated concept. Details of discretization and the solution method are available in [10] and will not be elaborated here. The temperature distribution is computed using the enthalpy approach (3). For solidification modeling, the volume fraction of solidified liquid is determined using a tabulated fraction solid (f_s) *vs.* temperature curve ($f_s(T)$). Latent heat L is released in proportion to the change in fraction solid ($S_Q = L\ \delta fs/\delta T$). Details of the method are presented in [12] and will not be repeated here.

Centrifugal and Coriolis forces are applied using the momentum source term

$$S_{rot} = -\rho\left[\frac{\partial\vec{w}}{\partial t}\times\vec{r} + 2\vec{w}\times\vec{v} + \vec{w}\times(\vec{w}\times\vec{r})\right], \tag{5}$$

where $\boldsymbol{w}$ is the angle velocity and $\boldsymbol{r}$ the vector to the rotation axis.

In order to calculate the surface force correctly, the method used to track the motion of the free surface must give a sharp interface. The Volume of Fluid (VOF) approach in combination with a High Resolution Interface Capturing (HRIC) scheme is used to tackle the problem: the entire fluid domain is considered to be filled with an effective fluid whose properties vary according to the distribution of volume fractions of melt C_m, solid C_s and gas C_g ($C_m + C_s + C_g = 1$). The transport of melt, solid and gas is computed by solving transport equations for their volume fractions (4) with a source term S_{Ci} for the phase change from melt to solid. In order to achieve sharpness of the interface a HRIC scheme [10, 16] is used, which typically resolves the interface within one cell.

The normal force due to surface tension is treated using the continuum surface force (CSF) model proposed by [15], which defines a volumetric source in the momentum equation that is expressed as:

$$S_\sigma = -\sigma \nabla \left(\frac{\nabla C_{nl}}{|\nabla C_{nl}|} \right) \nabla C_{nl}. \tag{6}$$

To simulate the escape of gas through the sand or shell molds and from other walls, we monitor pressure along non-wetted wall surfaces and compute the penetration velocity based on a pressure difference between the inside and outside wall surfaces and a resistance coefficient given by

$$p_w - p_{amb} = \rho_{scn}(a + b\vec{v}_n)\vec{v}_n, \tag{7}$$

where p_w is the pressure on the inside wall, p_{amb} is the specified reference pressure for the model (assumed to be the pressure outside the mold), ρ_{scn} is the density of the escaping fluid and v_n is the penetration velocity. a and b are free parameter, which need to be calibrated from experiment.

For a prediction of misruns the flow resistance in the mushy zone must be handled correctly. Assuming that the mushy zone acts like a porous media, the fluid velocity resistance in the mushy zone can be approximated as a pressure drop [13]:

$$-\frac{\partial p}{\partial x_i} = \frac{\mu}{K}\vec{v} + \frac{C_E}{\sqrt{K}}\rho|v|\vec{v}, \tag{8}$$

where μ is the viscosity, K the permeability and C_E the Ergun's coefficient. Permeability can be deduced from the Kozeny-Carman equation [14] as $K = ((1-f_s)^3 \lambda_2^2)/(180 f_s^2)$, where λ_2 is the secondary dendrite-arm spacing (SDAS).

Experimental Validation

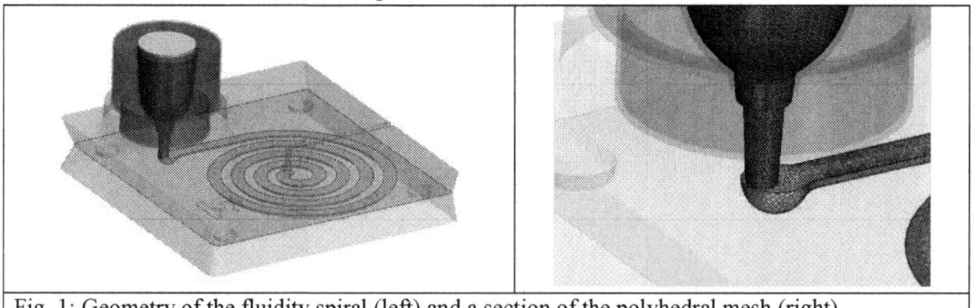

Fig. 1: Geometry of the fluidity spiral (left) and a section of the polyhedral mesh (right)

As test geometry for the validation of Aluminum sand casting a fluidity spiral made from coldbox sand and filled with A356 (AlSi7Mg0.3) alloy was used. The geometry (left) and a detail of the polyhedral mesh (right) are shown in Fig. 1. The thermo-physical data for A356 and cold box sand were taken from the STAR-Cast material database, giving T_{sol} = 555 °C and T_{liq} = 615°C for A356. The experiment was conducted with the pre-filled sprue cup equipped with a stopper. This was modeled in the simulation by keeping the sprue cup region filled using a pressure boundary condition at the top.

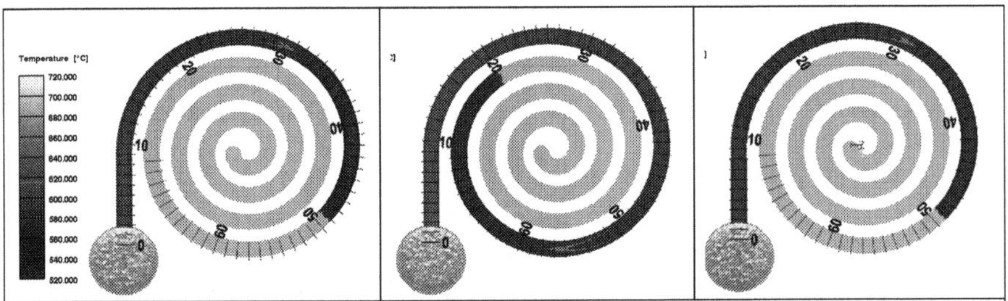

Fig. 2: Flow length in the fluidity spiral test for 3 different HTC settings:
left: $HTC_{contact}$ = 2500 W/m^{2}K, HTC_{gap} = 1500 W/m^{2}K,
middle: $HTC_{contact}$ = 1500 W/m^{2}K, HTC_{gap} = 500 W/m^{2}K,
right: $HTC_{contact}$ = 2500 W/m^{2}K, HTC_{gap} = 500 W/m^{2}K.

A critical parameter for misrun prediction is the heat transfer coefficient (*HTC*), which describes the heat exchange between melt and mold. The influence of surface roughness and gap formation during filling and solidification is difficult and numerical expensive to calculate. Here, a temperature dependent *HTC* was used to take into account the effect of gap formation. At temperatures above T_{liq} and for the heat exchange to the drag, where the spiral is pressed against the sand by gravity, a higher $HTC_{contact}$ was used to model a good contact between melt and mold. The heat exchange to the cope was reduced to HTC_{gap} below T_{sol} to take into account the gap formation. Fig. 2 shows the significant influence of the *HTC* values on the flow length. Simulation result displayed on the left was calculated using $HTC_{contact}$ = 2500 W/m^{2}K and HTC_{gap} = 1500 W/m^{2}K , while $HTC_{contact}$ = 1000 W/m^{2}K and HTC_{gap} = 500 W/m^{2}K were used to get the result displayed in the middle. The predicted flow length distance changes by 30 cm. A change of HTC_{gap} from 1500 W/m^{2}K to 500 W/m^{2}K while keeping $HTC_{contact}$ at 2500 W/m^{2}K has little effect on the flow length, as Fig. 3 right demonstrates. Case left with $HTC_{contact}$ = 2500 W/m^{2}K and HTC_{gap} = 1500 W/m^{2}K was found to be in good agreement with the experimental findings (see also Fig. 4) and is used throughout the rest of this publication.

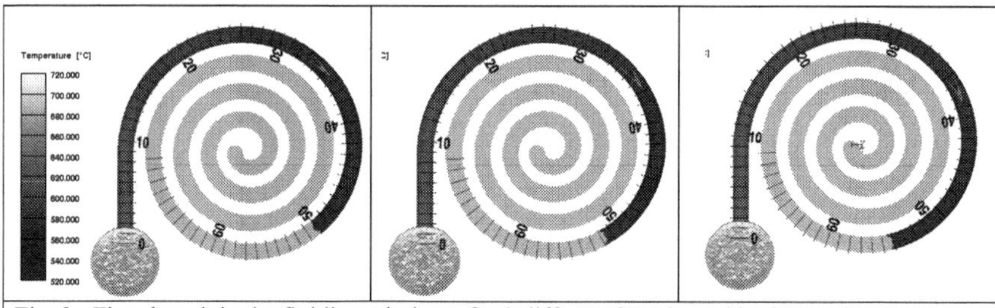

Fig. 3: Flow length in the fluidity spiral test for 3 different SDAS used in the mushy zone model: 7 μm (left), 14 μm (middle) and 30 μm (right).

Next the influence of the parameter of the mushy zone model, which describes the resistance of the dendrite network on the melt flow, was investigated. The SDAS in the middle of the spiral was found to be 14 μm, decreasing down to 7 μm near the sand mold. Fig. 3 shows the flow length obtained with an SDAS of 7 μm (left), 14 μm (middle) and 30 μm (right). The flow length increases from 50 cm to 51 cm and 54 cm, respectively. A larger SDAS leads to a larger permeability and, consequently, to a weaker resistance to the flow; as expected, the flow length increase with increasing SDAS. However, the influence is much smaller as in case of the *HTC* variation.

Finally, the fluidity spiral test simulation calibrated in this way was used to predict the influence of the casting temperature on the flow length as validation. Simulations for casting temperatures of 630, 645, 660, 675, 690, 705 and 720 °C were compared to the results of corresponding casting trials. Fig. 4 demonstrates that the flow lengths, predicted by the simulation, are in excellent agreement with the results of the casting trials.

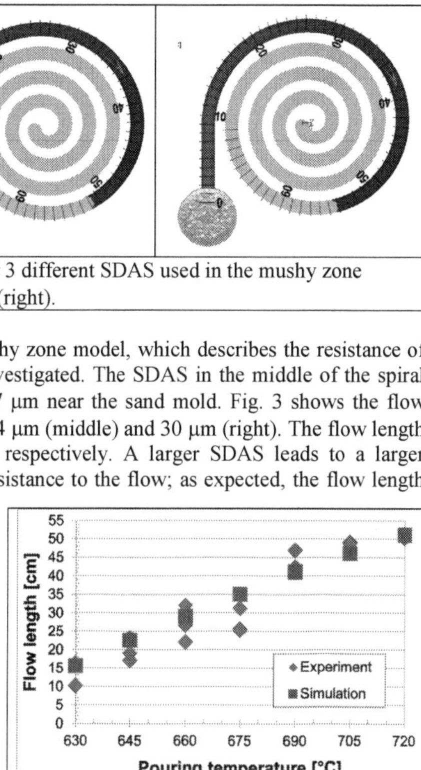

Fig. 4: Direct comparison of predicted and measured flow lengths of A356 as a function of the pouring temperature.

For TiAl centrifugal casting applications, the methodology was validated using simplified rectangular turbine blades of different thickness in a cluster set up with 12 blades. In the casting process, the TiAl alloy GE-48-2-2 (Ti-48Al-2Cr-2Nb) is melted under the influence of magnetic fields by induction skull melting (ISM). The liquid metal is transported through a funnel into the preheated shell mold, which is rotated at 250 rpm. The melt temperature was 1570 °C, the shell mold preheated to 1050 °C. The melt inflow time was 1.3 seconds and the simulation was continued until 1.5 s, when the flow in the blades was completely stopped. Material data for TiAl GE-48-2-2 and the ceramic shell mold was taken from the STAR-Cast material database. An average global *HTC* was applied, which was calibrated for such flow and heat transfer conditions (see [17] for details concerning the *HTC* calibration). Details of this validation experiment were published in [18]. In Fig. 5 two exemplary results for a 1 mm and 2 mm thin blade are shown and compared with experiments. Misrun predictions by the simulation were in excellent agreement with the experimental findings. Simplified blades having a thickness above 2 mm showed no misrun while thinner blades were not completely filled.

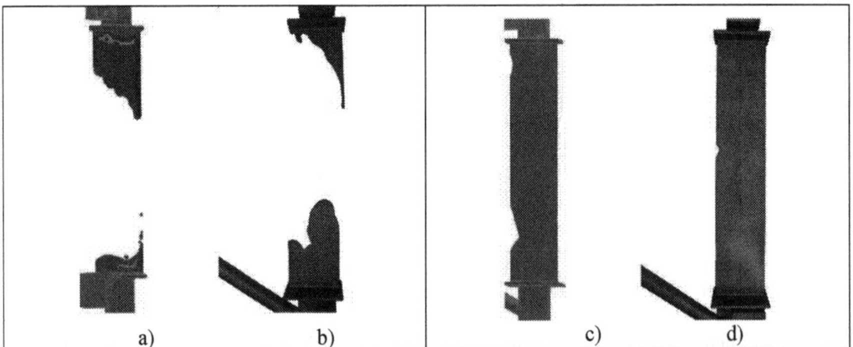

Fig. 5: Misrun prediction by simulation a) + c) compared to real castings b) + d) for a 1 mm a) + b) and 2 mm c) + d) thin simplified turbine blade. (Experiments by T. Stoyanov and R. Tiefers, simulation by S. Jana [18])

Application to industrial castings

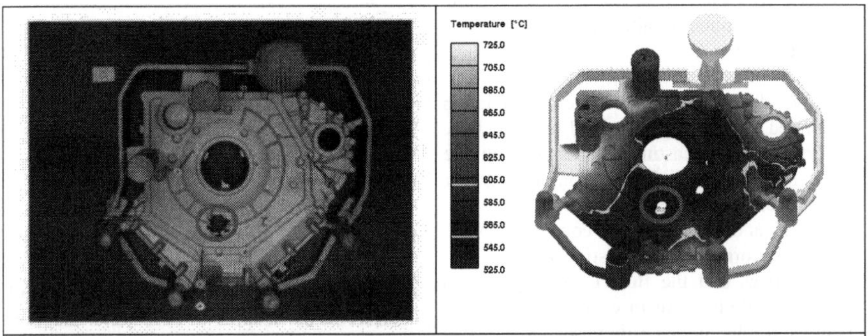

Fig. 6: Comparison of casting trial and numerical simulation of an automotive seal flange. The main misrun is nicely predicted by the simulation.

The application for aluminum sand casting was provided by the automotive industry. A seal flange is difficult to cast without misrun due to its large flat geometry. The part was cast with AlSi10Mg into a green sand mold. Material parameters were taken from the STAR-Cast material database, which gives a T_{liq} as 599 °C and a T_{sol} of 554 °C for $AlSi_{10}Mg$. The *HTC* values calibrated for the fluidity spiral using A356 and cold-box sand were used for the simulation, since no better data was available for the combination of AlSi10Mg and green sand. Fig. 6 left shows the casting with gating and feeding system. A large misrun is present below the central opening (see red circle). In the simulation a misrun is predicted at exactly same position, as Fig. 6 right shows. But the simulation predicts a second msirun location slightly right of the experimental misrun, which cannot be found in the casting. This deviation from the experimental findings can be explained by the missing calibration of the *HTC* for AlSi10Mg and green sand.

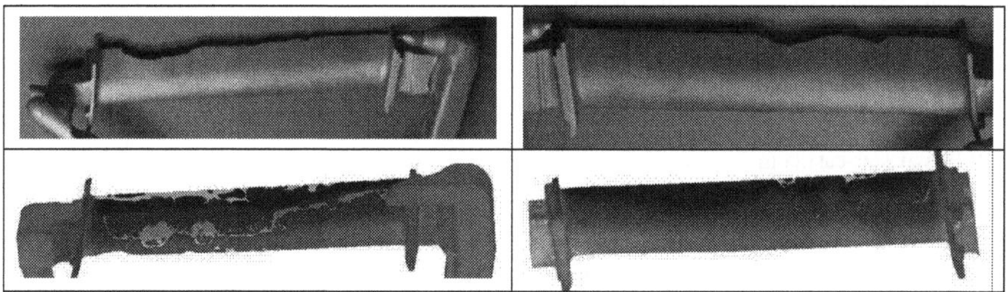

Fig. 7: Comparison of casting trials and simulation results of mold filling and solidification for a centrifugal investment casting of γ-TiAl turbine blades for two different casting setups. (Experiments by O. Kättlitz and J. Aguilar, simulation by S. Jana)

In case of γ-TiAl centrifugal investment casting a turbine blade was used as industrial application. Having a trailing edge thinner than 1 mm filling of the blade before solidification is only possible using centrifugal forces, which push the melt into the thin trailing edge. The casting process parameter, used for the validation experiment, were also used for the industrial application with a casting temperature of 1570 °C and the mold preheated to 1050 °C. The whole setup was rotated with 250 rpm. Fig. 7 compares casting and simulation result for two different casting setups. Since the geometry of sprue and runner system is confidential, only the blade geometry can be shown. Misruns at the trailing edge are predicted by the simulation in good agreement with the experimental findings.

Summary

A three-phase mold filling and solidification methodology has been presented and applied to the simulation of misruns in thin-walled castings in aluminum sand casting and TiAl centrifugal investment casting. The methodology was calibrated and validated using test geometries, namely a fluidity spiral and a simplified turbine blade. Finally, a comparison of the prediction of misruns by simulation to the experimental findings in industrial casting applications showed a good agreement with shape, size and position of the misruns. These results nicely demonstrate the ability to predict misruns in different casting applications by computer simulation.

Acknowledgement

The authors are grateful to Santhanu Jana for performing the centrifugal casting simulations and fruitful discussions, as well as to Rüdiger Tiefers, Todor Stoyanov, Oliver Kättlitz, Julio Aguilar and Stefan Zimmerman for the centrifugal investment casting experiments.

References

[1] Flemings M.C., Conrad H.F., Taylor H.F., 1959, AFS Trans., Vol. 93, p. 496-507

[2] Di Sabatino M., Arnberg L., 2004, Metall. Sci. and Tech., Teksid, vol.22, no.1, p.9-15

[3] Di Sabatino M., Arnberg L., Bonollo F., 2005, Metall. Sci. and Tech., Teksid, vol.23, no.1, p. 3-10

[4] Khan S.S., Hort N., Subasic E., Schmauder S., 2008, Proc. of 10th Asian Foundry Congress, Nagoya, Japan, p. 329-334

[5] Sung S. and Kim Y., 2007, Intermetallics 15, p. 465

[6] Wang H., Djambazov G., Pericleous K., Harding R.A., Wickins M., 2012, Int. J. Cast Metals Research 25, p. 65-74

[7] Ohnaka I., 2008, ASM Handbook vol.15, p. 462-467

[8] www.star-cast.com

[9] Jakumeit J., Laqua R., Peric M. and Schreck E., 2006, Proceedings of Modeling of Casting, Welding and Advanced Solidification Processes XI, p. 63

[10] Demirdzic I. and Muzaferija S., 1995, Comp. Met. in App. Mech. and Eng. 125, p. 235-255.

[11] Ferziger J.H. and Peric M., (ed.) (2002) Computational Methods for Fluid Dynamics, McGraw-Hill, ISBN 2-540-42074-6.

[12] Teskeredzic A., Demirdzic I., Muzaferija S., 2002, Num. Heat Trans., Part B 42, p. 437

[13] Sabau A.S. and Viswanath S., 2002, Met. Mat. Trans. B 33B, p. 131

[14] Poirier D.R., Yeum K. and Mapples A.L., 1987, Metall. Mater. Trans. A 18, p. 1979

[15] Brackbill J.U., Kothe D.B. and Zemach C., 1992, J. of Computational Physics 100, p. 335

[16] Muzaferija S. and Peric M., 1999, eds. Mahrenholtz and M. Markiewicz, WIP Press, Southampton, p. 59-100

[17] Jana S., Kättlitz O., Hediger F., Jakumeit J. and Aguilar J., 2012, MCWASP XIII, IOP Publishing, IOP Conf. Series: Materials Science and Engineering, p. 33

[18] Jana S., Jakumeit J., Tiefers R. and Stoyanov T., Proceedings of LMT-13, Trans. Tech. Publications, Old Windsor, 2013, p. 155

Shape Casting: 5th International Symposium 2014
Edited by: Murat Tiryakioğlu, John Campbell, and Glenn Byczynski
TMS (The Minerals, Metals & Materials Society), 2014

SHAPING OF METALS WITH MAGNETIC FIELDS

M.D. David[1], C.A. Monroe[1], J.A. Griffin[1]

[1]Department of Materials Science and Engineering, University of Alabama at Birmingham
Birmingham, AL 35294

Keywords: Lorentz force, coil design, electromagnetic levitation

Abstract

Electromagnetic fields offer significant potential in metals processing. The ability to shape using the Lorentz force that arises from the interaction between the magnetic field and the current can be controlled. Since this is a containerless technique, contamination in the processed part may be reduced especially when using reactive metals. The force has a rotational component which induces melt motion which may lead to a refined microstructure. However, free surface motion has been demonstrated to be unstable, leading to difficulty in permanent feature creation. Simulations were generated to predict the force and temperature gradient in the melt and the coil. These results can be linked to the surface stability and shaping ability. Parameters were varied to enable design of a system that optimizes the maximum force and minimum heat input. By de-coupling the forcing from the heating of the system, the widest range of operating conditions may be determined.

Introduction

There are many different electromagnetic forces which are already being used for a variety of applications such as maglev trains, magnetic bearings, and electromagnetic reactive armor on military tanks. This force, which is also referred to as the Lorentz force, is generated when currents are in the presence of magnetic fields. The resultant force is perpendicular to the magnetic field lines and the current flow. This relationship is mathematically expressed as:

$$F_L = J \times B \quad (1)$$

$$J = \sigma E \quad (2)$$

$$B = \mu H \quad (3)$$

where F_L is the volumetric Lorentz force (N/m^3), J is the current density (A/m^2), B is the magnetic flux density (T), σ is the electrical conductivity (S/m), E is the electric field (V/m), μ is magnetic permeability (N/A^2), and H is the magnetic intensity (A/m). Lorentz force has a conservative and a rotational term. The former creates a lifting force against gravity and produces an additional pressure, also known as magnetic pressure. The rotational term induces flow and causes electromagnetic stirring [1].

This electromagnetic force can be manipulated and has been investigated for a variety of applications [2-10]. One method of generating the fields necessary to maintain a high Lorentz force is through the use of induction coils. Induction fields are already commonly used in the metal casting industry to melt metal. One existing technique, used for highly reactive metals like titanium, even makes use of the Lorentz force to both melt the metal and levitate it away from the sides of the crucible to minimize contamination. However, the systems have always been optimized with the intent to maximize the heat input rather than the force. By maximizing the force and minimizing the heat input from an induction field, it is possible to create a force field that can be used to shape the metal during freezing.

In this study, induction circular coils are simulated by altering various factors to determine their effects on the resultant Lorentz force. For simplicity, the workpiece subjected to electromagnetic fields is at room temperature. The main objective is to design a coil that will generate sufficient force to levitate the workpiece with minimal heat input. A model is proposed as a theoretical evaluation of the circular coils' capacity to generate the required electromagnetic force to levitate the workpiece. Further work will involve levitating and shaping molten metals.

Methodology

Simulations were performed using COMSOL Multiphysics [11]. Using the induction heating (ih) interface and a time-dependent study, the Lorentz force and various heat transfer variables were calculated. Due to the symmetry of the problem, a 2D axisymmetric geometry was used. The horizontal axis of the modeling plane represents the r-axis while the vertical axis represents the z-axis. In the rz plane, the workpiece and air domains appear as a rectangle while the coil and water domains are concentric circles. Figure 1 shows a 3D view of the geometry.

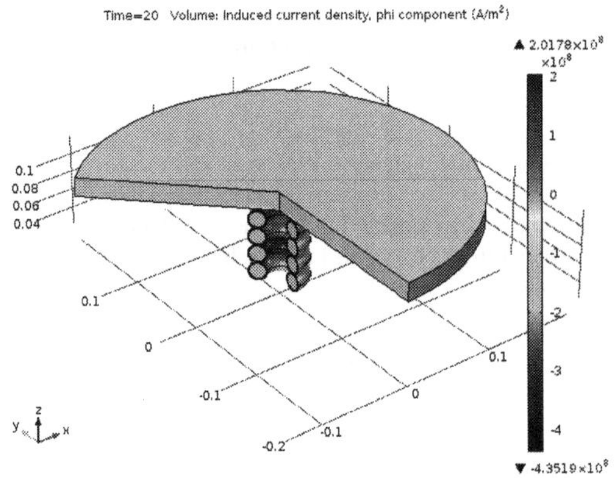

Figure 1. 2D axisymmetric diagram of the geometry showing a workpiece directly above a 4-turn coil surrounded by an air domain. The applied current in this time-dependent study is 500A at 125kHz.

The workpiece had a radius of 200mm and a height of 10mm and was placed directly above the coil. The workpiece material was aluminum A356 and its properties were imported from JMatPro [12]. The coil was made of copper and was water-cooled inside. The coil was tightly wound so each coil domain touched the outer boundary of an adjacent domain.

The resulting Lorentz force was calculated as a function of different factors. Table I summarizes all these variables and the range of values used. The tube dimensions were based on commercially available copper tubes. Each study was run for 20 seconds with a step size of 1s.

Table I. Variables Used to Simulate Electromagnetic Force of a Thin Aluminum Sample

Parameters	Unit	Values
Current	A	50, 100, 200, 500, 1000, 2000
Frequency	kHz	0.1, 05, 1, 2, 50, 125
Tube size: Outer Diameter (Inner Diameter)	mm	1.5 (0.8), 2.5 (0.9), 3.1(1.8), 6.4 (4.8), 9.5 (8.0), 12.7 (10.9), 19.1 (16.9)
Coil size: ID	mm	23.3, 40, 60, 100, 200, 400
Number of Turns	-	1, 2, 3, 4, 5, 6, 8, 10, 12
Number of Layers	-	1, 2, 3, 4, 5
Distance of coil from workpiece	mm	1.75

The turns mentioned above is the number of coil domain (each circle on the 2D plane revolved 360° around the z-axis) counted along the z-axis. Layers, on the other hand, can be visualized as the number of coil domain (revolved 360° around the z-axis) counted along the r-axis. Figure 2 shows 3D images of a 4-turn coil and a 4-layer coil, respectively. Take note that the geometry was modeled in 2D so the 3D image of the coils appears to show that each domain was revolved around the z-axis separately. However, the coil domains in both cases were defined to be connected in series in the simulations.

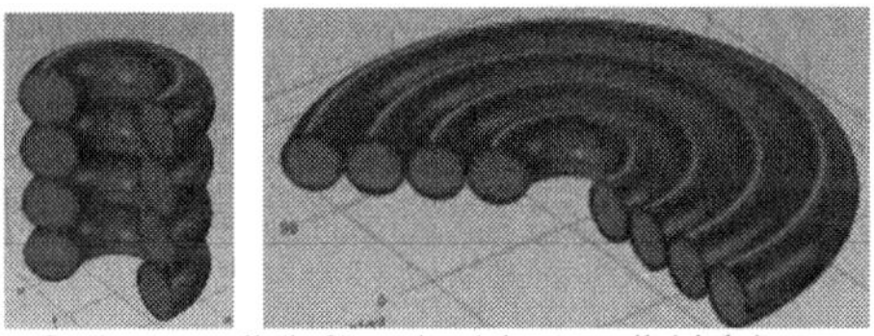

Figure 2. 3D diagram of a 4-turn coil (left) and a 4-layer coil (right) presented to distinguish the difference between the terminologies turn and layer used in this study

The heat source, Q, in the induction heating model was prescribed as:

$$Q = \frac{\frac{dM}{dt}C_p(T_{in}-T)}{2\pi r A} \tag{4}$$

where dM/dt is the water mass flow rate, T_{in} is the water inlet temperature, T is the actual temperature in the domain, r is the radial coordinate, and A is the cross section of the water domain. The electrical conductivity, σ, of the workpiece was defined as a function of temperature using the imported values from the external material database while the coil's conductivity was set to decrease linearly with temperature. Each turn of the coil was connected in series. The water entered the tube at 20°C with flow rates varying with tube size (see Table II). The values reported in the reference were scaled up by a factor of 260 based on flow rates used previously on related experiments which used a 6mm- and 5mm- tube OD. The surrounding air was set to room temperature. Predefined fine mesh size was used for all domains.

Table II. Water flow rates for different tube sizes [13]

Tube OD,mm	Tube ID, mm	flow rate, kg/s
1.5	0.8	0.018
2.5	0.9	0.019
3.2	1.8	0.033
6.4	4.8	0.066
9.5	8.0	0.132
12.7	10.9	0.198
19.1	16.9	0.412

Results and Discussion

The applied current in the coil domain was defined to flow into the page and the magnetic fields it generated move clockwise as shown in Figure 3. The Lorentz force created was mainly along the z axis and its radial component varied slightly. This force was higher on areas of the workpiece that were directly above the coil as represented by the length of the arrows.

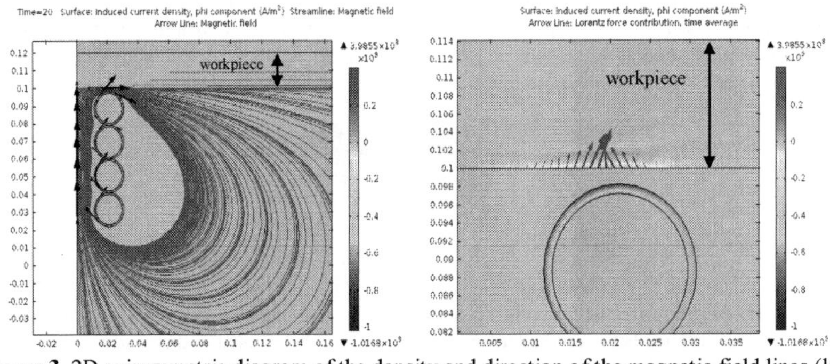

Figure 3. 2D axisymmetric diagram of the density and direction of the magnetic field lines (left) and the direction of the resultant Lorentz force for a magnitude of 1.5 N experienced by the workpiece (right) (tube OD: 19.1mm; 500A; 125kHz)

The mass of the aluminum workpiece was about 3.4kg (density is 2680 kg/m^3 and volume was about 1.26x10^{-3} mm^3); thus, a 33N upward force was needed to obtain a zero net force against gravity. The instability was neglected in the simulations discussed in this paper for the purpose of establishing the ideal relationships between various coil parameters and the resulting Lorentz force.

To determine the controlling parameters in force optimization, one factor was varied while fixing the remaining variables to a constant value. Tube size was varied for each condition. In each simulation, the coil's ID was fixed to 23.3mm except for the study of force as a function of coil size. The force significantly decreased as the coil moves farther below the sample; thus, the coil was constantly positioned at 1.75mm below the workpiece to maximize the force. Take note that Equations 5 to 9 that will be presented in the succeeding sections will predict the force only for the specified parameters used in the simulations. A generalized equation (see Equation 10) will be presented later on. Furthermore, these equations exclude the total heat dissipated and the final temperature on the coil. However, it will be briefly discussed at the end of this section.

Increasing the applied current significantly increased the electromagnetic force. Tube size was inversely proportional to the resultant force. Using a 4-turn coil and a frequency of 125kHz, a relationship between force, current, and tube size was established and is shown in Equation 5 where *I* is the current (A) and *tube OD* is the outer diameter of the tube (mm). Higher current in the inductor resulted in a higher induced current in the workpiece which increased the magnetic flux in the conductor and yielded a higher resultant force.

$$Force\ (I) = \left(\frac{0.00001}{e^{0.1(tube\ OD)}}\right) I^2 \tag{5}$$

Lorentz force was also directly related to the frequency of the alternating current. Increasing the frequency led to higher force for all tube sizes but it reached a point where the force plateaued. Thus, further increasing the frequency did not cause an apparent increase in force as compared to increasing the coil current. Equation 6 quantifies this effect where *f* is the frequency (Hz). The force decreased slightly as tube size decreased. This relationship was established from the simulation results using a 4-turn coil with an applied current of 500A.

$$Force\ (f) = \frac{0.5}{tube\ OD}\ln(f) + 0.3\ln(tube\ OD) - 1 \tag{6}$$

The force was also optimized by controlling the number of coil turns and layers. The current and frequency were fixed at 500A and 125kHz, respectively. Equations 7 and 8 summarize the effects of varying the turns (*tu*) and layers (*l*) of the coil on the resulting force. The force equation as a function of turns was obtained by modeling a single-layer coil and creating different number of turns for each simulation. A 1-turn coil with varying number of layers was used to define the force-layer relationship.

$$Force\ (tu) = \frac{1.4}{tube\ OD}(tu) - 0.004\,(tube\ OD)^2 + 0.1(tube\ OD) - 0.6 \tag{7}$$

$$Force\ (l) = \frac{0.3}{(tube\ OD)^2}\,l^2 + [0.0004(tube\ OD)^2 - 0.02(tube\ OD) + 0.1]\,l \tag{8}$$

Although increasing both variables led to higher electromagnetic force, adding more coil layers appeared to be more effective. There was a number of windings at which the resulting Lorentz force flattened out and additional turns had almost no effect on the force. This critical number increased as the tube size became smaller. Increasing the layers, on the other hand, continuously led to a much higher force. This increase grew with decreased tube size.

Coil size was also a critical factor to consider. The simulation results are summarized into Equation 9 where *ratio* corresponds to the coil's outer diameter divided by the total length of the workpiece. A 500A current at a frequency of 125kHz and a 400mm-long workpiece were used. As the coil became wider (i.e. *ratio* ≤ 1), the force increased and reached its maximum when the outer edge of the coil was aligned to the workpiece (i.e. *ratio* = 1). As the coil OD became larger than the length of the sample (i.e. *ratio* > 1), the force decreased again. The decrease in the force as the coil exceeded the length of the workpiece was more prominent for larger tubes. Equation 9, however, only works for *ratio* less than or equal to 1.

$$Force\,(ratio) = [3\ln(tube\ OD) + 11]\,ratio^2 + [-18\ln(tube\ OD) + 58]ratio \tag{9}$$

The established relationships between the resultant Lorentz force and the variables are summed up to form an equation that can predict the force for any circular coils (see Equation 10). The force as function of layer, *F(l),* is the main factor since it seemed to cause the most significant effect on force. The other variables were incorporated by multiplying the ratio of the function of each factor and the values of these functions for *tu=1, I=500, f=125000*, and *ratio= (23.3+ (tube OD x 2))/ 400*, respectively, to *F(l)*. Recall that these are the fixed values used in simulations performed to obtain *F(l)*. Take note that this equation is mainly for circular coils. It may work for conical coils; however, the force will depend on the angle between coil turns which was not considered in this model. In addition, the coil must be as close as possible to the workpiece.

$$Force = F(l) \times \frac{F(tu)}{F(tu=1)} \times \frac{F(I)}{F\,(I=500)} \times \frac{F(f)}{F\,(f=125000)} \times \frac{F(ratio)}{F(ratio=\frac{23.3+(tube\ OD\times 2)}{400})} \tag{10}$$

Figure 4 shows the comparison between the Lorentz force obtained from the simulations and the predicted value using Equation 10. In some cases, the calculated force is slightly lower than the one obtained from the simulations. Hence, the predictions made by the equation must not be treated as the sole basis of designing a coil. It may, however, be used as an initial tool in estimating the force that will be generated by a proposed coil configuration.

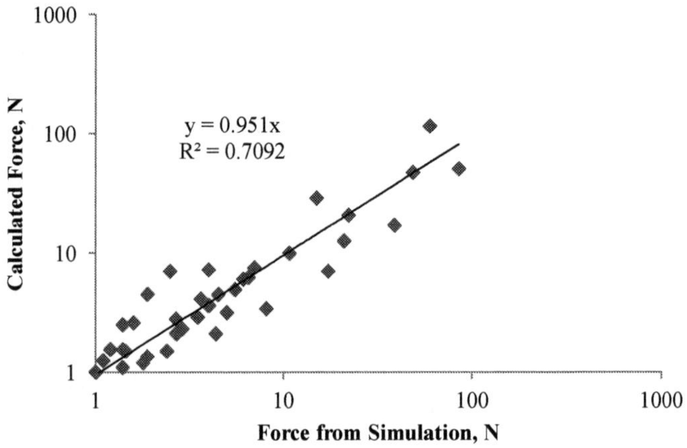

Figure 4. Correlation between the force obtained from simulations and the predicted force using the defined equation.

Based on the parameters examined, using a small tube and increasing the number of coil layers and the current will optimize the resultant Lorentz force. However, power supply requirements and heat input will limit the allowed coil designs in actual experiments. Figure 5 shows the force, power, and total heat source as a function of tube size for a coil with 4 turns and 4 layers. Decreasing the tube size and increasing the total coil length resulted in a high electromagnetic force but it also increased the heat and power generated by the coil. The coil power of a coil configuration must be within the operating limits of the available power supply. In addition, longer coils with smaller tubes will tend to heat up faster and may melt. Hence, designing the coil must be a balance between optimizing the force, minimizing the total heat that needs to be extracted from the coil, and meeting the power supply limits.

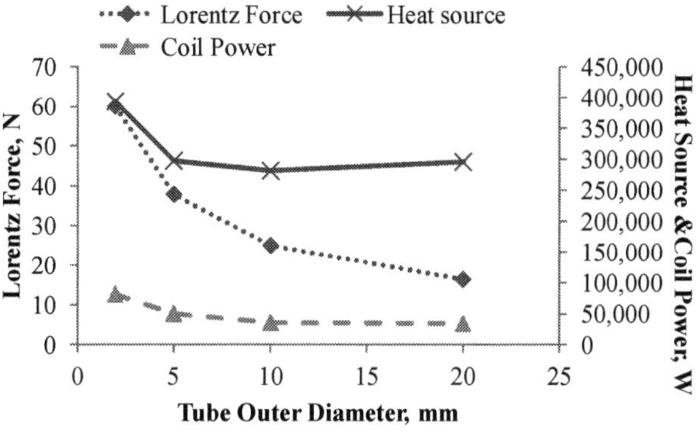

Figure 5. Lorentz force, Coil power, and heat input as a function of tube size for a 4x4 coil (4 turns and 4 layers).

Conclusions

It was shown that various factors can be controlled to optimize the force. Winding up a small tube into a multi-layer coil will result to a much higher force. Operating at a higher current will also significantly increase the electromagnetic force. Using the simulation results, a model was established that can predict the resultant Lorentz force on circular coils. However, the calculated force using the equation presented was lower compared to the force obtained from the simulations in some cases. Although various possible ways to maximize the force were presented, it must be noted that the total power and heat generated in the coil must also be considered. Depending on the available power supply, this will limit the coil configurations and process parameters that can be used.

Acknowledgements

"Research was sponsored by the U.S. Army Contracting Command-New Jersey, Emerging Technologies Contracting Center (ACC-NJ, ET), Benet Laboratories on behalf of the US Army Armament Research, Development & Engineering Command's (ARDEC) Benet Laboratories (BL) and the Defense Advanced Research Projects Agency (DARPA) and was accomplished under Grant Number W15QKN-12-1-0001. The views and conclusions contained in this document are those of the authors and should not be interpreted as representing the official policies, either expressed or implied, of the ACC-NJ, ET on behalf of the ARDEC/BL and

DARPA, or the U.S. Government. The U.S. Government is authorized to reproduce and distribute reprints for Government purposes notwithstanding any copyright notation hereon."

References

1. D.M. Matson and D.M. Herlach, *Solidification of Containerless Undercooled Melts,* (Hoboken: John Wiley & Songs, 2012).

2. R. L. Hollis and S. E. Salcudean, *Int'l Symposium for Robotics Research*, Hidden Valley, Pa., October 1-3, 1993.

3. I. Egry, A. Diefenbach, W. Dreier, and J. Piller, *International Journal of Thermophysics* 22, 2, 569-557 (2001).

4. H. Yasuda, I. Ohnaka, Y. Ninomiya, R. Ishii, S. Fujita, and K. Kishio, *Journal of Crystal Growth* 260, 475-485 (2004).

5. J. Szekely, E. Schwartz, and R. Hyers, *Journal of the Minerals, Metals, & Materials Society* 47, 5, 50-53 (1995).

6. H. Z. Fu, J. Shen, L. Liu, Q.T. Hao, S.M. Li, and J.S. Li, *Journal of Materials Processing Technology* 148, 25-29 (2004).

7. T. Campanella, C. Charbon, and M. Rappaz, *Metallurgical and Materials Transactions* 35A, 3201-3210 (2004).

8. V. Bojarevics and R.W. Hyers, *Journal of the Minerals, Metals, & Materials Society* 64, 9, 1089-1096 (2012).

9. C. Zhiqiang, J. Fei, Z. Xingguo, H. Hai, and J. Junze, *Materials Science and Engineering* A327, 133-37 (2002).

10. J.K.R. Weber, S. Krishnan, and P.C. Nordine, *Journal of the Minerals, Metals, & Materials Society* 43, 7, 8-14 (1991).

11. COMSOL, Inc., *COMSOL v4.2a*, 1 New England Executive Park, Burlington, MA 01803.

12. Sente Software Ltd., *JMatPro v6.2.1*, Guildford GU27YG: Surrey Technology Centre, 2013.

13. *Plumbing Supply*, http://www.plumbingsupply.com

Shape Casting: 5th International Symposium 2014
Edited by: Murat Tiryakioğlu, John Campbell, and Glenn Byczynski
TMS (The Minerals, Metals & Materials Society), 2014

The Formation of Hydrogen Related Porosity by Double Oxide Film Defects in Al Alloys

A.J. Gerrard[1] and W.D. Griffiths[2]

[1]Alcoa Europe, Kitts Green, Birmingham, UK, B33 9QR.
[2]School of Metallurgy and Materials, College of Engineering and Physical Sciences, University of Birmingham, Edgbaston, Birmingham, UK, B15 2TT.

Keywords: Oxide Films, Aluminium Alloys, Aluminium Oxide, Aluminium Nitride, Hydrogen Porosity.

Abstract

Liquid aluminium readily forms a surface oxide film, and during melting, casting and metal transfer operations double oxide film defects may be readily entrained into the liquid metal. Dissolved H is thought to diffuse into these defects and re-combine to form diatomic hydrogen molecules, leading to the formation of H porosity in the solid casting. However the diffusion of H into double oxide film defects may be considered unlikely as even thin Al oxide films have been reported to be effective barriers to the diffusion of H. In the experiments reported here, solid samples of commercial purity Al alloy were degassed by repeated use of a LECO hydrogen measurement device, then melted and exposed to an atmosphere of either oxygen or nitrogen. This created a surface containing Al_2O_3, or Al_2O_3 and AlN, respectively. When exposed to a H_2 gas atmosphere, the absorption of H was found to be greater in specimens containing AlN on the surface. The results suggest that AlN can form in cracks in an existing oxide layer and allowed greater diffusion of hydrogen when compared to samples with only alumina on their surface.

Introduction

An oxide film rapidly forms on the surface of molten aluminium and aluminium alloys due to their high chemical affinity for oxygen. These surface oxide films can fold over due to surface turbulence of the melt, such as that generated by a poor running system, resulting in the formation of double oxide film defects, (also referred to as bifilms), in a process initially proposed by Campbell [1]. Campbell also suggested that hydrogen dissolved in the melt, (which originates from the reaction between molten aluminium and water vapour), could diffuse into bifilms and inflate them, leading to the development of hydrogen gas porosity. This occurs because bifilms will encapsulate and entrap atmospheric gases during their entrainment (probably consisting principally of air) [2], and provide a pre-existing gas phase into which hydrogen can diffuse [3]. Diffusion of hydrogen into this gaseous envelope could cause expansion of the defect into spherical bubbles with dimensions far greater than the original bifilm [4]. Bifilms present a convenient solution to the problem of nucleation of hydrogen gas porosity during solidification, as both homogeneous and heterogeneous nucleation have been suggested to be unlikely to occur due to the high internal gas pressures required [1,5,6,7]. A pre-existing gas phase eliminates the requirement for nucleation of a gas pore.

Following entrainment and formation of a bifilm the gases within have been suggested to be consumed by reaction with the melt. Nyahumwa *et al.*[8] suggested that the continuous amorphous alumina formed on the surface of the melt is entrained, and then transforms to γ-Al_2O_3 and then to α-Al_2O_3, resulting in a 24% reduction in oxide volume. This transformation leads to rupture of the oxide film, allowing the gases within the bifilm to react with the surrounding melt. Raiszadeh and Griffiths [9] trapped an air bubble in an inert ceramic tube held in an Al melt, and observed a reaction between aluminium and oxygen, forming aluminium oxide, followed by a quicker consumption of gases after about 5 hours, perhaps attributable to break away oxidation and failure of the oxide film as described by Sleppy [10]. Once all the available oxygen was consumed the nitrogen contained within the bubble started to react with the melt (through the cracks in the oxide film) to form aluminium nitride.

For a bifilm to be a precursor to gas porosity, hydrogen must first be able to enter the bifilm envelope. Hydrogen in solution exists in the form of protons, diffusing interstitially throughout the liquid metal [7,11]. Protons would be expected to diffuse through or circumvent the ceramic bifilm and to recombine in the bifilm interior to form diatomic hydrogen molecules, which is an exothermic process [12]. H might also react with O to form H_2O. Aluminium oxide however, has been shown to be an effective barrier to the diffusion of hydrogen [13]. The diffusion coefficient of protons through alumina has been measured to be as little as $6.5x10^{-18}$ cm^2s^{-1} at 25°C [14]. Films of aluminium oxide as little as 20 nm thick are used as surface deposition barrier coatings in chemical and electrical engineering applications to control the diffusion of hydrogen. Relating this to double oxide film defects, aluminium oxide films that crack or tear should be quickly sealed by formation of aluminium oxide, which means that the theory that bifilms are a precursor to hydrogen gas porosity is undermined by the apparent inability of protons to diffuse into the defect through the oxide walls.

However, once all or most of the oxygen in the interior atmosphere has been consumed in this way, AlN should then form. It is unclear if hydrogen can diffuse through AlN structures, and this work is aimed at determining if hydrogen can diffuse into the bifilm defect interior atmosphere more readily through AlN than through alumina.

Experimental Procedure

Accurately assessing the H content of solid aluminium alloys is challenging because there are no standards to which measurements can be compared. The LECO hydrogen determination device heats solid specimens and extracts the hydrogen contained within into a nitrogen carrier gas. The hydrogen content is established by assessing the thermal conductivity of the resulting N_2 and H_2 gas mixture, however the extraction proces is not repeatable.

Commercially pure Al (>99.7% Al) and 5083 wrought alloy were used in these experiments. The composition of the alloy influences the oxide that forms on the sample surface; pure aluminium forms aluminium oxide (Al_2O_3) and 5083 (containing 4.0-4.9 wt.% Mg) forms magnesium oxide (MgO), respectively [1]. Several specimens were machined from each alloy of 8 mm OD and 50 mm length, for determination of H content using a LECO. In each instance the LECO was first prepared by heating it five times using a high temperature outgassing procedure (>700°C) that drives moisture from the furnace and the graphite crucible used in the measurement. The machine was then calibrated by dosing the thermal conductivity cell with pure hydrogen gas (>99.9995% H). Low oxygen nitrogen (>99.998% N) was used as a carrier gas to extract hydrogen from the samples.

In the first experiment the degassing potential of the LECO was evaluated by repeatedly analysing two 5083 alloy samples 10 times. After 10 analyses the two samples were left to stand exposed to air in the laboratory for 10 weeks, before being re-measured by the LECO, in order to determine the hydrogen absorption over this period.

In the second experiment analysed/degassed samples were placed in different environments to observe their influence on the absorption of hydrogen. Four 5083 alloy samples were analysed/degassed using the LECO (three times), and then placed in different atmospheric conditions for four days before their hydrogen absorption was determined using the LECO. A single sample was placed in each of the following environments; (1) submerged in a glass container filled with water, (2) placed in a glass jar open to the atmosphere, (3) sealed in a glass container (limiting the available water vapour), and (4) submerged in a glass container filled with acetone.

In a third experiment the composition of the surface film was modified by melting the alloy samples in different atmospheres. Pure aluminium and 5083 alloy samples were degassed using the LECO (three times), then placed in a furnace with a controlled atmosphere, with different experiments conducted as detailed in the following:

Experiment 3(a):
Two commercial purity aluminium samples were degassed and then sealed into the furnace at room temperature in an atmosphere of air. The furnace was heated to 700°C at a rate of 40 $Kmin^{-1}$ and the samples first expanded, cracking any surface oxide film, and then melted. The surface of these samples would therefore be oxidised during heating and melting in the air, forming an Al_2O_3 surface oxide film over the entire surface of the samples. They were held at 700°C for 15 minutes, after which the air was removed to a vacuum pressure of 1 Pa. The furnace was then filled with hydrogen gas to a pressure of 150 kPa and held at this temperature and pressure for one hour. The furnace was then cooled at a rate of 20 $Kmin^{-1}$ per minute to 50°C before the hydrogen was removed and the samples taken out. The samples were placed in sealed glass jars and their H content measured in the LECO.

Experiment 3(b):
Two degassed commercial purity aluminium samples were inserted into the furnace along with a ceramic dish that contained approximately 50g of Hf turnings, (to act as an oxygen scavenger). The atmosphere was reduced to a pressure of 1 Pa of air, and then filled with Ar (99.999% purity) to a pressure of 100 kPa. The Ar was then removed by vacuum until a residual pressure of 1 Pa remained within the furnace. The system was pressurised with Ar and removed by vacuum five times to try to ensure that minimal atmospheric oxygen or water vapour remained within the furnace. The evacuated furnace was then filled with low oxygen nitrogen (>99.998% N) to a pressure of 150 kPa, before being heated to 700°C and allowed to stand for 15 minutes in

a nitrogen atmosphere. It was expected that these samples would form aluminium nitride (AlN) on any exposed Al surface occurring during heating or melting. The furnace was then evacuated, removing the nitrogen gas, and filled with hydrogen gas to a pressure of 150 kPa and held at this temperature and pressure for one hour. The furnace was then cooled at a rate of 20 $Kmin^{-1}$ per minute to 50°C before the hydrogen was removed and the H content of the samples measured.

Experiment 3(c):
This was identical to the experiment described in 3(a) but used 5083 alloy samples instead of commercial purity aluminium.

Results

In experiment one, the repeated H measurements showed that about 85% of the initial hydrogen content was removed from each sample during the first analysis in the LECO. Nearly all of the hydrogen was removed from the sample after three analyses, after which an average of 0.005 ppm H remained, (see Figure 1), (the accuracy of the device is claimed to be ± 0.005 ppm). Ten weeks later, after resting uncovered in a laboratory environment, both samples had absorbed hydrogen. 5083 alloy would be expected to have a porous magnesium oxide surface layer, which is not protective, and which would allow continued reaction between the surface and atmospheric water vapour. The two samples were shown to have picked up almost as much hydrogen as their initial content, suggesting that this was the limit to which the solid alloy could absorb hydrogen.

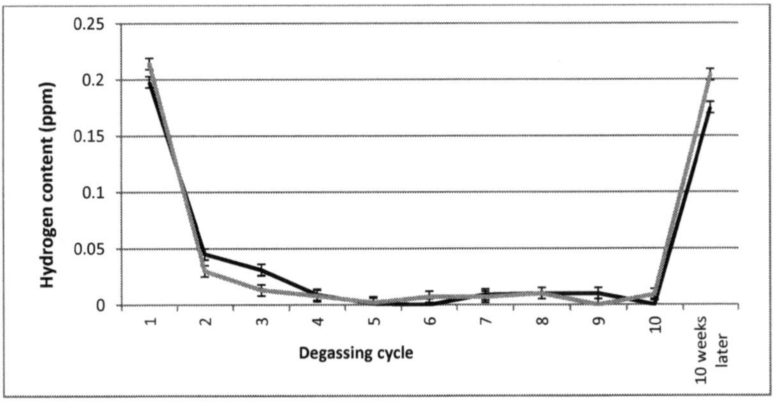

Figure 1: Two 5083 alloy samples measured using the LECO ten consecutive times. The absorption of hydrogen was evaluated ten weeks later.

Experiment two showed that the 5083 alloys placed in specific environments absorbed varying amounts of hydrogen. Figure 2 shows that the sample submerged in water and the sample in laboratory air absorbed 0.16 and 0.15 ppm of H respectively. In both cases moisture was readily available for reaction with the alloy. The sample that was placed in a sealed container (i.e. limited water vapour) was observed to have absorbed only 0.024 ppm of hydrogen after four days. The sample submerged in acetone also absorbed very little hydrogen, only 0.038 ppm.

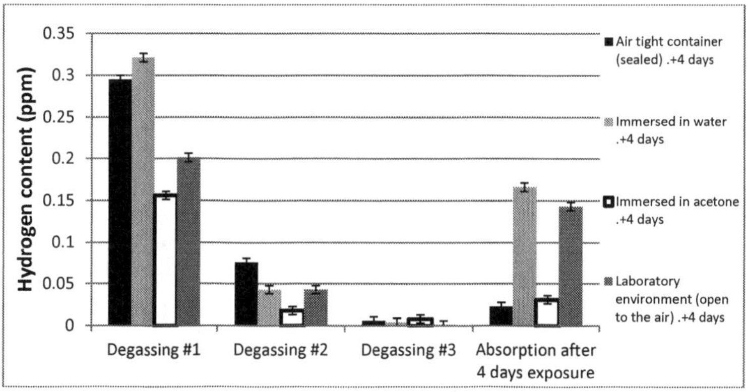

Figure 2: The degassing of 5083 alloy samples (three times) and the absorption of hydrogen into those samples after exposure to various environments.

In experiment 3, the hydrogen absorbed during each experiment, in air and in nitrogen, was determined by analysing each sample three times. The results are shown in Figure 3. The measured mean absorption figures were, for Experiment 3(a) an increase of 0.04 ppm for the pure Al heated in air, for Experiment 3(b) an increase of 0.08 ppm for the pure Al heated in N_2 and, for Experiment 3(c) an increase of 0.279 ppm for the 5083 alloy heated in air.

SEM analysis of the commercial purity Al samples heated in nitrogen revealed that cracks that had developed in the original aluminium oxide film during expansion and melting were decorated with AlN crystals, (established by EDX). These crystals, shown in Figures 4(a) to 4(c), may have nucleated from the edges of the aluminium oxide layer and grown upon the aluminium exposed underneath,

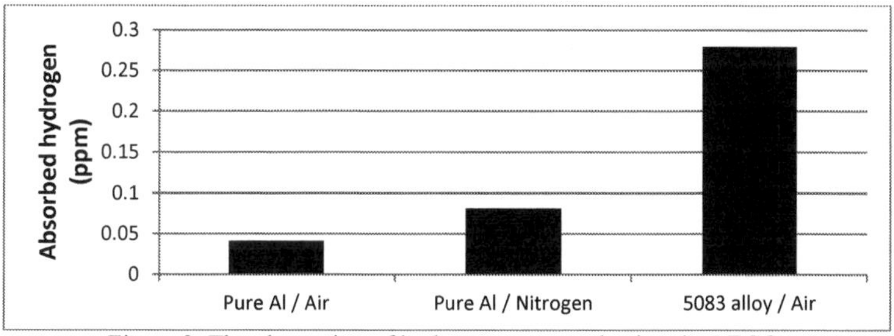

Figure 3: The absorption of hydrogen measured using the LECO.

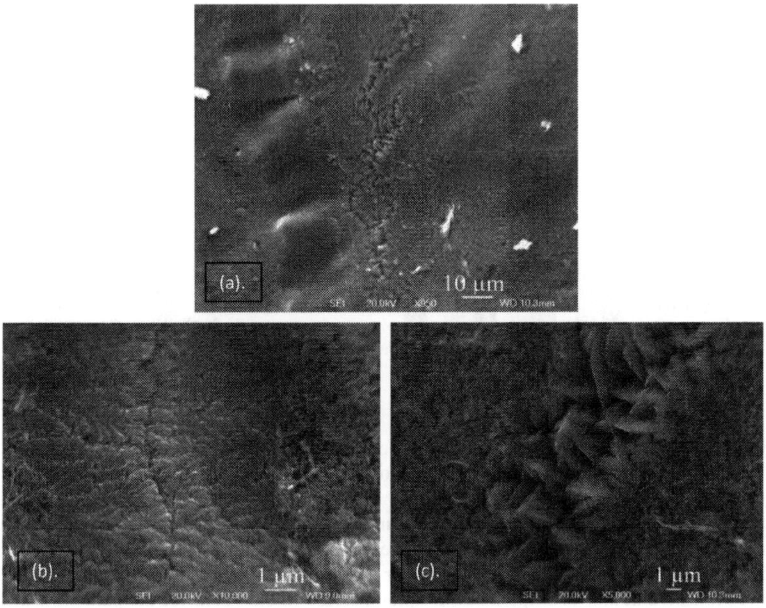

Figure 4: The surface of a commercial purity aluminium sample heated in nitrogen, showing cracks in the original oxide film (a). The cracks were observed to be covered with a mixture of (b) fine or (c) coarse AlN crystals.

Discussion

In experiment 3 the pure aluminium samples heated in both air and nitrogen absorbed measurable amounts of H. This demonstrated that hydrogen was able to diffuse through the aluminium oxide and nitride films on the surface of the specimens. But the hydrogen absorption of the pure aluminium samples heated in nitrogen was greater, by 0.04 ppm, or about double, a difference demonstrated to be statistically significant at the 95% level.

The original aluminium oxide film present on the surface of the commercial purity aluminium should have cracked during heating, due to thermal expansion. As the oxide film cracked the aluminium beneath was exposed to air, and instantaneous oxidation of the aluminium would occur, effectively healing the cracks as soon as they formed. When they were exposed to hydrogen, the gas must diffuse through the oxide film into the sample.

When the commercial purity aluminium samples were heated in nitrogen, however, the cracks in the oxide film that formed would allow the aluminium to react to form AlN, as shown in Figure 4. The AlN crystals may have nucleated and grew from both edges of the original aluminium oxide film, until they met and completely covered the cracks. The permeability of hydrogen through the nitride crystals was evidently greater (compared to the oxide) and these samples absorbed significantly more hydrogen. The coarse AlN crystals shown in figure 4(c) appear interlocked, and it is proposed that the crystal structure has greater permeability for hydrogen because of this morphology. The experiments here demonstrate that hydrogen can diffuse through both aluminium oxide and aluminium nitride layers, but that the thickness and structure of the films likely influences the rate at which hydrogen can diffuse into the envelope over time, and that AlN offers less of a barrier to the passage of hydrogen than does alumina.

The 5083 alloy samples also readily absorbed hydrogen. This occurred because the magnesium oxide that formed on the surface of the sample was permeable, allowing hydrogen to diffuse through the MgO. The internal atmosphere of a bifilm composed of MgO would be expected to react with the surrounding melt relatively quickly (compared to pure Al) because when the film cracks or tears, the oxygen contained within reacts with the melt to form a discontinuous oxide, which would easily tear again. Cracks that form in both MgO and any nitride that formed would also allow protons to diffuse into the internal atmosphere of the bifilm for continued reaction.

Conclusions

1. Repeated use of the LECO was used to degas aluminium samples to low H levels (<0.01 ppm).
2. Degassed 5083 alloy specimens absorbed hydrogen by reaction with atmospheric water vapour, as did specimens that were immersed in water. Specimens that were degassed and sealed in an air tight container absorbed very little hydrogen over a period of four days.
3. Degassed 5083 alloy specimens that were melted in air and exposed to hydrogen gas readily absorbed hydrogen as the porous magnesium oxide surface film did not impede the diffusion of hydrogen.
4. In a low oxygen atmosphere, (such as might exist in a double oxide film defect), nitrogen will react with aluminium to form aluminium nitride at cracks or tears in oxide film.
5. Commercial purity aluminium specimens that were heated in air or nitrogen (forming Al_2O_3 or AlN on the surface, respectively) were both observed to absorb hydrogen. Absorption into samples that were heated in nitrogen was significantly greater, and about double, probably due to the formation of porous AlN crystals that grew within cracks in the original Al_2O_3 film.

Acknowledgements

The authors wish to thank EPSRC and Alcoa Europe for financial support.

References

[1] J. Campbell, Castings, 2nd ed., Butterworth-Heinemann, pp. 17-69, 2003.

[2] J. Campbell, "An overview of the effects of bifilms on the structure and properties of cast alloys," Metallurgical and Materials Transactions B, vol. 37 A, pp. 857-863, 2006.

[3] D. Dispinar and J. Campbell, "Porosity, hydrogen and bifilm content in Al alloy castings," Materials Science and Engineering A, vol. 528, pp. 3860–3865, 2011.

[4] J. Campbell, "Entrainment defects," Materials Science and Technology, vol. 22, pp. 127-145, 2006.

[5] K. J. Brondyke and P. D. Hess, "Interpretation of vacuum gas test results for aluminium alloys," Transactions of the Metallurgical Society of AIME, vol. 230, pp. 1542 - 1546, 1964.

[6] X. G. Chen and J. E. Gruzleski, "Influence of melt cleanliness on pore formation in aluminium-silicon alloys," International Journal of Cast Metals Research, vol. 9, pp. 17-26, 1996.

[7] S. N. Tiwari and J. Beech, "Origin of gas bubbles in aluminium," Metal Science, vol. 12, No 8, pp. 356-362, 1978.

[8] C. Nyahumwa, N. R. Green and J. Campbell, "Effect of mould filling on fatigue properties of cast aluminium alloys," AFS Transactions, vol. 58, pp. 215-223, 1998.

[9] R. Raiszadeh and W. D. Griffiths, "A method to study the history of a double oxide film defect in liquid aluminium alloys," Shape Casting: The John Campbell Symposium, TMS (The Minerals, Metals & Materials Society), pp. 865-871, 2005.

[10] W. C. Sleppy, "Oxidation of molten high purity aluminium in dry oxygen," Journal of the Electrochemical Society, vol. 108, pp. 1097-1102, 1961.

[11] D. J. Talbot, The Effect of Hydrogen in Aluminium and its Alloys, Leeds: Maney, pp. 8-64, 2004.

[12] T. L. Cottrell, The Strengths of Chemical Bonds, 2d edition., London: Butterworth, 1958.

[13] T. H. Yamada, "Hydrogen permeation barrier performance characterization of vapor deposited amorphous aluminum oxide films using coloration of tungsten oxide," Surface and Coatings Technology, Vols. 153(2-3), pp. 114-118, 2002.

[14] S. Aggarwal, "Use of amorphous aluminium oxide on a capacitor sidewall for use as a hydrogen barrier," Patent , no. US6876021 B2, 2005.

[15] N. B. Pilling and R. E. Bedworth, "The oxidation of metals at high temperatures," Journal of the Institute of Metals, vol. 29, pp. 529-582, 1923.

Shape Casting: 5th International Symposium 2014
Edited by: Murat Tiryakioğlu, John Campbell, and Glenn Byczynski
TMS (The Minerals, Metals & Materials Society), 2014

THE INFLUENCE OF Fe, Mn AND Cr ADDITIONS ON THE FORMATION OF IRON-RICH INTERMETALLIC PHASES IN AN Al-Si DIE-CASTING ALLOY

Alberto Fabrizi, Stefano Ferraro, Giulio Timelli

University of Padova – Dep. of Management and Engineering,
Strad. S. Nicola 3; Vicenza, 36100, Italy

Keywords: Al-Si die-casting alloy, primary α-Fe phase, intermetallic morphology

Abstract

The microstructure evolution of a high-pressure die-cast AlSi9Cu3(Fe) alloy is studied over different Fe, Mn and Cr content. Fe-rich intermetallic phases were characterized by using optical microscope, image analysis, scanning electron microscopy (SEM), energy dispersive spectroscopy (EDS). The results revealed that the amount of Fe-rich phases as well their size increase by increasing the elemental content. Needle-like particles appear in the alloy with higher Fe:Mn ratio, while coarse α-phases assume polygonal, stark-like or Chinese script morphology according to the Fe:Mn:Cr balance as revealed by EDS analysis. The 3D-SEM investigation suggests that both the polygonal and stark-like morphology can be associated to a rhombic dodecahedron structure. The morphology evolution of the Fe-bearing particles as a function of the chemical composition and casting process is proposed and discussed.

Introduction

It is widely known that Fe-rich intermetallic phases represent one of the most critical issue in the Al foundry because of their intrinsic harmful effects on the integrity and the mechanical properties of casting components [1-3]. Furthermore, Fe-rich intermetallics are abrasive and increase bulk hardness with a resulting decrease of alloy machinability [4]. Among various Fe-rich intermetallic phases in Al-Si alloys, the β-Al_5FeSi platelets are considered the most deleterious compounds since their edges act as stress concentration sites, facilitating crack initiation in the matrix [5-7]. A common technological action to counteract the undesirable presence of β phases consists in modifying the initial chemical composition of the alloy in order to form intermetallic compounds with more compact morphology.

Among various chemical treatments, manganese and chromium addition has been largely demonstrated to be effective in neutralizing the deleterious effects of the β particles by promoting the formation of compact α-Al(Fe,Mn,Cr)Si phase with polyhedral, star-like and Chinese script morphologies. Furthermore, the addition of such modifying elements results more effective for high cooling rate [6], which is a key feature on controlling the morphology and chemical composition of Fe-rich compounds. Indeed, the α-Fe intermetallic phase generally occurs when the cooling rate is sufficiently high, i.e. of the order of 10°C/s: polyhedral α-Fe phases were firstly detected in Al die-casting alloys where the cooling rate is significantly high (50-500°C/s [8]).

The beneficial amount of Mn and Cr needed to induce the phase transition from the β- to the α-Fe varies with the cooling rate [9]. A common rule to encourage α-Fe phase precipitation is to keep the Mn/Fe ratio > 0.5 [10]; nevertheless, it is reported that high Mn addition does not provide a completely inhibition of β-Fe needles formation into α-Fe Chinese scripts even for

Mn/Fe ratio up to 2 because of the solidification conditions and/or the presence of other elements in the alloys (e.g. Cu and Mg), which reduce the effectiveness of Mn to modify the Fe intermetallics [2,7]. Furthermore, an excessive Mn concentration could cause severe sludge problems. These coarse compact α-Fe particles can induce furnace-wall damage and hot tearing of the casting products. Sludge particles nucleate at temperature higher than crystallization of Al dendrites [6] and, due to their higher density, sediment to the bottom of the furnace and increase in size. These primary α-Fe particles were identified as complex intermetallic phases due to their chemical variability, besides aluminum, silicon and iron they can contain, manganese, chromium, copper, nickel and other elements [11]. Sludge also reduce the fluidity and machinability of the alloys [4], change the chemical composition of the melt and also reduce the capacity of the furnace.

Concerning the Cr amount, some works have demonstrated how a relatively high Cr content facilitates the formation of the α-Fe phase instead of the coarse platelet compounds [6,12], and a Cr/Fe ratio higher than 0.25 is recommended for α-Fe phase precipitation. In conclusion, iron in Al alloys is generally considered very harmful; however, an economical process to remove it from aluminum-silicon alloys has not been found yet. At the same time, a high amount of Fe is required to prevent die-soldering, especially in die-casting processes. It results therefore extremely important to control the concentration of Fe and *modifying elements* (such as Cr and Mn) especially in secondary Al alloys, where sludge is a problem of increasing importance due to the increasing concentration of Fe, Mn and Cr in the scrap cycle.

The present study aims to investigate the influence of Fe, Cr and Mn addition on the α-Fe intermetallic phases in a secondary Al-Si die-casting alloy in order to clarify the interactions of these elements on size, amount, chemical compositions and morphology of Fe-rich compounds. Furthermore, an explanation of morphological evolution and growth mechanism of primary α-Fe particles is proposed, supported by detailed 3D morphology observations.

Experimental procedure

A secondary AlSi9Cu3(Fe) cast alloy (EN AB-46000, equivalent to the US designation A380) was supplied by Raffineria Metalli Capra as commercial ingots and was selected as base alloy. The experimental alloys were prepared from the base material by progressively addition of Fe, Cr and Mn in the form of commercial Al-25Fe, Al-10Cr and Al-25Mn master alloys. In Table I, the chemical composition of the resulting alloys are reported.

Table. I. Chemical composition of the experimental alloys (wt.%); the sludge factor, SF, and the Mn/Fe ratio are also reported.

Alloy No.	Si	Fe	Mn	Cr	Cu	Mg	Zn	Ni	Ti	Pb	Al	SF	Mn/Fe
1-base	8.40	0.72	0.22	0.06	2.43	0.19	1.05	0.05	0.04	0.09	bal.	1.3	0.31
2	8.11	1.22	0.21	0.06	2.35	0.19	1.03	0.05	0.04	0.09	bal.	1.8	0.17
3	8.32	1.38	0.22	0.10	2.42	0.20	1.05	0.05	0.04	0.09	bal.	2.1	0.16
4	8.20	1.42	0.59	0.10	2.39	0.21	1.02	0.04	0.04	0.09	bal.	2.9	0.42

After the addition of the master alloys, the melt was heated in the furnace at 760 ± 5°C and held for 30 min. The temperature was then gradually decreased by following the furnace inertia up to 690 ± 5°C, which is the holding temperature commonly used for AlSi9Cu3(Fe) type alloys. Cast-to-shape specimens were produced using a multicavity die in a cold chamber die-casting machine with a locking force of 2.9 MN. A more detailed description of high-pressure die-casting machine, process procedure and multi-specimen casting is published elsewhere [13].

The specimens for the investigation have been drawn from the cross section of the 6mm-thick plate of the multi-specimen castings. The samples for microstructure examinations were polished according the standard metallographic procedures and etched by a 10% heated sulphuric acid solution. Microstructural investigations were carried out by means of an optical microscope (OM) and a field emission gun-environmental scanning electron microscope (FEG-ESEM) equipped with an energy-dispersive spectrometer (EDS). In order to compute the primary α-Fe compounds, the etched samples were processed by image analysis software. Only primary compact α-Fe particles (polyhedral and star-like) greater than 50 μm^2 were considered for the statistical counts of sludge, and all the resulting data were taken from the average measurements of approximately total area of 20 mm^2. The chemical micro-analysis of α-Fe particles was performed by EDS at 10 kV using a point count of 40 seconds for each reading. Some samples were also deeply etched in a concentrated NaOH solution in order to reveal the 3D morphology of α-Fe intermetallic phases.

Results and Discussion

The microstructure of the experimental alloys are compared in Fig. 1. The eutectic Al-Si structure and Fe-rich intermetallic particles with polyhedral and star-like morphologies are recognizable. The number of primary α-Fe compact particles as well as their size and roundness were estimated by image analysis after chemical etching and the resulting data are plotted in Fig. 2. In the base alloy, referred here to Alloy 1, primary polyhedral and star-like α-Fe particles with a mean diameter of about 10 μm appear, while needle-like β-Fe seems to be mostly suppressed and only few and very small ones have been detected. In Alloy 2 the amount of primary α-Fe particles with polyhedral and faceted star-like morphologies significantly increase as well as the β-Fe needles. The increase of Fe amount (1.22 wt.%) has consequently reduced the Mn:Fe ratio (Tab. I), if compared to Alloy 1, and it has promoted the nucleation of β-Fe compounds. A comparison of the particle size to the α-Al cell size of the alloys evidences how these compounds can be labeled as primary or pro-dendritic phases. As the Cr content is enhanced up to 0.1 %wt. (Alloy 3), the morphology of α-Fe particles seems to degenerate and more irregular morphologies appear, as it is shown by comparing the roundness distributions in Fig. 2b. The presence of β-Fe phases is also observed, although they result to be less and smaller in comparison to Alloy 2.

Finally, the increase of Mn content up to 0.59 wt.% (Alloy 4) promotes a large precipitation of compact Fe-rich particles with higher mean size, as revealed by Fig. 2a. In addition, the amount of β-Fe needles is considerable reduced as the Mn/Fe ratio enhances compared to Alloy 3 (Tab. I). The size range of primary α-Fe particles is consistent with the typically data of casting process with very high cooling rates (e.g. 10 - 50 μm for high-pressure die-casting [18]). Therefore, from the OM investigations and image analysis, it results that, by increasing the elemental concentration in the alloys, an increase in size and area fraction of primary α-Fe intermetallics is noticed. According to its definition (Eq. 1), sludge factor, SF, was used to evaluate the tendency of primary α-Fe particles (*sludge*) to form as a function of Fe, Mn, and Cr amount [14]:

$$\text{Sludge Factor} = (1\times \text{wt.\%Fe}) + (2\times \text{wt.\%Mn}) + (3\times \text{wt.\%Cr}) \quad (1).$$

The addition of Fe, Cr and Mn in the experimental alloys causes an increase of the SF values (Tab. I) and consequently a larger α-Fe intermetallic precipitation. These results are in agreement with Shabestari [19] who found that no intermetallics form in alloys with a sludge factor less than 1.20 and their volume percent increases linearly with SF. From the investigations, it appears that the increase of Cr and Mn content promotes also the modification of α-Fe particles morphology from polyhedral to a more branched or star-like form. These results agree with Šerák *et al.* [11]. The chemical compositions of α-Fe intermetallics were studied by EDS and the concentrations of Al, Si, Fe, Mn and Cr expressed in at.% were correlated to the different α-Fe morphologies, as reported in Fig. 3. Considering that Fe, Mn and Cr can substitute to each other in the α-Fe crystal structure [15], the stoichiometry of the polyhedral, star-like and Chinese script morphologies are close to α-$Al_{15}(Fe,Mn,Cr)_3Si_2$ and α-$Al_{12}(Fe,Mn,Cr)_3Si_2$ for all the alloys. In addition, gradual changes in chemical composition are recorded when α-Fe compounds' morphology degenerates from compact structures (polygonal or star-like) to Chinese script: a reduction of Al, Mn and Cr levels with a corresponding increase of Si and Fe occurs for all the α-Fe phases in all the alloys.

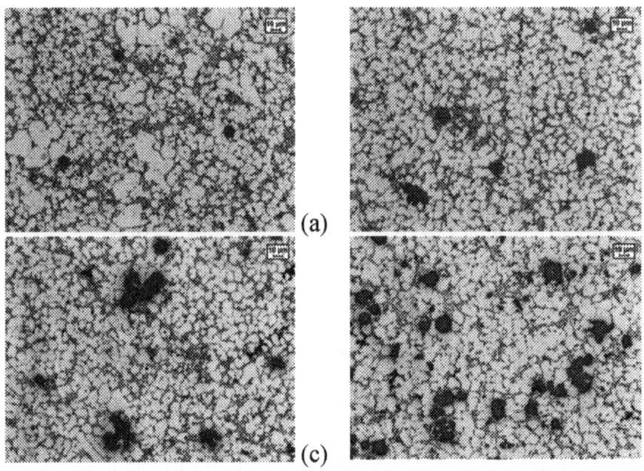

Figure 1. OM microstructure of: (a) Alloy 1, (b) Alloy 2, (c) Alloy 3 and (d) Alloy 4.

On the other hand, a low difference of elemental concentration is observed for polygonal and star-like structures. It is also important to note that by increasing the concentration of Fe, Cr or Mn in the alloy, the level of the same element in all the α-Fe morphologies increases. Thus, the chemical composition of primary α-Fe particles is strictly related to the initial composition of the melt, as found also by Šerák *et al.* [11].

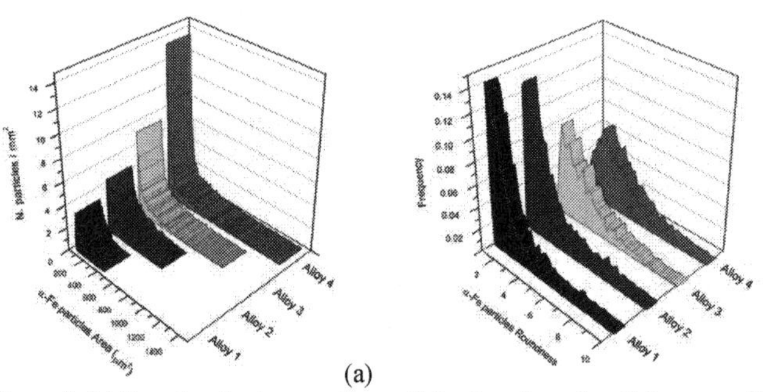

Figure 2. (a) Density of primary α-Fe particles (number of particles per unit area) as function of their area; (b) roundness distribution of primary α-Fe particles.

Since Fe, Mn and Cr are interchangeable to each other as mentioned above, the increase of an element level causes a gradual decreased of the other two ones in the chemical composition of the α-Fe morphologies, while the corresponding phase stoichiometry is kept mostly constant.

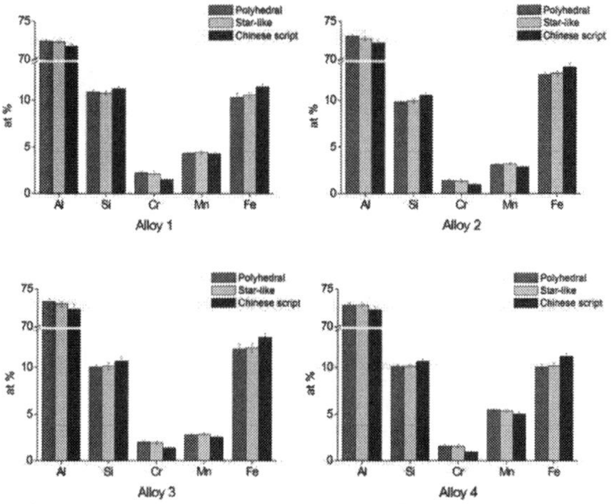

Figure 3. Chemical composition (at.%) of coarse α-Fe phases as function of their morphology for the experimental alloys.

After the dissolution of α-Al matrix by deep etching, SEM investigations reveal the 3D morphology of the polyhedral α-Fe particles, which correspond to a regular rhombic dodecahedron (Fig. 4). This result is agreement with Kim *et al.* [12], who revealed how the primary α-Fe phase nucleates and grows with the same polyhedral structure by using time-resolved radiography in a synchrotron radiation facility. In Fig. 5, typical SEM micrographs of various polyhedral shapes are reported. It appears evident how such shapes can be obtained from different cross-sections of the rhombic dodecahedron, as illustrated by the pictures in the insets. In some cases, the rhombic dodecahedron structures show hollows in the center of the 12 rhombic facets, as shown in Fig. 6.

Figure 4. 3D SEM micrograph of a deep etched sample with α-Fe particles having a regular rhombic dodecahedron structure, as indicated in the insets.

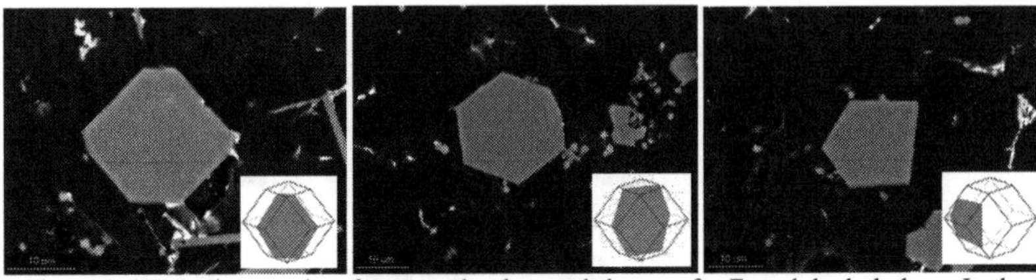

Figure 5. SEM micrographs of commonly observed shapes of α-Fe polyhedral phase. In the insets, the corresponding cross-sections of rhombic dodecahedron structure.

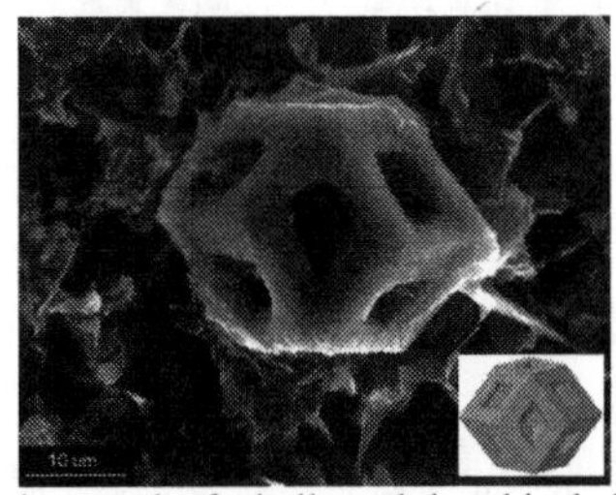

Figure 6. 3D SEM micrograph of a hollowed rhombic dodecahedron structure; in the inset, a reproduction of the polyhedron by using a 3D modeling software.

By using a 3D modeling software, a reproduction of the hollowed rhombic dodecahedron has been proposed and the various cross-sections seem to well approximate the star-like and branched morphologies commonly appearing in the OM micrographs (Fig. 7).

It is worth mentioning that primary Mg_2Si particles in Mg and Al-Mg alloys [16-18] and Silicon crystals in Al-Si alloys [19] can exhibit similar hollowed polyhedral structures, called *hopper-like crystals*. The growth mechanism of hollowed α-Fe dodecahedron and the forming of cavities are yet under discussion and further studies will be performed in order to understand more clearly such phenomena. Anyway, the authors suggest that the growth mechanism of Mg_2Si particles and Si crystals proposed in Ref. [19,20], can been adopted to explain the morphology evolution of compact α-Fe particles too. Generally, the micro-segregations forming in the alloy melt constitute the preferred nucleation site for the seed crystal of primary particles.

The initial growth of the seed crystal is diffusion controlled, so that radial growth rates of an isolated crystal floating in the melt are essentially isotropic. As the spherical seed crystals grow and exceed a critical size, the growth manner will proceed according to the inherent crystal structure (determining the relative growth rates of different crystal planes) and external growth conditions such as solidification conditions, solute concentration in melt, the direction of heat [16], melt superheating [17]. Thus, the final morphology of the crystal results from the competition of internal and external factors.

In some cases, impurities in the melt or high concentration of Al atoms limit the growth of the interior of the facets resulting in a relative higher growth rate of the edges/corners. This growth gap leads to form hollows in the center of polyhedron facets. If the concentration of Al atoms and impurities cannot be reduced effectively by diffusion or convection, these cavities are then definitely filled by the alloy melt appearing in the final solidification structure.

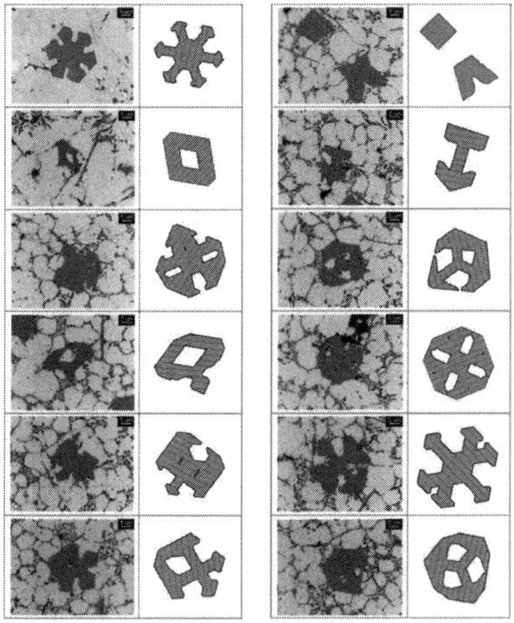

Figure 7. Micrographs of typical star-like and branched morphologies of primary α-Fe phases and cross-sections of the hollowed rhombic dodecahedron simulated by a 3D modeling software.

Conclusions

The increase of Fe, Cr and Mn content in an AlSi9Cu3(Fe) die-casting alloy induces an increase of size and number density of primary α-Fe particles. β-Fe needle-like particles appear only in the alloy with higher Fe:Mn ratio.

The α-Fe Chinese script phases show a higher level of Fe and Si in comparison with compact morphologies. The polyhedral α-Fe phases are associated to a regular rhombic dodecahedron while the star-like morphology originate from hollowed rhombic dodecahedron.

Acknowledgements

The authors are grateful to Dr. Aldo Dalla Via who carried out the reproduction of the hollowed rhombic dodecahedron by using 3D modeling software.

References

1. J.A. Taylor, "Iron-containing intermetallic phases in Al-Si based casting alloys", *Procedia Materials Science*, 1 (2012), 19–33.
2. S. Seifeddine, S. Johansso and I. L. Svensson, "The influence of cooling rate and manganese content on the β-Al_5FeSi phase formation and mechanical properties of Al–Si-based alloys", *Materials Science and Engineering A*, 490 (2008), 385–390.

3. G. Gustafsson, T. Thorvaldsson and G.L. Dunlop, "The Influence of Fe and Cr on the Microstructure of Cast Al-Si-Mg Alloys", *Metallurgical Transactions A*, 17 (1986), 45-52.
4. Y. Zedan, F.H. Samuel, A.M.Samuel, H.W. Doty, "Effects of Fe intermetallics on the achinability of heat-treated Al–(7–11)% Si alloys", *Journal of Materials Processing Technology, 210*(2) (2010), 245-257.
5. L. Lu and A.K. Dahle, "Iron-Rich Intermetallic Phases and Their Role in Casting Defect Formation in Hypoeutectic Al-Si Alloys, Metallurgical and Materials Transactions, A36 (2005), 819-835.
6. W. Eidhed, "Modification of β-Al_5FeSi Compound in Recycled Al-Si-Fe Cast Alloy by Using Sr, Mg and Cr Additions", *Journal of Materials Science and Technology*, 24 (2008), 45-47.
7. W. Khalifa, A. M. Samuel, F. H. Samuel, H. W. Doty and S. Valtierra, "Metallographic observations of β-AlFeSi phase and its role in porosity formation in Al–7%Si alloys", *International Journal of Cast Metals Research*, 19 (2006), 156-166.
8. J.R. Davis, *Aluminum and Aluminum Alloys* (Ohio: ASM Specially Handbook, International Materials Park, 1993), 98-99.
9. S. Belmares-Perales, M. Castro-Román, M. Herrera-Trejo and L. E. Ramírez-Vidaurri, "Effect of cooling rate and Fe/Mn weight ratio on volume fractions of α-AlFeSi and β-AlFeSi phases in Al–7.3Si–3.5Cu alloy", *Metals and Materials International*, 14 (2008), 307-314.
10. S. Ji, W. Yang, F. Gao, D. Watson and Z. Fan, "Effect of iron on the microstructure and mechanical property of Al–Mg–Si–Mn and Al–Mg–Si diecast alloys", *Materials Science and Engineering A*, 564 (2013), 130–139.
11. J. Šerák and D. Vojtěch, "Influence of (AlSiFeMnCr) intermetallic phases on the casting properties of AlSi9Cu2FeMnCr alloys", *Aluminium*, 78 (2002), 384–387.
12. B.Kim, S. Lee and H. Yasuda, "Morphological Variation of the Fe/Cr-containing Intermetallic Phase in the Al-Si casting Alloy as a Function of Cooling Rate: Time-Resolved Radiography", *Materials Science Forum*, 654-656 (2010), 974-977.
13. G. Timelli, S. Ferraro, F. Grosselle, F. Bonollo, F. Voltazza and L. Capra, "Caratterizzazione meccanica e microstrutturale di leghe di alluminio pressocolate", *La Metallurgia Italiana*, 1 (2011), 5-17.
14. J.L. Jorstad, "Understanding Sludge", *Die Casting Engineer*, 30 (1986), 30-36.
15. S.G. Shabestari, "The effect of iron and manganese on the formation of intermetallic compounds in aluminum–silicon alloys", *Materials Science and Engineering A*, 383 (2004) 289–298.
16. H. Kang, X Li, Y. Su, D. Liu, J. Guo and H. Fu, "3-D morphology and growth mechanism of primary Al_6Mn intermetallic compound in directionally solidified Al-3at.%Mn alloy", *Intermetallics*, 23 (2012), 32-38.
17. Q.D. Qin, Y.G. Zhao, W. Zhou and P.J. Cong, "Effect of phosphorus on microstructure and growth manner of primary Mg2Si crystal in Mg2Si/Al composite, Materials Science and Engineering A, 447 (2007),186–191.
18. M.F. Ourfali, I. Todd and H. Jones, "Effect of Solidification Cooling Rate on the Morphology and Number per Unit Volume of Primary Mg2Si Particles in a Hypereutectic Al-Mg-Si Alloy, Metallurgical and Materials Transaction A, 36 (2005), 1368–1372.
19. R.-Y. Wang, W.-H. Lu and L.M. Hogan, "Faceted Growth of Silicon Crystals in Al-Si Alloys", Metallurgical and Materials Transactions A, 28 (1997), 1233-1243.
20. C. Li, Y.Y. Wu, H. Li and X.F. Liu, "Morphological evolution and growth mechanism of primary Mg2Si phase in Al–Mg2Si alloys", Acta Materialia, 59 (2011), 1058–1067.

Shape Casting: 5th International Symposium 2014
Edited by: Murat Tiryakioğlu, John Campbell, and Glenn Byczynski
TMS (The Minerals, Metals & Materials Society), 2014

THERMAL ANALYSIS OF SOLIDIFYING SECONDARY PHASES IN THE EUTECTIC Al-Si ALLOY AND THE EFFECT OF Sr AND Fe

Jeyakumar Manickaraj, Anton Gorny, Sumanth Shankar

Light Metal Casting Research Centre (LMCRC), Department of Mechanical Engineering, McMaster University, Hamilton, ON, Canada L8S 4L7

Keywords: Al-Si eutectic alloy, Al-Fe-Si intermetallic, Thermal analysis, microstructure; Sr modification

Abstract

The effect of Fe and Sr on the evolution of the Fe intermetallic and eutectic phases in the Al-Si eutectic (Al-12.6 wt%Si) alloy was studied. Two different amounts of Sr (0.02 and 0.1 wt%) and three levels of Fe (0.5, 1 and 2 wt%) were introduced in the alloy; thermal analysis during solidification coupled with optical microscopy was carried out. It was observed, from the thermal data, that the nucleation temperature of the Fe intermetallic phase was not affected by the addition of Sr, but there is a significant variation in the heat balance during nucleation and growth of these phases during solidification. Additionally, the order of evolution of Al, Fe intermetallic phases and Si was significantly altered by the addition of Sr to the alloy. The analyses of the alloys showed that Fe intermetallic phase acted as nucleation sites for the Si phase while this was not the case in alloys with Sr addition. There is significant reduction in the size and alteration of the morphology of the Fe intermetallic phases when Sr was added to the alloy.

Introduction

Al-Si alloy is one of the important Al alloys used in automobile industries due to its excellent castability, and high strength to weight ratio [1-6]. To improve the mechanical properties of the Al-Si alloy trace levels (50 to 250 ppm) of Sr have been added to alter the morphology of the eutectic Al and Si such that the eutectic Al is grain refined and Si phase changes from a coarse plate like to a fine fibrous structure; this structure in Al-Si alloy is commonly referred to as modified structure whereas the coarse structure found without the Sr addition is called the unmodified structure. Initially, it was believed that Sr modified the Si morphology during its growth [7], but later it was observed that Sr addition played a more significant role in altering the nucleation environment in the liquid alloy during the solidification of the secondary phases including the eutectics; the Sr addition has been shown to significantly alter the atomic arrangements and clusters in the liquid prior to the solidification event of a secondary phase [8].

Thermal analysis using cooling curve is a powerful and efficient tool to detect any miniscule change in the thermal energy of the system during solidification and reflects evolution of even minor secondary phases in an entire bulk alloy melt [9]. Thermal study of the Al-Si alloy during solidification, shows that Sr addition increases the undercooling of eutectic nucleation, however, there are only a few research work that correlates undercooling to other elemental additions such as Fe, Mn, Cu, Zn, etc, and not merely to the effect of Sr alone. Fe is one of the main impurities in Al-Si alloy and a recent study conducted by our group [10] shows Fe playing a major role in Sr modification, however further study is required to understand the importance of the Al-Si-Fe phase to the modification of the eutectic Al and Si phases. Therefore, the present study is focused to understand the eutectic nucleation of the Al-Si alloy with the effect of Sr and Fe in the alloy.

Materials and Experimental Procedure

Raw materials used for this study were 99.99 % pure Al, 99.9999% pure Si, Al-10Sr alloy and Al-25Fe alloy and the compositions were verified by ICP* spectrometry. Solidification was carried out at approximately 0.02 $°C.s^{-1}$ in a furnace cooling environment from an initial alloy melt at 200 °C above the liquidus temperature and maintained isothermally for 30 min; to avoid the effect of high cooling rate and maintain a near-equilibrium solidification condition. In all the experiments, three thermocouples were placed linearly and the distance between them was 10 mm as shown in Figure 1. The cooling rates were evaluated from the thermal data (T versus t) acquired during the solidification experiments with a K-type ungrounded thermocouple of about 1.6 mm diameter and the same thermocouple was used for all solidification experiments in order to avoid variations due to calibration issues.

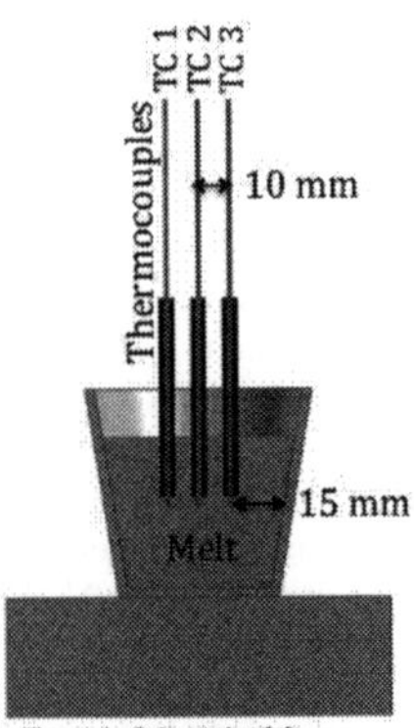

Figure 1. Schematic of the experimental setup for thermal study.

The standard metallographic sample preparation techniques were employed for the microstructure analysis, which was carried out using the Light optical microscopy (Nikon Eclipse LV 100). A total of 16 alloy compositions with varying levels of Fe and 9 alloy compositions with varying levels of Sr and Fe in them were solidified and analyzed to achieve the objectives of this study. All compositions in this publication will be in weight percentage unless otherwise mentioned.

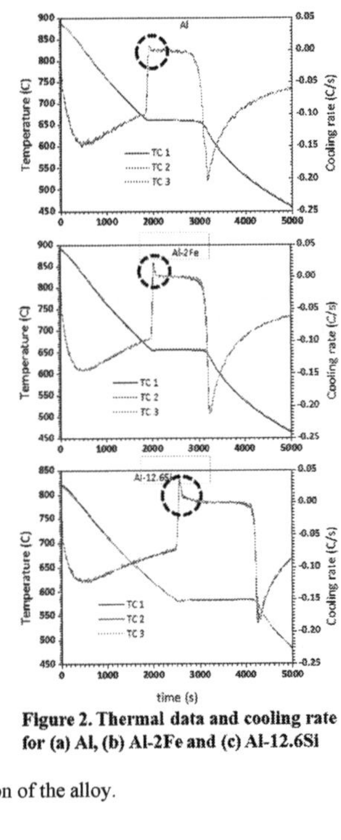

Figure 2. Thermal data and cooling rate for (a) Al, (b) Al-2Fe and (c) Al-12.6Si

Results and discussion

Thermal data and the first derivate (cooling rate) for pure Al are shown in Figure 2a, thermal data and cooling rate for Al-2Fe and Al-12.6Si are shown in Figure 2 (b) and (c) respectively. From Figure 2, it is observed that the nucleation temperature of the only solidifying event in the pure Al, Al-2Fe and Al-12.6Si were 666.45, 656 and 578 °C respectively. Also, in all the experiments, the variation in the temperature across the melt volume was not negligible at any given time during solidification. As in Figure 2, the expression 'TC' in the legends of all the figures on thermal data represents the thermocouple.

* Inductively Coupled Plasma Spectrometry to measure chemical composition of the alloy.

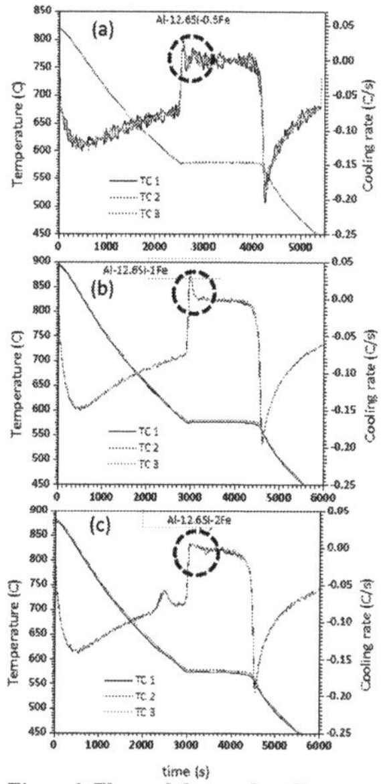

Figure 3. Thermal data and cooling rate for (a) Al-12.6Si-0.5Fe, (b) Al-12.6Si-1Fe and (C) Al-12.6Si-2Fe

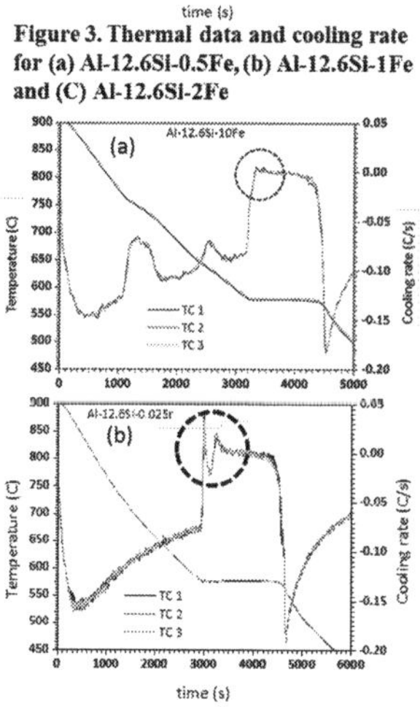

Figure 4. Thermal data and cooling rate for (a) Al-12.6Si-10Fe and (b) Al-12.6Si-0.02Sr

Thermal data and the first derivative for Al-12.6Si-0.5Fe, Al-12.6 Si-1Fe and Al-12.6Si-2 Fe are shown in Figure 3(a), (b) and (c) respectively. In the eutectic alloy when 0.5Fe was added nucleation of two phases at the eutectic temperature (576.13 and 576.41 °C) were observed, both phases nucleated at nearly the same temperature but at different times. In figure 3(b) nucleation of only one phase occurred at 574.25 °C, which clearly indicates Al-12.6 Si-1 Fe is the eutectic composition of this alloy. When the Fe amount was increased to 2 wt% in the eutectic alloy a secondary Al-Fe-Si intermetallic phase nucleated at 623.54 °C while at the eutectic temperature nucleation of only one phase evolved at 574.55 °C.

The thermal data and the cooling rate for Al-12.6Si-10Fe are shown in Figure 4(a) and it is observed that prior to the eutectic solidification, two types of Al-Fe-Si intermetallic phases formed, however, as per the phase diagram [10], three types (θ-$Al_{13}Fe_4$, γ-Al_8Fe_2Si and β-$Al_9Fe_2Si_2$) of nucleation had to occur prior the eutectic nucleation. The thermal data and the first derivative of the thermal data for the addition of 0.02 wt% Sr in the eutectic alloy are shown in Figure 4(b); contrary to the alloy without Sr (Figure 2(c)), the nucleation of two phases occurred at the eutectic temperature when Sr was added to the eutectic alloy.

The thermal data and cooling rate for Al-12.6Si-0.5Fe-0.02Sr, Al-12.6Si-1Fe-0.02Sr and Al-12.6Si-2Fe-0.02Sr are shown in Figure 5(a), (b) and (c) respectively. Also, the thermal data and first derivative for Al-12.6Si-0.5Fe-0.1Sr, Al-12.6Si-1Fe-0.1Sr and Al-12.6Si-2Fe-2Sr are shown in Figure 5(d), (e) and (f) respectively. Addition of Sr in the Al-12.6Sr-0.5Fe did not produce any difference in the nucleation at the eutectic temperature (compare Figures 3(a) to 5(a) and 5(d)) but the time difference between the formation of the two phases at the eutectic temperature was affected by Sr. In eutectic Al-Si alloy with 1 and 2 wt% Fe, the addition of Sr resulted in two nucleation events at the eutectic temperature contrary to the one observed in alloys without Sr (as in Figures 3(b) and 3(c)). In Al-12.6Si-2Fe-2Sr alloy the Al-Si-Sr intermetallic phase formed before the nucleation of eutectic phase but this did not affect the two nucleation events at the eutectic temperature.

The microstructures for Al-2Fe, Al-12.6Si, Al-12.6Si-0.02Sr, Al-12.6Si-2Fe, Al-12.6Si-2Fe-0.1Sr and Al-12.6Si-2Fe-2Sr are as shown in Figure 6(a), (b), (c), (d), (e) and (f) respectively. From the microstructure it is clear that 0.02Sr modified the Al-Si eutectic alloy completely and adding 2 wt%Fe in the eutectic alloy, the Al-Si-Fe intermetallic phase evolved prior the eutectic temperature and clearly shows the formation of primary blocky Si on the intermetallic phases indicating that the intermetallic phases acted as a nucleation site for the Si phase.

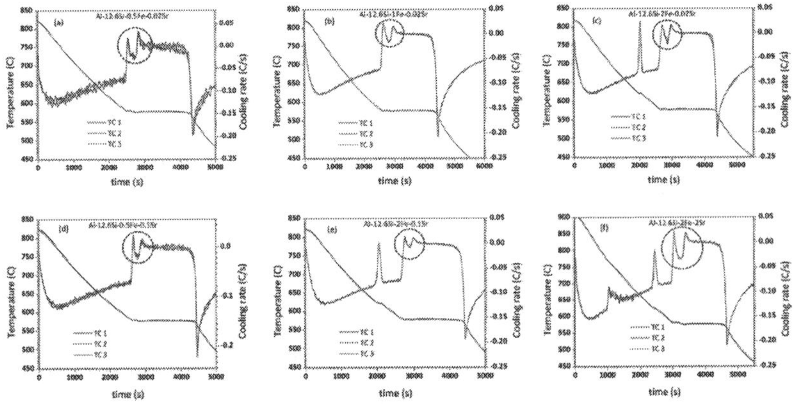

Figure 5. Thermal data and cooling rate for (a) Al-12.6Si-0.5Fe-0.02Sr, (b) Al-12.6Si-1Fe-0.02Sr, (C) Al-12.6Si-2Fe-0.02Sr, (d) Al-12.6Si-0.5Fe-0.1Sr, (e) Al-12.6Si-2Fe-0.1Sr and (f) Al-12.6Si-2Fe-2Sr

In Al-12.6Si-2Fe alloy, the Al-Si-Fe phase denoted by the first derivative of the thermal data prior to the eutectic temperature for with and without Sr alloy were different; for the alloy without Sr the peak was a broad hump (Figure 3(c)) but that for with Sr was sharp and narrow (Figure 5(c)), thus indicating that without Sr, the number of nucleation sites was low and growth contributed more to the broad hump and with Sr, the peak was narrow and sharp indicating that there were more nucleation sites but less growth. Therefore, the size of Al-Si-Fe phase was large in the alloy without Sr as compared with Sr added alloy. Also, the density of Al-Si-Fe intermetallic phase distribution in the microstructure was high in Sr added alloy (size of the intermetallic phases were less) compared to without Sr as shown in Figure 6 (d) and (e). When Sr was added in the Al-12.6Si-2Fe alloy, the Si phase was well modified and primary Si was not observed (Figure 6(e)), and in the alloy without Sr alloy (Figure 6(d)) primary Si was observed and eutectic Si was unmodified. If the Sr amount was increased from 0.02 to 0.1, a few Al_2Si_2Sr phase were observed attached to the Al-Si-Fe intermetallic phase (Figure 6(e)) and when the Sr increased to 2 wt%, the primary blocky Al_2Si_2Sr phase was well distributed in the microstructure (Figure 6(f)).

In the eutectic alloys with Fe and Sr, the split in the cooling rate curve showing two separate nucleation events of phases at the eutectic temperature was further analyzed by carrying out similar solidification experiments with Al-7Si alloys to differentiate the order of nucleation events of Al and Al-Fe-Si intermetallic phases in this system. The thermal data and its first derivative for Al-7Si, Al-7Si-1Fe and Al-7Si-1.8Fe alloy without and with Sr are shown in Figure 7(a) to (f). Contrary to the Al-12.6Si alloys, in the Al-7Si alloy, irrespective of the addition of Fe and/or Sr, only one nucleation event was observed at the eutectic temperature. When 1Fe was added to the Al-7Si alloy, the Al-Si-Fe intermetallic phase evolved between evolution of the primary Al and eutectic phases; while addition of 1.8 wt%Fe resulted in the co-evolution of the primary Al and Al-Si-Fe intermetallic phases at around the same temperature.

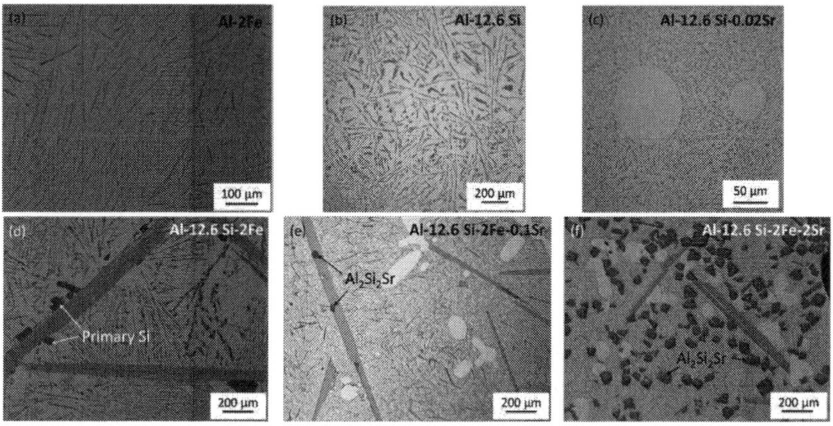

Figure 6. Microstructure of (a) Al-2Fe, (b) Al-12.6Si, (c) Al-12.6Si-0.02Sr, (d) Al-12.6Si-2Fe (e) Al-12.6Si-2Fe-0.1Sr (f) Al-12.6Si-2Fe-2Sr

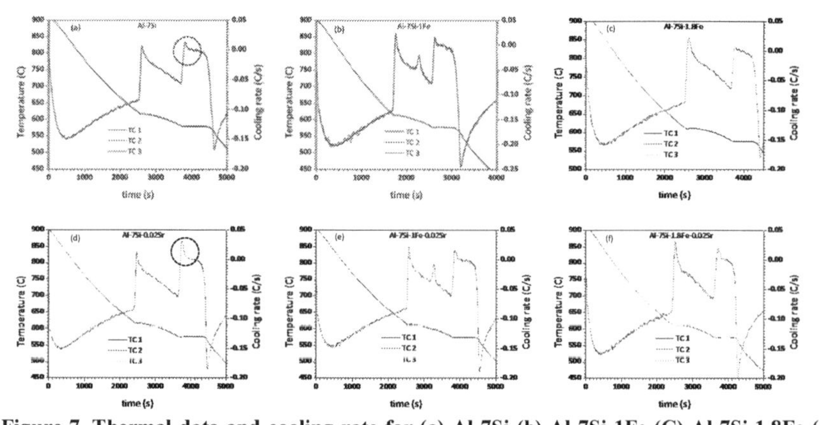

Figure 7. Thermal data and cooling rate for (a) Al-7Si (b) Al-7Si-1Fe (C) Al-7Si-1.8Fe (d) Al-7Si-0.0.02Sr (e) Al-7Si-1Fe-0.0.2Sr and (f) Al-7Si-1.8Fe-0.0.2Sr

The profile of the encircled region in the first derivative of the thermal data at around the eutectic temperature for all the compositional variations (Fe and /or Sr addition) of the Al-12.6Si and Al-7Si alloys in this study are shown in Figure 8 (a) and (b), respectively. In these figures, left side of the vertical line denote the Al-Si-Fe intermetallic phase formation along at the eutectic temperature and right side of the vertical line denote the Al-Si-Fe intermetallic phase formation prior to the eutectic temperature. The maximum value of the first derivative at the eutectic temperature is shown for each profile. In Figure 8(a), the addition of <1Fe in alloys with and without Sr resulted in two nucleation events at the eutectic temperature, wherein, the addition of Sr resulted in a significant increase in the time lapse between these two nucleation events. In

Figure 8(a), the addition of ≥1Fe to the alloys showed only one nucleation event in alloys without Sr, whereas, two nucleation events persisted in alloys with Sr at the eutectic temperature.

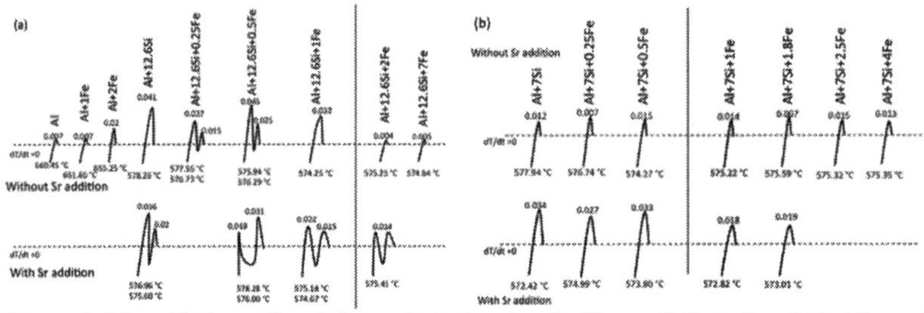

Figure 8. Magnified profile of the encircled region in Figure 2, 3, 4, 5 and 7 at the eutectic temperature for alloys with and without Sr. (a) Al-12.6Si and (b) Al-7Si alloys. The left side of the vertical line within each figure represents the alloys where Fe intermetallic phase evolved at around the eutectic temperature and right side of the vertical line represents the alloys where Fe intermetallic phase formed prior to the eutectic temperature. The horizontal dotted line denotes the zero cooling rate and the maximum cooling rate is noted on top of the respective curves along with the alloy composition. The nucleation temperatures are noted at the bottom of the respective curves.

In Figure 8(b), for the Al-7Si alloys, the Sr addition did not result in two nucleation events unlike the Al-12.6Si alloys in Figure 8(a), at the eutectic temperature but the maximum value of the cooling rate attained during phase evolution increased significantly with Sr addition. From Figure 8(a) and (b), the highest cooling rate attached by each nucleation event represents the net energy released due to the competition between the latent heat of fusion released by the nucleation event and the heat extracted from the bulk liquid alloy during solidification; the latter is constant for all the experiments in this study; hence, the higher value of maximum cooling rate for a nucleation event denotes higher energy released from the latent heat of fusion. Notably, the sharp and narrow nature of the derivative curve during solidification at any nucleation event denotes high degree of nucleation and low growth, while the short and broad nature of the derivative curve at the nucleation event denotes low degree of nucleation and large growth.

In Figures 2(a) and 2(b), also shown in Figure 8(a), the heat evolution for the nucleation and growth of the primary Al phase in pure Al is nearly zero and increases to 0.01 and 0.02 for the Al-1Fe and Al-2Fe alloys, respectively; indicating that the heat released by nucleation of Al is negligible and Fe amount increased in these alloys, the co-nucleation of Al-Fe phase with Al phase added incrementally higher heat during the nucleation events of these alloys; it is notable in these alloys, that $Al_{13}Fe_4$ phase only nucleated on the Al phase as there is no independent nucleation of the $Al_{13}Fe_4$ phase prior to the Al phase observed during solidification of these alloys albeit being predicted by the phase diagram as in the case of Al-2Fe alloys, as shown in Figure 9. This is an critical information that will help to explain the sequence of event in the Al-12.6Fe and Al-7Fe alloy systems. Adding 0.05Fe with Al-12.6Si alloy resulted in two nucleation events, in Figure 8(a), analyzing the maximum cooling rate (energy of fusion released) at the eutectic temperature shows that the Al and Al-Fe-Si intermetallic phase co-nucleates first with the former leading the event while the second event which is 153 s after the first is that of the eutectic Si on the Al-Fe-Si phase.

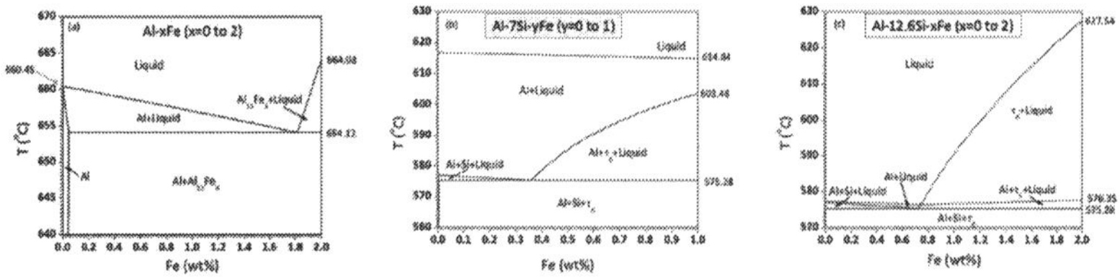

Figure 9. Equilibrium phase diagram of the (a) Al-Fe system between 0 and 2 Fe, (b) Al-7Si-xFe (x= 0 to 1 Fe) and (c) Al-12.6Si-xFe (x= 0 to 2 Fe).

In Al-12.6Si for ≥2Fe, the Al-Si-Fe intermetallic phase formed prior to the eutectic temperature (Figure 3(c)) and hence at the eutectic temperature, the Si phase could nucleate on this Fe intermetallic phase and trigger the initiation of the eutectic reaction, thereby, causing an insignificant value of the maximum cooling rate (low heat of fusion) at the nucleation event as shown by Figures 3(c) and 8(a). In the Al-12.6Si alloys with Sr, there are two distinct nucleation events, irrespective of the Fe content in the alloy: for the alloys with <1Fe, as shown in Figure 5, the eutectic reaction is triggered by the co-nucleation of the Al and Al-Si-Fe intermetallic phases and subsequently, the Si phase nucleates on the Fe intermetallic phase after a significant time lag (far greater than the alloys without Sr). Recent research [11,12] has suggested that the Al-Fe-Si phase formed when Sr is added to the Al-Si alloy is not the predicted τ_6-$Al_9Fe_2Si_2$ as per the phase diagram shown in Figure 9, but rather, another phase κ-Al_5Fe_2Si which has a different composition and crystal structure when compared to the τ_6-$Al_9Fe_2Si_2$ phase. This is elucidated by comparison of the Figures 3(c) and 5 (c), wherein, a marked difference is observed in the heat evolved during the nucleation of the Al-Si-Fe phase prior to the eutectic temperature in both the cases of Al-12.6Si-2Fe alloys with and without Sr addition, respectively; the broad and shallow peak in the temperature derivative curve for the alloy without Sr (Figure 3(c)) is in sharp contrast to the sharp, high and narrow peak in the alloy with Sr (Figure 5(c)). All the experiments in this study have been repeated to assure accuracy of the data. In Figure 8(a), for the Al-12.6Si alloys with Sr, two nucleation events are always observed at the eutectic temperature, albeit the nucleation of the Fe intermetallic phase prior to this temperature in alloys with >1Fe, which is contrary to the alloys without Sr addition. Further, the time lapse between the two nucleation events in the Al-12.6 Si alloy with Sr addition (Figure 8(a)) is much higher than those in the no Sr counterparts

In Al-7Si alloy, the addition of Sr did not change the number of nucleation at the eutectic temperature, only one phase nucleated as shown in Figure 8(b), because the primary Al nucleated before eutectic nucleation. The Al-Fe-Si phase in the Al-7Si alloy for any composition of Fe, will evolve prior to the eutectic reaction presenting the favorable τ_6-$Al_9Fe_2Si_2$ on which the eutectic Si nucleates at the eutectic temperature [13]; the eutectic Al phase nucleates on the Si and primary Al, alike. When Sr was added to the Al-7Si alloy, there is a marked undercooling of the eutectic temperature coupled with a significant increase in the peak height of the recalescence from eutectic nucleation as shown by Figure 7 and Figure 8(b). Further, Gorny et al [10,11] has shown that Sr addition to these alloys alter the Fe intermetallic phase from τ_6-$Al_9Fe_2Si_2$ to a new κ-Al_5Fe_2Si phase which does not enable the nucleation of the eutectic Si on itself. Hence, in Al-7Si alloy with Sr, the eutectic Si phase does not nucleate at the predicted temperature and results in an undercooling of the eutectic melt during solidification during which, the primary Al phase continues to grow, enriching the liquid with Si until super saturation. At an adequate super saturation of the Si in the eutectic liquid, the Si phase crystallizes on the primary Al phase triggering the evolution of highly grain refined eutectic Al phase in the remaining liquid through which the Si grows by multiple mechanical twinning process [14].

Summary

The results show that there are changes to the liquid structure causing significant differences in the nucleation of Al and secondary phases in the Al-Si hypoeutectic alloys due to the changes in the concentrations of Fe and addition of trace levels of Sr. In alloys without the Sr addition, the eutectic Si phase requires the presence of the Al-Fe-Si phase in the liquid to nucleate on; this Al-Fe-Si phase requires the presence of primary Al phase to nucleate on. In Al-Si alloys with Sr, the Sr addition alters the type of terminal intermetallic phase that evolves prior to or at the eutectic temperature from τ_6-$Al_9Fe_2Si_2$ to a new κ-Al_5Fe_2Si phase which eliminates the inoculant for the eutectic Si and results in a significant undercooling of the inter-dendritic eutectic alloy during solidification prior to the evolution of the blocky Si phase on the primary Al phase which results in a highly refined grain structure of the eutectic Al and refined morphology of the eutectic Si phases.

References

[1] Joseph R D, Davis J R. & Associates, ASM International. Handbook Committee "Aluminum and aluminum alloys", ASM International, (1993) 22.
[2] Stevens R H., ed.: "Aluminum alloys", ASM International, Materials Park, OH, USA, (1985).
[3] Hatch J E, ed.: "Aluminum properties and physical metallurgy", ASM Metals Park, Ohio (1984).
[4] Ye H, "An overview of the development of Al-Si-alloy based material for engine applications", J. Mater.Eng. and performance, 12 (2003) 288.
[5] Donahue R. and Fabiyi P, "Manufacturing feasibility of all-aluminum automotive engines via application of high silicon aluminum alloy", Society of Automotive Engineers, (2000).
[6] Miller W, Zhuang L, and Bottema J, "Recent development in aluminum alloys for the automotive industry", Mater. Sci. and Eng., A280 (2000) 37.
[7] Lu S and Hellawell A, "The mechanism of silicon modification in aluminum-silicon alloys: impurity induced twinning", Met. Trans, 18a (1987) 1721.
[8] Srirangam P, Kramer M J and Shankar S, "Effect of strontium on liquid structure of Al–Si hypoeutectic alloys using high-energy X-ray diffraction", Acta Mater., 59 (2011) 503.
[9] Loizaga A, Niklas A, Fernandez-Calvo A I, Lacaze J, "Thermal analysis applied to estimation of solidification kinetics of Al-Si aluminium alloys", Int. J. Cast Metals Res., 22(2009) 345.
[10] Gorny A, Manickaraj J, Cai Z, Shankar S, "Evolution of Fe based intermetallic phases in Al–Si hypoeutectic casting alloys: Influence of the Si and Fe concentrations, and solidification rate", J. Alloys and Comp., 577 (2013) 103.
[11] Anton Gorny PhD dissertation "Characterization of Major Intermetallic Phases in solidified Al-xSi-yFe-zSr (x=2 to 12.5wt%, y=0 to 0.5wt% and z=0 and 0.02wt%) alloys" McMaster University 2012.
[12].Gorny A, Manickaraj J, Shankar S, "Effect of Fe and Sr in the evolution of Iron based intermetallic phases in Al-Si hypoeutectic alloys", Materials Science and Technology (MS&T) (2013) 1203.
[13] Sumanth S, Yancy W R, Makhlouf M M, "Nucleation mechanism of the eutectic phases in aluminum-silicon hypoeutectic alloys",. Acta Mater., 52 (2004) 4447.
[14] Makhlouf M M, Shankar S and Riddle Y W" "Chemical modification of the eutectic phases in hypoeutectic aluminium-silicon alloy" AFS Trans. 05-88(02) (2005) 18.

Shape Casting: 5th International Symposium 2014
Edited by: Murat Tiryakioğlu, John Campbell, and Glenn Byczynski
TMS (The Minerals, Metals & Materials Society), 2014

Thin-wall Model for Use in Multiple Casting Conditions

A.T. Noble[1], C.A. Monroe[1], A.K. Monroe[2]

[1]School of Engineering, University of Alabama at Birmingham, Birmingham, Alabama
[2]MAGMA Foundry Technologies, Schaumberg, Illinois

Keywords: thin-wall, casting, simulation, modeling, flow, filling

Abstract

Predicting the misruns in thin-wall castings is required to produce lightweight high performance components. A general 1D heat transfer model for predicting the heat extraction of thin-wall castings is proposed. Temperature dependent properties calculated from a thermodynamic database are used to incorporate latent heat. A stopping criterion, related to a critical solid fraction, has been introduced to determine when the fluid solidifies and obstructs the flow. The model has been used to predict cast lengths in several conditions found in literature including die casting, sand casting and Ragone tests with good agreement to reported lengths. This model shows potential for use in 3D simulations of thin-wall castings.

Introduction

Thin-wall castings are a focus of foundries looking to satisfy the demand for improved casting performance and cost efficiency. For sand casting processes, castings with walls thinner than ¼ inch (6 mm) are considered thin-walled castings whereas high pressure die casting processes can achieve sections down to 0.03 inch of 0.76 mm. Successful casting designs are created based on well-established design rules and procedures such as Chvorinov's Rule, which relates solidification time to the casting geometric modulus. Other factors in addition to solidification time are used to determine the fluidity of cast meals including metal composition, velocity, and others as described extensively by Campbell [1]. The filling of thin-walled castings can be modeled using a heat transfer analysis of the heat extraction from the cast metal to the mold. According to Dantzig and Tucker, there are two methods to evaluate fluidity using heat extraction models [2]. The first method is called mold control, and assumes heat extraction is limited by the mold. This is the method assumed in Chvorinov's rule. Mold control methods work well for solidification of filled molds, but their use during filling results in over-predictions of fill length. The second method, solid control, more closely predicts fill lengths. The model described in this paper uses solid control assumptions to predict fluidity in thin-wall castings.

The derivation assumes the metal flow front has a constant velocity with a sufficiently short fill time such that the mold ahead of the metal front does not significantly heat up, which allows the mold surface temperature to be treated as constant. Justification for the constant velocity assumption is found from the Prandtl number Pr of the metal flow. Prandtl numbers much less than one indicate that a fully developed temperature profile is achieved much more quickly than a fully developed velocity profile [2]. Many liquid metals do have $Pr \ll 1$, and so a uniform velocity may be assumed across the flow cross section. The derivation of the model is given by Dantzig and Tucker and reiterated by Monroe and Monroe is as follows [2,3].

$$\frac{\partial\theta}{\partial t^*} = \frac{\partial^2\theta}{\partial y^{*2}} \tag{1}$$

Equation 1 can be used to determine the temperature at any point in the thin-wall channel using the scaled boundary conditions 2a and b

$$\pm\frac{\partial\theta}{\partial y^*} = Nu\theta \qquad \theta(t=0) = 1 \tag{2a,b}$$

where Nu is the Nusselt number. The Nusselt number is a useful dimensionless quantity that is found from the ratio of the HTC and the characteristic width to the thermal conductivity of the metal. It is often referred to as the dimensionless HTC. Rohsenow et al. demonstrated that the Nusselt number is a function of the Péclet number for cases of negligible compressibility, as can be assumed of liquid metal flow [4]. Several correlations between the two dimensionless numbers are provided in literature, and one correlation that is often cited as useful in metal castings analysis is the Lubarsky-Kaufman correlation.

$$Nu = 0.625 * Pe^{0.4} \tag{3}$$

Carbajo found the Lubarsky-Kaufman correlation to be the most applicable of five correlations to the study of HTC in liquid metal heat convection [5]. Long et al. also used the Lubarsky-Kaufman correlation to justify the use of large HTCs in the modeling of high pressure die casting [6]. The advantage of the correlation is that the often approximated HTC model parameter can be calculated as a function of velocity, section thickness, and material properties of the metal. Equation 7 is then solved to determine the temperature distribution by separation of variables. This is discussed in more detail in the methods section.

The model described thus far applies only to castings of constant cross section, i.e. Ragone tests, fluidity spirals, and others. Using three dimensional (3D) filling simulation results as input, a new method has been demonstrated in the authors' previous work with potential for modeling geometry of varying thicknesses [7]. The difficulties discussed earlier with high density of thermal boundary layers associated with high Pe are magnified in 3D cases. In more complex geometries, it is not feasible to fully resolve the thermal boundary layers by mesh refinement. In previous studies, it was found that in some cases the minimum mesh requirement of three elements across the thin-wall section is sufficient to produce reasonable results. This method form the previous work and the solution to equation 1 are described in the methods.

Methods

To calculate fill predictions, equation 1 was solved in the method described by Monroe and Monroe [3]. Figure 1 shows the flow chart for the method of solving the thin-wall filling equations. Using the nine parameters, the Nusselt and Péclet numbers, and the dimensionless stop temperature, were calculated. In Figure 1, the first eight λ_n terms from equation 10a were used to calculate the summation terms that add to approximate the dimensionless stop temperature. The difference between the dimensionless stop temperature and the summation results was minimized by varying the dimensionless time. The dimensionless time was then multiplied by both the Péclet number and the modulus to find the predicted fill length.

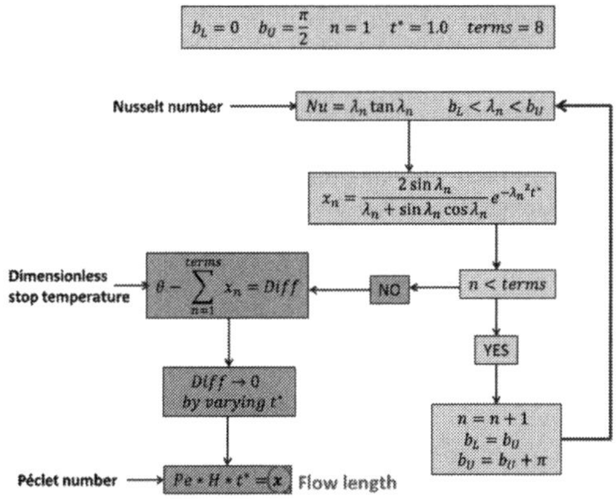

Figure 1. This shows a flow chart of the method of calculating flow length in thin-wall castings described by Monroe and Monroe [3]. The Nusselt, Péclet, and dimensionless stop temperature are calculated from the nine parameters. The first eight terms are used to match the dimensionless stop temperature. The dimensionless time was found from minimizing the difference between the summation of the eight terms and the stop temperature. Dimensionless time is then converted to the flow length prediction using the Péclet number and the channel half-thickness.

The one dimensional (1D) model may also be applied to more complex 3D geometries using the method described in the authors' most recent work [7]. Filling simulation of 3D CAD castings was performed using commercially available software. The filling results yield two of the primary parameters, temperature and velocity, which are calculated at sufficiently fine time intervals to capture the details of filling. The thin-wall model described previously was then used to calculate a fill length prediction for each element in the meshed CAD model at each filling time-step using the temperature and velocity results as well as the local section thickness. The result of this calculation was stored and can be seen as the current fill length field plotted on the casting geometry just as any other result is visualized. By comparing these results to characteristic filling lengths, described later through example, areas of fill risk can be visualized in 3D parts. This is the primary industrial interest in this study and work is ongoing to experimentally evaluate the validity of this approach. Recently, the 1D solution has been altered to include temperature dependent material properties, which accounts for the latent heat during filling. To include temperature dependent properties, the temperature solution throughout the channel thickness was needed rather than solving for centerline temperatures alone. The calculations were performed using an implicit solving method, described by Özişik [8], using the same governing equations and boundary conditions as in the previous method. The fill length calculation was written into a Java application for quick calculations and cross-platform capabilities. The 1D results presented here were obtained using the latent heat method.

Results

The model described has been applied to study various landmark fluidity cases from literature with both simple and complex geometries. The processes employed in the case studies include sand casting, Ragone fluidity tests, and die castings. The full results of the various case studies will be available soon in a Metallurgical and Materials Transactions B publication. Here a summary of the results and some select examples are presented. The most significant finding in the 1D case studies is the HTC sensitivity to mold properties including surface roughness, thickness, material, and the presence of a coating. These properties are accounted for with correction factors to the Lubarsky-Kaufman correlation. In general, it was found that the Lubarsky-Kaufman correlation was most applicable with metal flowing in molds made of material with high thermal conductivity, and a smooth surface, and with coatings that do not significantly insulate the metal. These results seem to indicate the model may apply only to die castings, yet this trend was shown to be independent of metal velocity. A Ragone case study using a copper mold rather than the traditional quartz or stainless steel tube required no correction factor to approximate the fill lengths found in literature. A replication of the findings is shown in Table 1 with the mold property responsible for the correction is highlighted.

Table I.Summary of the Lubarsky-Kaufman correlation correction factors used in the case studies along with the conditions at the metal-mold interface.

	Case Study 4	Case Study 5	Case Study 2	Case Study 1		Case Study 3
correction factor	0.03	0.07	0.75	0.5	1	1
mold material	copper	sand	pyrex/air	steel	steel	Copper
mold coating	sand	none	none	oil	oil diluted 500 times with kerosene	None
mold thickness	N/A	N/A	~0.1 in	N/A	N/A	N/A
surface roughness	rough	rough	smooth	smooth	smooth	Smooth
cast metal	AlCu & AlSi	medium carbon steel	Al4.3Cu	ZDC2	ZDC2	AlCu

One of the more well-documented case studies from Table 1 is the work performed by Taylor, Rominski, and Briggs on steel fluidity spirals (case study 5) [9]. A section of this extensive work is dedicated to producing a "standard curve" for fluidity of steels most familiar to foundrymen as a function of initial pouring temperature. The filling lengths of five steel alloys cast at various pour temperatures were recorded using a robust fluidity spiral geometry. The cross section of the fluidity spiral had a hydraulic diameter of 0.328 inches which is equivalent to a modulus of 4.17 mm. Through simulation work, Monroe and Huff [10] estimated the velocity of metal at the spiral inlet to be 0.3 m/s. A commercially available thermodynamic database software calculated the temperature dependent material properties for use in the model calculation. Figure 2 displays the comparison between the experimental standard curve and the model prediction.

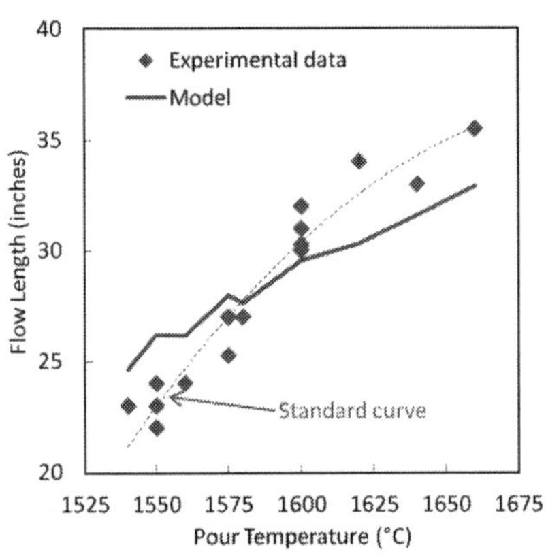

Figure 2. Comparison of the standard curve developed from experimental results and the model prediction. The local peaks in the prediction curve are caused by averages of multiple values at 1550, 1575, and 1600°C.

The model predictions mimic the experimental standard curve though with a reduced slope. A correction factor of 0.07 was employed to account for the mold being made of sand and having a rough surface. The insulating property of the rough mold surface is due to low thermal conductivity of sand molds and also the molten metal's high surface tension. The surface tension prevents the metal from taking advantage of enhanced heat transfer from more surface area in a rough mold surface. Instead the rough surface holds air gaps which insulate the metal. The study included multiple fill length results at three pouring temperatures, 1550, 1575, and 1600°C. In Figure 2, multiple model results for the three temperatures were averaged to produce the model curve resulting in the observed local peaks. Nevertheless, the 1D thin-wall calculations gives a reasonable approximation to the standard fluidity curve.

The die casting case study from Nishi is also well-documented (case study 1) [11]. The study focused on the effects of injection velocity, section thickness, die and pour temperatures, and die lubricant on the filling of the mold. The most interesting of these was the casting of 100 x 70 mm zinc plates with 0.2 mm thicknesses. Plates with 0.4 mm thickness also were studied between 1 and 1.6 m/s, where filling defects were observed at the slower speeds. Injection velocities ranged from 0.7 to 2.0 m/s (piston speed). The plates of interest were cast with a die temperature of 180°C and a pour temperature of 430°C. The fluidity of plates cast with undiluted lubricant and lubricant diluted at a ratio of 500:1 with kerosene were also compared. The results of the 1D model calculation were compared to the average length along the cast plate which was obtained from the reported volume percent fill, as shown in Figure 3.

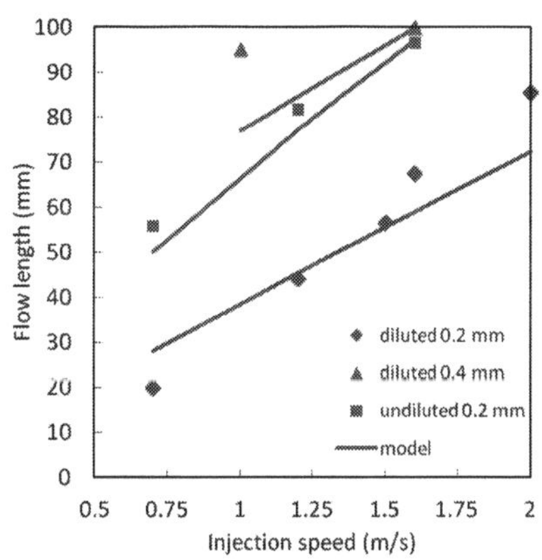

Figure 3. The model predictions of the die casting plate filling are overlaid onto the results of the three main casting conditions of interest in the study.

The model prediction curves show reasonable approximation of the experimental fluidity results. In all but two cases, the model is conservative in the fill length predictions. The diluted cases of 0.2 and 0.4 mm thicknesses needed no correction factor. Experimental data showed that 100% filling is achieved in the 0.4 mm diluted plate between injection speeds of 1 and 1.6 m/s which the model also predicts. A correction factor of 0.5 was introduced for the 0.2 mm plate with undiluted lubricant to account for the increased insulation of the molten zinc. The use of the correction factor was found to increase the slope of the model prediction in addition to increasing flow length based on lowered HTC. This should be expected based on the 0.4 exponent in the Lubarsky-Kaufman correlation.

The other case study results will be available in a future Metallurgical and Materials Transactions publication. Now the 3D implementation methodology will be demonstrated with a continuation of the Nishi zinc plate case study. The recreated plate geometry and resulting mesh are shown in the authors' previous work [7]. Filling results were sufficiently captured in the thin plate section with three mesh elements across the thickness. Further refinement of the mesh would better resolve the boundary layers but at the cost of calculation time. In addition, the filling length predictions are made using the entrance conditions of the flow rather than the conditions of the flow front, therefore the mesh refinement is less significant. The plate with the 0.7 m/s injection speed was chosen to represent the method. Fill length predictions were obtained using the model calculations for every metal cell during every time step. During filling, it was determined that the temperature drop of the metal is relatively small at any fixed point and the velocity at the front is unstable but becomes steady at a maximum value as the bulk flow passes. The filling length was determined to mimic the velocity result which led to the conclusion of describing the local fill conditions with the maximum filling length calculated during the simulation. The maximum fill length is shown in Figure 4 for the 0.7 m/s injected plate.

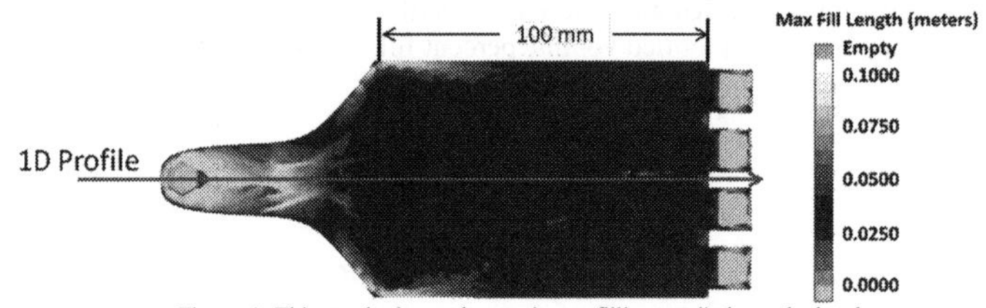

Figure 4. This result shows the maximum filling prediction calculated at every mesh element over the course of the simulated filling process.

The plate shown in Figure 4 fills only 20% of the volume during experimental casting. Yet, temperature and velocity results during filling permit the calculation of fill predictions for elements that should not contain metal. A suggested solution is for the casting software to include solidification in the filling process to point out where metal flow should stop. This would require reliable knowledge of the flow front during filling which, as discussed previously is not feasible given the need of a fine mesh and the associated calculation time. Instead, a characteristic fill length may be chosen for comparison to the fill predictions. A good characteristic length for comparison is the total length traveled by the metal in each element along its flow path. This result will be called accumulated length and is provided in commercial casting software. In geometries with relatively simple flow patterns, like the Nishi plate, the accumulated length will increase linearly with the part length, because most flow paths are along the length of the plate. Figure 5 shows the filling results collected along the centerline drawn in Figure 4 with the accumulated length overlaid.

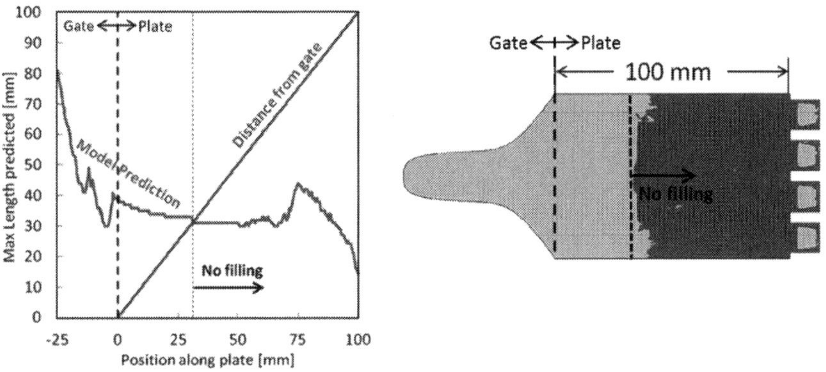

Figure 5. The plot shows the model prediction overlaid onto the accumulated length which for this unidirectional flow casting is the same as the distance from the gate. Using the overlay, the areas of fill risk can be determined and applied to the rest of the geometry as shown in the plate figure where the red areas indicate filling risk.

Closer to the gate, the fill prediction is greater than the accumulated length, indicating areas that should fill. Areas of fill risk occur when the predicted length is less than the accumulated length. The plate in Figure 5 shows the application of this comparison where areas in red indicate areas of fill risk. The filled section of the plate more closely resembles the experimental plate. The other injection velocities were similarly modeled with reasonable correlation to the experimental plates.

Conclusions

The thin-wall model has the potential to predict fluidity in thin-wall castings produced with various processes and conditions. Nine casting parameters were used to calculate the flow predictions: velocity; pouring and mold temperatures; casting modulus; density, specific heat, and thermal conductivity of the metal; stopping temperature; and the HTC at the mold-metal interface. The last two parameters must be chosen carefully for calculating thin-wall filling. While the stopping temperature is bounded by the solidification range of the alloy, the HTC has no bounds. According to literature, the Lubarsky-Kaufman correlation has been used with some success to predict the HTC in terms of the velocity, local section thickness, and material properties. Accounting for the effect on HTC of the mold conditions can be achieved with correction factors to the Lubarsky-Kaufman correlation. In general, a mold with a rough surface and lower thermal conductivity will require a more severe correction factor. The 1D model using constant thickness castings has been extended for use in predicting the fill of 3D geometries. In the example shown, the maximum filling length was representative of the steady state flow which may not be the case in other casting geometries. Future work will include evaluating the filling predictions at steady state conditions with more complex flow patterns. Subtracting the accumulated lengths from the filling length predictions yields a visualization of the filling risks areas in the casting. This approach is a quick method for evaluating the local fill risk for a part in order to focus on the root cause of non-fill defects for a particular casting. Further work on heat transfer coefficient, latent heat, and metal stop temperature should improve these results.

References

1. John Campbell, *Complete Casting Handbook Metal Casting Processes, Metallurgy, Techniques and Design* (Oxford, UK: Elsevier Butterworth-Heinemann, 2011).

2. Jonathan A. Dantzig and Charles L. Tucker III, *Modeling in Materials Processing* (Cambridge, England New York: Cambridge University Press 2001).

3. C.A. Monroe and A.K. Monroe, "Predicting flow lengths in die casting including heat advection" (Paper presented at the 2012 Die Casting Congress & Exposition, Indianapolis, Indiana, 10 October 2012).

4. Warren Rohsenow, James Hartnett, and Ejup Ganić, *Handbook of Heat Transfer Fundamentals* (New York: McGraw-Hill 1985).

5. J.J. Carbajo, "Applicability of Heat Transfer Coefficient Correlations to Single-Phase Convection in Liquid Metals" (Pub9924, Oak Ridge National Laboratory, 2011).

6. A. Long et al. "Determination of the heat transfer coefficient at the metal-die interface for high pressure die cast AlSi9Cu3Fe," *Applied Thermal Engineering*, 31 (2011), 3996-4006.

7. A.T. Noble, C.A. Monroe, and A.K. Monroe, "Predicting flow lengths in die casting in 3D" (Paper presented at the 2013 Die Casting Congress & Exposition, Louisville, Kentucky, 18 September 2013).

8. M. Necati Özişik, *Heat Conduction, 2nd ed* (New York: John Wiley & Sons 1993)

9. H.F. Taylor, E.A. Rominski, and C.W. Briggs, "The Fluidity of Ingot Iron and Carbon and Alloy Cast Steels," *Transactions of the American Foundrymen's Association*, 49 (1941), 1-93.

10. C.A. Monroe and R. Huff, "Filling Thin Wall Steel Castings" (Paper presented at the SFSA 64th National Technical and Operating Conference, Chicago, Illinois, December 2010).

11. N. Nishi, "Challenge to Ultra Thin-Wall Zinc Die Casting under 0.5 mm Thickness," *Die Casting Engineer*, 2007, September, 42-46.

Shape Casting: 5th International Symposium 2014
Edited by: Murat Tiryakioğlu, John Campbell, and Glenn Byczynski
TMS (The Minerals, Metals & Materials Society), 2014

ELECTROCHEMICAL BEHAVIOR OF Al-1wt.%Cu AND Al-4.5wt.%Cu ALLOYS

Alejandra S. Roman[1], Claudia M. Méndez[1], Carlos E. Schvezov[1,2] , Alicia E. Ares[1,2]

[1] Materials Institute of Misiones, IMAM (CONICET-UNaM), University of Misiones; 1552 Azara Street, Posadas, Misiones, 3300 Argentina.
[2] Member of CIC of the National Research Council (CONICET) of Argentina.

Keywords: Al-Cu alloys, structural parameters, corrosion resistance.

Abstract

The objective of this work was to study the corrosion behavior of dendritic diluted Al-Cu alloys, i.e., Al-1wt.%Cu and Al-4.5wt.%Cu (weight percent), directionally solidified vertically upward. During the solidification of the alloys, the phenomenon of columnar-to-equiaxed transition (CET) was obtained, and then, the properties are assessed according to the type of structure (columnar, equiaxed or CET) and composition. To determine the susceptibility to corrosion, working electrodes were prepared with different types of structures and cyclic potentiodynamic polarization measurements were conducted, as well as assays, electrochemical impedance spectroscopy (EIS) using solution of 3% NaCl at room temperature. It was observed that increasing the percentage of copper in the alloy, the corrosion potential moves to more anodic values. Impedance diagrams of the corrosion potential were obtained and then through a simulated equivalent circuit model. It is further noted that the resistance values decrease with increasing copper content in the alloys.

Introduction

The first and most widely used aluminum alloys are those containing between 4 and 10 wt% Cu. Metallurgical factors that may affect the corrosion of an alloy include: crystallography, size, shape and heterogeneity of the grain, impurities, inclusions and residual stresses due to cold working. Alloys (Al-Cu) have a high tensile strength as used in aircraft structural parts, car and bus bodies, and fuel tanks [1].
As was mentioned by Campbell [2] when metal is solidifying, central regions which may form hot spots will cause some local shrinkage that possibly will take the form of porosity showing the existence of bifilms. In absence of bifilms, local shrinkage can cause porosity like surface sink. To avoid this problem, it is necessary using a feeder or use a chill (block or cold metal forming part of the mold). Directional solidification refers to the process of controlled feeding of the molten metal into a temperature-controlled mold to produce a part that is free of hollow spots, called shrink defects. By controlling the rate of flow for the molten metal feed and introducing thermal variations in the mold, shrink defects can be eliminated, because the liquid metal will naturally run into these dips and vacant areas. By this process is possible to obtain the columnar-to-equiaxed transition phenomenon [3].
Osorio et al. [4] studied the effect of macrosegregation and secondary dendritic spacing on the corrosion resistance of alloys Al-4.5% Cu. Found two factors that influence the corrosion resistance in this alloy: 1) the difference in corrosion potential between the Al-rich phase and 2) intermetallic particles Al_2Cu. Conclude that the first factor is leading to localized corrosion and the second is influenced by the cooling rate imposed during solidification.

In another study, they analyzed the corrosion resistance in alloys Al-5%Cu and Al-8%Cu, directionally solidified with CET. They found that both morphologies (columnar and equiaxed) have similar results in trials of experimental EIS and polarization curves. In addition, the corrosion resistance decreases with increasing Cu content in it and that this behavior and may be associated with a decrease in the secondary dendritic spacing [5].
The objective of this work was to study the corrosion behavior of dendritic diluted Al-Cu alloys, i.e., Al-1wt.%Cu and Al-4.5wt.%Cu (weight percent), directionally solidified vertically upward in samples with the phenomenon of columnar-to-equiaxed transition (CET) occurring.

Experimental Procedure

The samples of Al-Cu alloys (Al-1wt.%Cu and Al-4wt.%Cu) were unidirectionally solidified. To obtain the transition from columnar to equiaxed structure (CET) in the samples, directional solidification was performed in the same way as in previous work [3, 6], that is, the alloys were melted and directionally solidified in alumina molds in a directional solidification furnace comprising a heating unit and control systems and acquisition of temperature, which was added a directional heat extraction system. After directional solidification, the samples were cut longitudinally and grounded using SiC papers of different grain sizes, from grade 60 to grade 1200, and then polished with diamond paste of 1 micron.
The etching was performed with 15 ml HF, 4.5 ml HNO_3, 9.0 ml HCl and 271.5 ml H_2O at room temperature (25 ° C), what was appropriate to reveal the macrostructure [7]. The position of the CET was found by visual observation and optical microscopy, and the distance from the base of the samples was measured with a ruler. The positions where the CET occurred in the samples were between 2.5 cm and 4.9 cm from the base. In Figure 1 (a) and (b) shows the macrograph of an experience with Al-1wt.%Cu and Al-4.5wt.%Cu, respectively, showing the three areas in the samples, columnar (base), CET (middle) and equiaxed (above). To analyze the microstructure, the samples were etched with a solution containing 1 g NaOH in 100 ml of distilled H_2O, during a time of 5 to 15 seconds [7, 8]. The micrograph of different zones of structures can be seen in Figures 1 (c) to (e) for Al-1wt.% Cu and (f) to (h) for Al-4.5wt.%Cu alloy.
Equiaxed grain size was measured according to ASTM standard E112 [1], using equally spaced at equal intervals in the specimens. Similarly, the columnar region was divided into intervals and directly measured the width and length of the columnar grains. Dendritic spacing measurements were made using the linear interception technique, preferably in regions close to the positions of the thermocouples during directional solidification in order to make correlations with the parameters of solidification. Using an Arcano® optical microscope (MO), and with the help of image processing program TSView the values of secondary dendritic spacing, λ_2, were obtained.
For the electrochemical tests (samples were prepared, which are used as working electrodes, approximately 2 cm long, each of the three zones (columnar, equiaxed and CET) and for each concentration from sections specimens cut longitudinally were sanded to # 1200 SiC particle size, washed in distilled water and dried by natural air flow.
Cyclic potentiodynamic polarization curves were performed, following the Standard G61-86, using a potentiostat model M7 LYP® Electronics Argentina. An electrochemical Pyrex® glass cell with conventional three electrodes (ASTM G-5), with saturated calomel (SCE) as reference electrode and a platinum foil as counter electrode.

The scanning of polarization curve was started from a lower potential of 300 mV at open circuit potential to more anodic potentials. Upon completion of these experiences, the working electrodes were examined using Nikon® metallographic microscope to visualize the effects of corrosion on the microstructure.

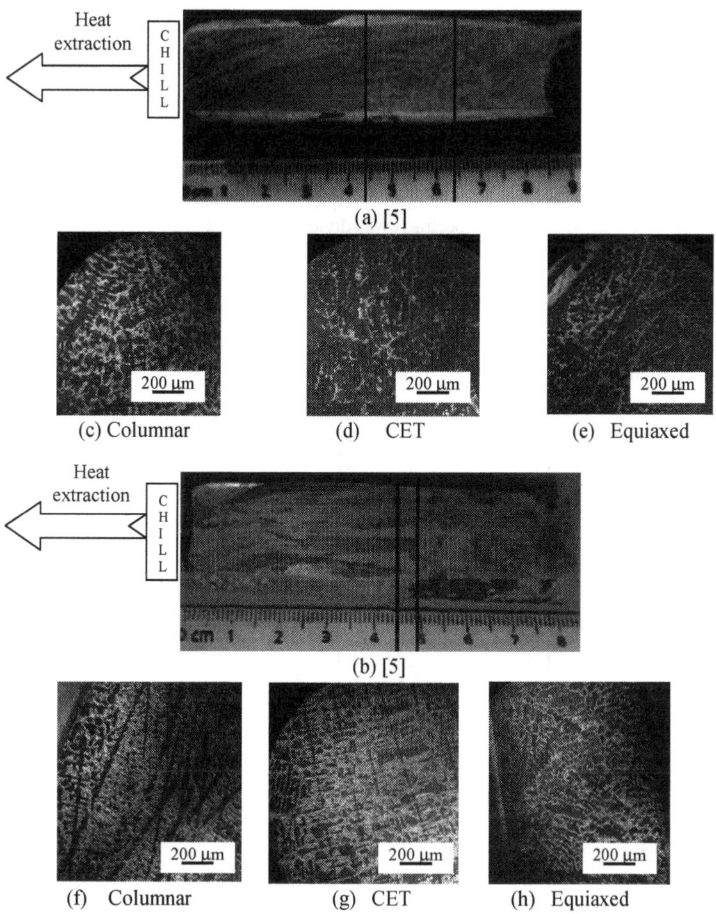

Figure 1. Macrostructures of Al-Cu alloys. (a) Al-1wt.%Cu y (b) Al-4.5wt.%Cu. (c) to (e) Microstructures of Al-1wt.%Cu and (f) to (h) Microstructures of Al-4.5wt.%Cu.

Measurements were also performed using Electrochemical Impedance Spectroscopy (EIS) technique, using a Gamry® Instruments, in combination with M7 LYP potentiostat. Potential amplitude of 10 mV/s was used at open circuit potential with an initial period of stabilization system of 600 seconds. For adjustment of the results the method of nonlinear least squares designed by Bouckamp [9] was used.
For all electrochemical tests, a 300 ml solution of 3% NaCl was employed as electrolyte, pH = 5, at room temperature, which previously removing oxygen by bubbling nitrogen during 20 minutes.

Results and Discussion

Grain size (Gs)
To determine the size of equiaxed grains a typical frequency histogram of equiaxed grain size (g) for each of the intervals in which the specimen was divided was used. From these histograms the equiaxed grain size was determined [3, 6].
The results are shown in Figure 2, which includes the width of columnar grains, and the size of the grains is plotted as a function of position in the samples. As shown, the equiaxed grain size varies from 0.5 mm in the CET zone and increases to 0.8 mm in the equiaxed zone, located at the end of the sample, which is also the last part to solidify. In the case of the columnar grain width, it maintain constant (0.5 mm) until to reach the zone of the CET.

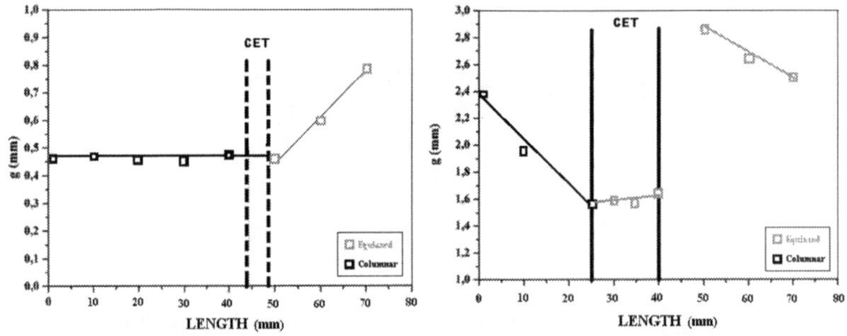

(a) (b)

Figure 2. Variation of grain size versus distance from the base of the sample [5].
(a) Al-1wt.%Cu and (b) Al-4.5wt.%Cu. The vertical lines are indicating the CET zone in the sample.

Secondary dendritic spacing (λ_2)
Measurements of the secondary dendritic arm spacings include active and inactive branches. In Figure 3 shows that 1) the structure changes from columnar (base of the sample) to equiaxed (top of the sample) and 2) that it increases with distance from the base.

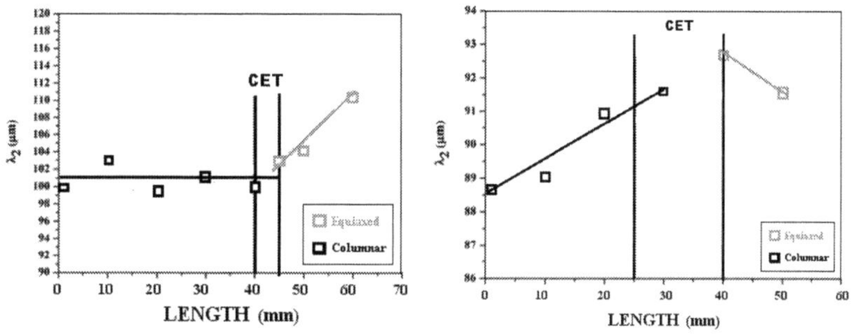

(a) (b)

Figure 3. Secondary dendritic arm spacing variation versus distance from the base of the sample [5]. (a) Al-1wt.%Cu and (b) Al-4.5wt.%Cu.

Corrosion resistance

Potentiodynamic measurements

For Al-1wt.%Cu alloy, in each of the areas analyzed showed passivation, where lower currents remained virtually constant over a range of 100 mV (Figure 4). Then, there was a rapid increase in current and a rapid dissolution of the metal, with the highest pitting potential (E_P) in the columnar zone equal to -707 mV. Table 2 shows the measured potential for such pitting sample.

In the case of Al-4.5wt.%Cu alloy sample, passivation was only seen on the CET sample, commencing the increase in dissolution rate of the sample at E_P = -672.6 mV.

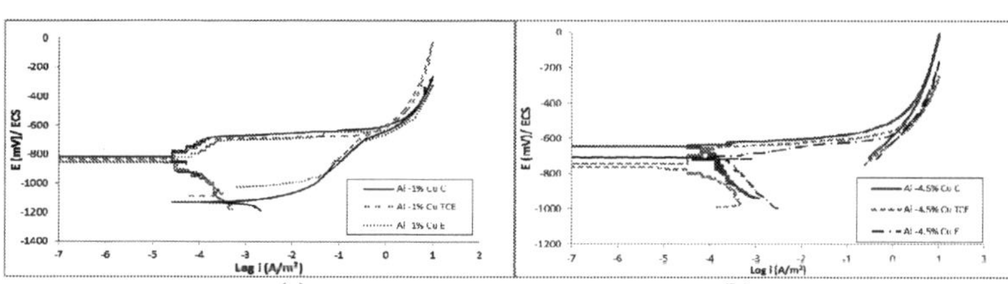

(a) **(b)**

Figure 4. (a) Al-1wt.%Cu and (b) Al-4.5wt.%Cu potentiodynamic curves.

However, the E_P for the equiaxed zone was equal to -703.6 mV, which is nobler than the E_P in the columnar zone, equal to -643.5 mV. By evaluating the relationship between the copper content of alloys and potentials, it was found that in columnar and equiaxed zones, the corrosion potential (E_C) appears nobler to Al-4.5wt.% Cu alloy. This composition is the nearest to commercial Al-Cu alloys. So too, the lowest E_C was evident for samples with lower copper content. In the case of the CET region was found that the corrosion potential increases with increasing solute content in the alloy. Table 3 presents the E_C measured for each composition analyzed.

These results are consistent with the observations that the columnar zones of the samples are free of the presence of shrinkage and therefore bifilms [2] and in the CET and equiaxed zones were the final contraction of solidification process may occur.

Table 2. Pitting potentials in columnar, equiaxed and CET zones.

Alloy	**Zone**	**E_P (mV)**
Al-1wt.%Cu	Columnar	-707
	CET	-731
	Equiaxed	-733
Al-4.5wt.%Cu	Columnar	-
	CET	-670
	Equiaxed	-

Table 3. Corrosion potentials for the three alloys studied.

	E_C (mV)		
Alloy	Columnar	CET	Equiaxed
Al -1wt.% Cu	-820	-833	-854
Al -4.5wt.% Cu	-645	-744	-707

Electrochemical Impedance Spectroscopy (EIS)

The impedance response for samples obtained for various zones of Al-1wt.%Cu alloy can be simulated by the circuit shown in Figure 6 (a). The parameters corresponding to EIS can be observed in Table 4. R_Ω corresponds to the resistance of the electrolyte and R_1 resistance corresponds to the charge transfer that accompanies the double layer. Given that the values of n_1 are close to 0.8, which can be attributed to this constant phase element, CPE_1 is the double layer capacity.

For the three regions of Al-4.5wt.%Cu alloy, a proposed model was used to represent the impedance response of the electrodes having a coverage of porous surface film, where the reactions occur only in the exposed surface, i.e., the end of the pore. Within the pore, concentrations of different species differ from the concentration in the bulk solution [10] (Figure 6 (b)). Setting parameters are presented in Table 4. R_2 is considered the resistance associated with the porous oxides and constant phase element, CPE_2, correspond to porous oxides capabilities.

Observing the impedance setting parameters in the various zones for a given composition of the sample, shows that the equiaxed zone have the lowest values of the charge transfer resistance,R_1, while the CET zone have the highest valued, assumed resistor values in the three alloys tested. Figure 7 shows the Nyquist plots for the three zones of the samples of Al-4.5wt.%Cu alloy.

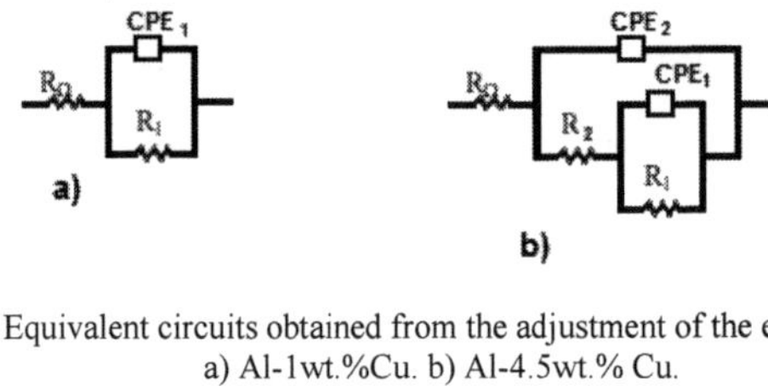

Figure 6. Equivalent circuits obtained from the adjustment of the experimental data. a) Al-1wt.%Cu. b) Al-4.5wt.% Cu.

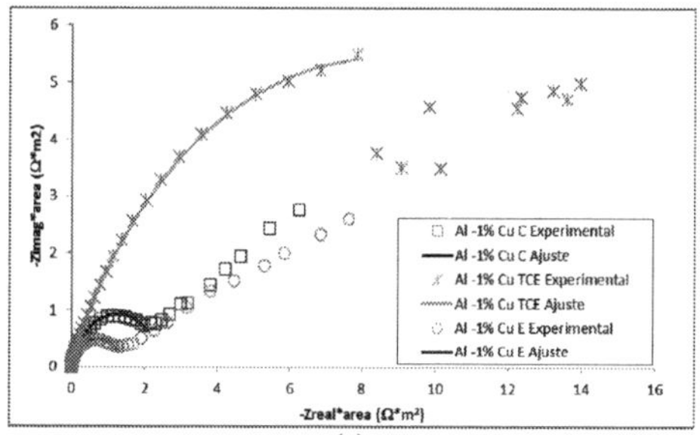

(a)

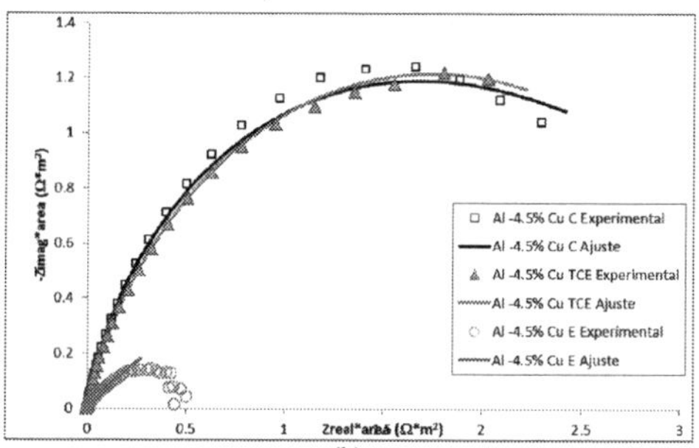

(b)

Fig. 7. Nyquist diagram for (a) Al-1wt.%Cu and (b) Al-4.5wt.%Cu.

Table 4.Parameters obtained from EIS adjustment.

Alloy	Zone	R_{omh} (Ω*m^2)	CPE_1 (Ω/m^2)	n_1	R_1 (Ω*m^2)	R_2 (Ω*m^2)	CPE_2 (Ω/m^2)	n_2	Circuit
	Columnar	0.067	7.0E-03	0.82	251	-	-	-	Fig.6. (a)
Al-1wt.%Cu	CET	0.039	2.6E-03	0.82	1139	-	-	-	Fig.6. (a)
	Equiaxed	0.040	6.8E-02	0.80	139	-	-	-	Fig.6. (a)
	Columnar	0.049	5.8E-04	1.00	334	47.3	$2.4E^{-05}$	1.00	Fig. 6. (b)
Al-4.5wt.%Cu	CET	0.039	1.8E-03	0.53	336	38.26	$2.2E^{-03}$	0.88	Fig.6. (b)
	Equiaxed	0.019	1.8E-02	0.51	64	0.12	$6.6E^{-04}$	0.99	Fig.6. (b)

Conclusions

The main conclusions of this investigation are:

Electrochemical parameters do not indicate a relationship between the structure of the alloy and corrosion resistance. However, microscopic observations show that the interdendritic region, rich in copper, acting as a cathode favoring the dissolution of the metal matrix formed in a higher percentage of aluminum.

Given the presence of a passive region in the cyclic potentiodynamic polarization curves and high resistance values in the equivalent circuits of the electrochemical impedance spectra can be concluded that the alloy Al-1wt.% Cu exhibits good corrosion resistance.

Analysis of electrochemical impedance spectroscopy revealed that in general, for each of the alloy compositions, the equiaxed zone has the lowest values of resistance.

The impedance response of the three zones of Al-4.5wt.%Cu alloy has contributions from a porous oxide layer.

Acknowledgments

This work was partially supported by CONICET (Consejo Nacional de Investigaciones Científicas y Técnicas, Argentina).

References

1. H. E. Boyer and T.L. Gall, *Metals Handbook*, (Metals Park, Ohio: American Society for Metals, 1984), 35-18 -35-19.

2. J. Campbell, *Complete Casting Handbook*, (Elsevier Ltd., 2011), 1013-1018.

3. A.E. Ares, S.F. Gueijman, C.E. Schvezov, "An experimental investigation of the columnar-to-equiaxed grain transition in aluminum hypoeutectic and eutectic alloys", *Journal of crystal Growth*, 312 (2010), 2154-2170.

4. W.R. Osório, J.E. Spinelli, I.L. Ferreira and A. Garcia, "The rol of macrosegregation and of dendritic array spacings on the electrochemical behavior of an Al-4.5wt%Cu alloy", *Electrochimica Acta*, 52 (2007), 3265-3273.

5. W.R. Osório, J.E. Spinelli, I.L. Ferreira and A. Garcia, "Experimental analysis of corrosion resistance on columnar to equiaxed transition region of as cast structures of Al-Cu alloys", *Materials Science and Technology*, 24 (2008), 1433-1437.

6. C.M. Rodriguez, "Relación entre la microestructura y la microdureza de las aleaciones Al-Cu solidificadas direccionalmente", Master Thesis, Instituto Sábato, CNEA- UNSaM, 2013.

7. Alicia E. Ares, Liliana M. Gassa, Carlos E. Schvezov, Sergio F. Gueijman, Relationship Between Structure and Properties of Al-Cu Alloys, *Shape Casting: The 4° International Symposium*, Murat Tiryakioglu, John Campbell and Paul Crepeau Editors (2011) 207-214.

8. G. F. Vander Voort, *Metallographic Principles and Practice*, (Metals Park, Ohio: ASM International, 2001) 525-661.

9. B. A. Boukamp, *Equivalent Circuit*, University of Twente-Holand, Enschede, 1989.

10. J. Esteban, M. Lowry, M.E. Orazem, Correction of experimental data for the ohmic potential drop corresponding to a secondary current distribution on a disk electrode, in: L.L. Scribner, S.R. Taylor (Eds.), The Measurement and Correction of Electrolyte Resistance in Electrochemical Tests, ASTM STP, 1056, American Society for Testing and Materials, Philadelphia, (1990) 127.

AUTHOR INDEX
Shape Casting: 5th International Symposium 2014

A

B

C

D

E

F

G

H

J

K

L

M

SUBJECT INDEX
Shape Casting: 5th International Symposium 2014

MIX
Papier aus verantwortungsvollen Quellen
Paper from responsible sources
FSC® C105338

If you have any concerns about our products,
you can contact us on
ProductSafety@springernature.com

In case Publisher is established outside the EU,
the EU authorized representative is:
Springer Nature Customer Service Center GmbH
Europaplatz 3, 69115 Heidelberg, Germany

Printed by Libri Plureos GmbH
in Hamburg, Germany